Advanced Video Coding: Principles and Techniques

Series Editor: J. Biemond, Delft University of Technology, The Netherlands

Volume 1 Three-Dimensional Object Recognition Systems (edited by A.K. Jain and P.J. Flynn)
Volume 2 VLSI Implementations for Image Communications (edited by P. Pirsch)
Volume 3 Digital Moving Pictures – Coding and Transmission on ATM Networks (J.-P. Leduc)
Volume 4 Motion Analysis for Image Sequence Coding (G.Tziritas and C. Labit)
Volume 5 Wavelets in Image Communication (edited by M. Barlaud)
Volume 6 Subband Compression of Images: Principles and Examples (T.A. Ramstad, S.O. Aase and J.H. Husøy)
Volume 7 Advanced Video Coding: Principles and Techniques (K.N. Ngan, T. Meier and D. Chai)

ADVANCES IN IMAGE COMMUNICATION 7

Advanced Video Coding: Principles and Techniques

King N. Ngan, Thomas Meier and Douglas Chai

University of Western Australia,
Dept. of Electrical and Electronic Engineering,
Visual Communications Research Group,
Nedlands, Western Australia 6907

1999

Elsevier
Amsterdam – Lausanne – New York – Oxford – Shannon – Singapore - Tokyo

ELSEVIER SCIENCE B.V.
Sara Burgerhartstraat 25
P.O. Box 211, 1000 AE Amsterdam, The Netherlands

First edition 1999

Library of Congress Cataloging in Publication Data
A catalog record from the Library of Congress has been applied for.

ISBN: 0 444 82667 X

Transferred to Digital Printing 2007

To *Nerissa, Xixiang, Simin, Siqi*

To *Elena*

To *June*

Preface

The rapid advancement in computer and telecommunication technologies is affecting every aspects of our daily lives. It is changing the way we interact with each other, the way we conduct business and has profound impact on the environment in which we live. Increasingly, we see the boundaries between computer, telecommunication and entertainment are blurring as the three industries become more integrated with each other. Nowadays, one no longer uses the computer solely as a computing tool, but often as a console for video games, movies and increasingly as a telecommunication terminal for fax, voice or videoconferencing. Similarly, the traditional telephone network now supports a diverse range of applications such as video-on-demand, videoconferencing, Internet, etc.

One of the main driving forces behind the explosion in information traffic across the globe is the ability to move large chunks of data over the existing telecommunication infrastructure. This is made possible largely due to the tremendous progress achieved by researchers around the world in data compression technology, in particular for video data. This means that for the first time in human history, moving images can be transmitted over long distances in real-time, i.e., the same time as the event unfolds over at the sender's end.

Since the invention of image and video compression using DPCM (differential pulse-code-modulation), followed by transform coding, vector quantization, subband/wavelet coding, fractal coding, object-oreinted coding and model-based coding, the technology has matured to a stage that various coding standards had been promulgated to enable interoperability of different equipment manufacturers implementing the standards. This promotes the adoption of the standards by the equipment manufacturers and popularizes the use of the standards in consumer products.

JPEG is an image coding standard for compressing still images according to a compression/quality trade-off. It is a popular standard for image exchange over the Internet. For video, MPEG-1 caters for storage media up to a bit rate of 1.5 Mbits/s; MPEG-2 is aimed at video transmission of typically 4-10 Mbits/s but it also can go beyond that range to include HDTV (high-definition TV) images. At the lower end of the bit rate spectrum, there are H.261 for videoconferencing applications at $p \times 64$ Kbits/s, where $p = 1, 2, \ldots, 30$; and H.263, which can transmit at bit rates of less than 64 Kbits/s, clearly aiming at the videophony market.

The standards above have a number of commonalities: firstly, they are based on predictive/transform coder architecture, and secondly, they process video images as rectangular frames. These place severe constraints as demand for greater variety and access of video content increases. Multimedia including sound, video, graphics, text, and animation is contained in many of the information content encountered in daily life. Standards have to evolve to integrate and code the multimedia content. The concept of video as a sequence of rectangular frames displayed in time is outdated since video nowadays can be captured in different locations and composed as a composite scene. Furthermore, video can be mixed with graphics and animation to form a new video, and so on. The new paradigm is to view video content as audiovisual object which as an entity can be coded, manipulated and composed in whatever way an application requires.

MPEG-4 is the emerging standard for the coding of multimedia content. It defines a syntax for a set of content-based functionalities, namely, content-based interactivity, compression and universal access. However, it does not specify how the video content is to be generated. The process of video generation is difficult and under active research. One simple way is to capture the visual objects separately, as it is done in TV weather reports, where the weather reporter stands in front of a weather map captured separately and then composed together with the reporter. The problem is this is not always possible as in the case of outdoor live broadcasts. Therefore, automatic segmentation has to be employed to generate the visual content in real-time for encoding. Visual content is segmented as semantically meaningful object known as video object plane. The video object plane is then tracked making use of the temporal correlation between frames so that its location is known in subsequent frames. Encoding can then be carried out using MPEG-4.

This book addresses the more advanced topics in video coding not included in most of the video coding books in the market. The focus of the book is on coding of arbitrarily shaped visual objects and its associated topics.

It is organized into six chapters: Image and Video Segmentation (Chapter 1), Face Segmentation (Chapter 2), Foreground/Background Coding

(Chapter 3), Model-based Coding (Chapter 4), Video Object Plane Extraction and Tracking (Chapter 5), and MPEG-4 Video Coding Standard (Chapter 6).

Chapter 1 deals with image and video segmentation. It begins with a review of Bayesian inference and Markov random fields, which are used in the various techniques discussed throughout the chapter. An important component of many segmentation algorithms is edge detection. Hence, an overview of some edge detection techniques is given. The next section deals with low level image segmentation involving morphological operations and Bayesian approaches. Motion is one of the key parameters used in video segmentation and its representation is introduced in Section 1.4. Motion estimation and some of its associated problems like occlusion are dealt with in the following section. In the last section, video segmentation based on motion information is discussed in detail.

Chapter 2 focuses on the specific problem of face segmentation and its applications in videoconferencing. The chapter begins by defining the face segmentation problem followed by a discussion of the various approaches along with a literature review. The next section discusses a particular face segmentation algorithm based on a skin color map. Results showed that this particular approach is capable of segmenting facial images regardless of the facial color and it presents a fast and reliable method for face segmentation suitable for real-time applications. The face segmentation information is exploited in a video coding scheme to be described in the next chapter where the facial region is coded with a higher image quality than the background region.

Chapter 3 describes the foreground/background (F/B) coding scheme where the facial region (the foreground) is coded with more bits than the background region. The objective is to achieve an improvement in the perceptual quality of the region of interest, i.e., the face, in the encoded image. The F/B coding algorithm is integrated into the H.261 coder with full compatibility, and into the H.263 coder with slight modifications of its syntax. Rate control in the foreground and background regions is also investigated using the concept of joint bit assignment. Lastly, the MPEG-4 coding standard in the context of foreground/background coding scheme is studied.

As mentioned above, multimedia content can contain synthetic objects or objects which can be represented by synthetic models. One such model is the 3-D wire-frame model (WFM) consisting of 500 triangles commonly used to model human head and body. Model-based coding is the technique used to code the synthetic wire-frame models. Chapter 4 describes the procedure involved in model-based coding for a human head. In model-based coding, the most difficult problem is the automatic location of the object in the image. The object location is crucial for accurate fitting of the 3-D WFM onto the physical object to be coded. The techniques employed for automatic facial feature contours extraction are active contours (or snakes) for face profile and eyebrow extraction, and deformable templates for eye and mouth extraction. For synthesis of the facial image sequence, head motion parameters and facial expression parameters need to be estimated. At the decoder, the facial image sequence is synthesized using the facial structure deformation method which deforms the structure of the 3-D WFM to stimulate facial expressions. Facial expressions can be represented by 44 action units and the deformation of the WFM is done through the movement of vertices according to the deformation rules defined by the action units. Facial texture is then updated to improve the quality of the synthesized images.

Chapter 5 addresses the extraction of video object planes (VOPs) and their tracking thereafter. An intrinsic problem of video object plane extraction is that objects of interest are not homogeneous with respect to low-level features such as color, intensity, or optical flow. Hence, conventional segmentation techniques will fail to obtain semantically meaningful partitions. The most important cue exploited by most of the VOP extraction algorithms is motion. In this chapter, an algorithm which makes use of motion information in successive frames to perform a separation of foreground objects from the background and to track them subsequently is described in detail. The main hypothesis underlying this approach is the existence of a dominant global motion that can be assigned to the background. Areas in the frame that do not follow this background motion then indicate the presence of independently moving physical objects which can be characterized by a motion that is different from the dominant global motion. The algorithm consists of the following stages: global motion estimation, object motion detection, model initialization, object tracking, model update and VOP extraction. Two versions of the algorithm are presented where the main difference is in the object motion detection stage. Version I uses morphological motion filtering whilst Version II employs change detection masks to detect the object motion. Results will be shown to illustrate the effectiveness of the algorithm.

The last chapter of the book, Chapter 6, contains a description of the MPEG-4 standard. It begins with an explanation of the MPEG-4 development process, followed by a brief description of the salient features of MPEG-4 and an outline of the technical description. Coding of audio objects including natural sound and synthesized sound coding is detailed in Section 6.5. The next section containing the main part of the chapter, Coding of Natural Textures, Images And Video, is extracted from the MPEG-4 Video Verification Model 11. This section gives a succinct explanation of the various techniques employed in the coding of natural images and video including shape coding, motion estimation and compensation, prediction, texture coding, scalable coding, sprite coding and still image coding. The following section gives an overview of the coding of synthetic objects. The approach adopted here is similar to that described in Chapter 4. In order to handle video transmission in error-prone environment such as the mobile channels, MPEG-4 has incorporated error resilience functionality into the standard. The last section of the chapter describes the error resilient techniques used in MPEG-4 for video transmission over mobile communication networks.

King N. Ngan
Thomas Meier
Douglas Chai

June 1999

Acknowledgments

The authors would like to thank Professor K. Aizawa of University of Tokyo, Japan, for the use of the "Makeface" 3-D wireframe synthesis software package, from which some of the images in Chapter 4 are obtained.

Table of Contents

Chapter 1

Image and Video Segmentation

Segmentation plays a crucial role in second-generation image and video coding schemes, as well as in content-based video coding. It is one of the most difficult tasks in image processing, and it often determines the eventual success or failure of a system.

Broadly speaking, segmentation seeks to subdivide images into regions of similar attribute. Some of the most fundamental attributes are luminance, color, and optical flow. They result in a so-called low-level segmentation, because the partitions consist of primitive regions that usually do not have a one-to-one correspondence with physical objects.

Sometimes, images must be divided into physical objects so that each region constitutes a semantically meaningful entity. This higher-level segmentation is generally more difficult, and it requires contextual information or some form of artificial intelligence. Compared to low-level segmentation, far less research has been undertaken in this field.

Both low-level and higher-level segmentation are becoming increasingly important in image and video coding. The level at which the partitioning is carried out depends on the application. So-called second generation coding schemes [1, 2] employ fairly sophisticated source models that take into account the characteristics of the human visual system. Images are first partitioned into regions of similar intensity, color, or motion characteristics. Each region is then separately and efficiently encoded, leading to less artifacts than systems based on the discrete cosine transform (DCT) [3, 4, 5]. The second-generation approach has initiated the development of a significant number of segmentation and coding algorithms [6, 7, 8, 9, 10], which are based on a low-level segmentation.

The new video coding standard MPEG-4 [11, 12], on the other hand, targets more than just large coding gains. To provide new functionalities for future multimedia applications, such as content-based interactivity and content-based scalability, it introduces a content-based representation. Scenes are treated as compositions of several semantically meaningful objects, which are separately encoded and decoded. Obviously, MPEG-4 requires a prior decomposition of the scene into physical objects or so-called video object planes (VOPs). This corresponds to a higher-level partition.

As opposed to the intensity or motion-based segmentation for the second-generation techniques, there does not exist a low-level feature that can be utilized for grouping pixels into semantically meaningful objects. As a consequence, VOP segmentation is generally far more difficult than low-level segmentation. Furthermore, VOP extraction for content-based interactivity functionalities is an unforgiving task. Even small errors in the contour can render a VOP useless for such applications.

This chapter starts with a review of Bayesian inference and Markov random fields (MRFs), which will be needed throughout this chapter. A brief discussion of edge detection is given in Section 1.2, and Section 1.3 deals with low-level still image segmentation. The remaining three sections are devoted to video segmentation. First, an introduction to motion and motion estimation is given in Sections 1.4 and 1.5, before video segmentation techniques are examined in Sections 1.6 and 5.1. For a review of VOP segmentation algorithms, we refer the reader to Chapter 5.

1.1 Bayesian Inference and Markov Random Fields

Bayesian inference is among the most popular and powerful tools in image processing and computer vision [13, 14, 15]. The basis of Bayesian techniques is the famous inversion formula

$$P(X|O) = \frac{P(O|X)P(X)}{P(O)}. \tag{1.1}$$

Although equation (1.1) is trivial to derive using the axioms of probability theory, it represents a major concept. To understand this better, let X denote an unknown parameter and O an observation that provides some information about X. In the context of decision making, X and O are sometimes referred to as hypothesis and evidence, respectively.

$P(X|O)$ can now be viewed as the likelihood of the unknown parameter X, given the observation O. The inversion formula (1.1) enables us to express $P(X|O)$ in terms of $P(O|X)$ and $P(X)$. In contrast to the *posterior probability* $P(X|O)$, which is normally very difficult to establish, $P(O|X)$ and the *prior probability* $P(X)$ are intuitively easier to understand and can usually be determined on a theoretical, experimental, or subjective basis [13, 14]. Bayes' theorem (1.1) can also be seen as an updating of the probability of X from $P(X)$ to $P(X|O)$ after observing the evidence O [14].

1.1.1 MAP Estimation

Undoubtedly, the maximum a posteriori (MAP) estimator is the most important Bayesian tool. It aims at maximizing $P(X|O)$ with respect to X, which is equivalent to maximizing the numerator on the right-hand side of (1.1), because $P(O)$ does not depend on X. Hence, we can write

$$P(X|O) \propto P(O|X)P(X). \tag{1.2}$$

For the purpose of a simplified notation, it is often more convenient to minimize the negative logarithm of $P(X|O)$ instead of maximizing $P(X|O)$ directly. However, this has no effect on the outcome of the estimation. The MAP estimate of X is now given by

$$\begin{aligned} X_{MAP} &= \arg\max_X \{P(O|X)P(X)\} \\ &= \arg\min_X \{-\log P(O|X) - \log P(X)\}. \end{aligned} \tag{1.3}$$

From (1.3) it can be seen that the knowledge of two probability functions is required. The likelihood $P(X)$ contains the information that is available a priori, that is, it describes our prior expectation on X before knowing O. While it is often possible to determine $P(X)$ from theoretical or experimental knowledge, subjective experience sometimes plays an important role. As we will see later, Gibbs distributions are by far the most popular choice for $P(X)$ in image processing, which means that X is assumed to be a sample of a Markov random field (MRF).

The conditional probability $P(O|X)$, on the other hand, defines how well X explains the observation O and can therefore be viewed as an observation model. It updates the a priori information contained in $P(X)$ and is often derived from theoretical or experimental knowledge. For example, assume we wanted to recover the unknown original image X from a blurred image O. The probability $P(O|X)$, which describes the degradation process leading to O, could be determined based on theoretical considerations. To this end, a suitable mathematical model for blurring would be needed.

The major conceptual step introduced by Bayesian inference, besides the inversion principle, is to model uncertainty about the unknown parameter X

by probabilities and combining them according to the axioms of probability theory. Indeed, the language of probabilities has proven to be a powerful tool to allow a quantitative treatment of uncertainty that conforms well with human intuition. The resulting distribution $P(X|O)$, after combining prior knowledge and observations, is then the a posteriori belief in X and forms the basis for inferences.

To summarize, by combining $P(X)$ and $P(O|X)$ the MAP estimator incorporates both the a priori information on the unknown parameter X that is available from knowledge and experience and the information brought in by the observation O [16].

Estimation problems are frequently encountered in image processing and computer vision. Applications include image and video segmentation [16, 17, 18, 19], where O represents an image or a video sequence and X is the segmentation label field to be estimated. In image restoration [20, 21, 22], X is the unknown original image we would like to recover and O the degraded image. Bayesian inference is also popular in motion estimation [23, 24, 25, 26], with X denoting the unknown optical flow field and O containing two or more frames of a video sequence. In all these examples, the unknown parameter X is modeled by a random field.

1.1.2 Markov Random Fields (MRFs)

Without doubt the most important statistical signal models in image processing and computer vision are based on Markov processes [27, 20, 28, 29]. Due to their ability to represent the spatial continuity that is inherent in natural images, they have been successfully applied in various applications to determine the prior distribution $P(X)$. Examples of such Markov random fields include region processes or label fields in segmentation problems [16, 17, 18, 30], models for texture or image intensity [20, 21, 30, 31], and optical flow fields [23, 26].

First, some definitions will be introduced with focus on discrete 2-D random fields. We denote by $L = \{(i,j)|1 \leq i \leq M, 1 \leq j \leq N\}$ a finite $M \times N$ rectangular lattice of sites or pixels. A neighborhood system $\mathcal{N}$ is then defined as any collection of subsets $\mathcal{N}_{i,j}$ of L,

$$\mathcal{N} = \{\mathcal{N}_{i,j}|(i,j) \in L \text{ and } \mathcal{N}_{i,j} \subseteq L\}, \tag{1.4}$$

such that for any pixel (i,j)

$$\begin{aligned} &1) \quad (i,j) \notin \mathcal{N}_{i,j} \text{ and} \\ &2) \quad (k,l) \in \mathcal{N}_{i,j} - (i,j) \in \mathcal{N}_{l,k}. \end{aligned} \tag{1.5}$$

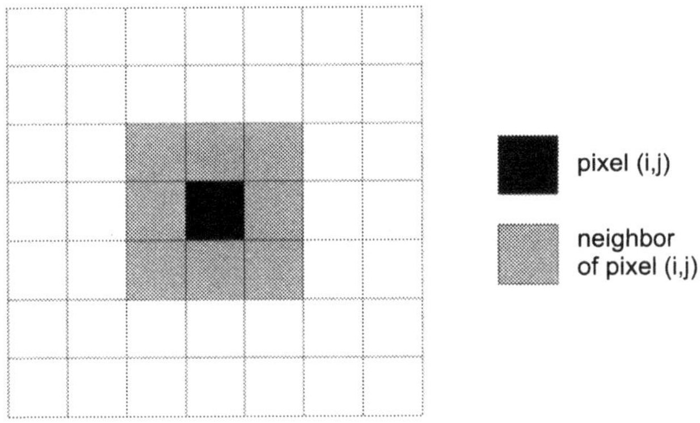

Figure 1.1: Eight-point neighborhood system: pixels belonging to the neighborhood $\mathcal{N}_{i,j}$ of pixel (i, j) are marked in gray.

Generally speaking, $\mathcal{N}_{i,j}$ is the set of neighbor pixels of (i, j).

A very popular neighborhood system is the one consisting of the eight nearest pixels, as depicted in Fig. 1.1. The neighborhood $\mathcal{N}_{i,j}$ for this system can be written as

$$\mathcal{N}_{i,j} = \{(i+h, j+v) \mid -1 \leq h, v \leq 1 \text{ and } (h,v) \neq (0,0)\}, \tag{1.6}$$

whereby boundary pixels and the four corner pixels have only five and three neighbors, respectively. The eight-point neighborhood system is also known as the second-order neighborhood system. In contrast, the first-order system is a four-point neighborhood system consisting of the horizontal and vertical neighbor pixels only.

Now let X be a two-dimensional random field defined on L. Further, let Ω denote the set of all possible realizations of X, the so-called sample or configuration space. Then, X is a Markov random field (MRF) with respect to $\mathcal{N}$ if [20]

$$\begin{aligned} &1) \quad P(X(i,j) \mid X(k,l), \text{ all } (k,l) \neq (i,j)) \\ &\qquad\quad = P(X(i,j) \mid X(k,l), \ (k,l) \in \mathcal{N}_{i,j}) \\ &2) \quad P(X = x) > 0 \text{ for all } x \in \Omega \end{aligned} \tag{1.7}$$

for every $(i, j) \in L$.

The first condition is the well-known Markovian property. It restricts the statistical dependency of pixel (i, j) to its neighbors and thereby significantly reduces the complexity of the model. It is interesting to notice that this condition is satisfied by any random field defined on a finite lattice if the neighborhood is chosen large enough [29]. Such a neighborhood system would, however, not benefit from a reduction in complexity like, for example, a second-order system. The second condition in (1.7), the so-called positivity condition, requires all realizations $x \in \Omega$ of the MRF to have positive probabilities. It is not always included into the definition of MRFs, but it must be satisfied for the Hammersley-Clifford theorem below.

The definition (1.7) is not directly suitable to specify an MRF, but fortunately the Hammersley-Clifford theorem [27] greatly simplifies the specification. It states that a random field X is an MRF if and only if $P(X)$ can be written as a Gibbs distribution[1]. That is,

$$P(X = x) = \frac{1}{Z} \exp\left(-\frac{1}{T} U(x)\right), \qquad \forall x \in \Omega. \tag{1.8}$$

The Gibbs distribution was first used in physics and statistical mechanics. Best known is the Ising Model, which was proposed to model the magnetic properties of ferromagnetic materials [33].

Due to the analogy with physical systems, $U(x)$ is called the energy function and the constant T corresponds to temperature. For high temperatures T, the system is "melted" and all realizations $x \in \Omega$ are more or less equally probable. At low temperatures, on the other hand, the system is forced to be in a state of low energy. Thus, in accordance with physical systems, low energy levels correspond to a high likelihood and vice versa. The so-called partition function Z is a normalizing constant and usually does not have to be evaluated.

The energy function $U(x)$ in (1.8) can be written as a sum of potential functions $V_C(x)$:

$$U(x) = \sum_{\text{all cliques } C} V_C(x). \tag{1.9}$$

A clique C is defined as a subset $C \subset L$ that contains either a single pixel or several pixels that are all neighbors of each other. Note that the neighborhood system $\mathcal{N}$ determines exactly what types of cliques exist. For example, all possible types of cliques for the eight-point neighborhood system in Fig. 1.1 are illustrated in Fig. 1.2.

The clique potential $V_C(x)$ in (1.9) represents the potential contributed by clique C to the total energy $U(x)$ and depends only on the pixels belonging to C. It follows that the energy function $U(x)$, and therefore the

[1] sometimes called a Boltzmann-Gibbs distribution [32]

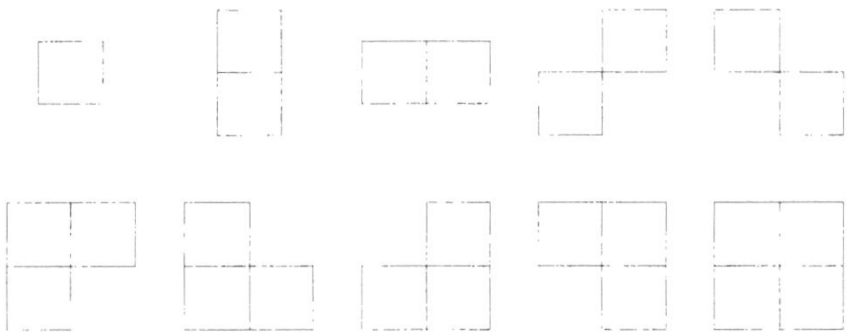

Figure 1.2: All possible types of cliques C associated with the eight-point neighborhood system $\mathcal{N}$ shown in Fig. 1.1.

likelihood $P(X)$, consists of contributions from local interactions within cliques. This conforms with the Markovian property of X in (1.7), where pixels are statistically depending only on their neighbors.

This section is concluded with an example of a simple but very popular clique potential function [17]. Consider a segmentation label field X such that $X(i,j) = q$ means pixel (i,j) is assigned to region q. In this example, only the two-point cliques in Fig. 1.2 are used, consisting of pairs of horizontally, vertically, and diagonally adjacent pixels. Our intuition tells us that such two adjacent pixels are very likely to carry the same label q. Hence, the two-point clique potential $V_C(x)$ could be defined as

$$V_C(x) = \begin{cases} -\beta, & \text{if } x(i,j) = x(k,l) \text{ and } (i,j),\ (k,l) \in C \\ +\beta, & \text{if } x(i,j) \neq x(k,l) \text{ and } (i,j),\ (k,l) \in C \end{cases} \tag{1.10}$$

By choosing a positive value for β, a large potential or low probability is assigned to two neighbor pixels (i,j) and (k,l) if they belong to different regions. On the other hand, neighbor pixels that are member of the same region correspond to a high probability.

This example demonstrates how easily clique potentials can be specified, guaranteeing that the resulting likelihood $P(X)$ is a Gibbs distribution and therefore X is a Markov random field.

1.1.3 Numerical Approximations

Finding the MAP estimate X_{MAP} in (1.3) can be viewed as a combinatorial optimization problem [34]. Let Ω be the set of all possible realizations of X, the so-called configuration space. The function $-\log P(O|X) - \log P(X)$

in (1.3) then defines a cost function of many variables that must be minimized, i.e., we would like to find the configuration $x_{opt} \in \Omega$ for which the cost takes its minimum value. In other words, once the distributions $P(O|X)$ and $P(X)$ are defined, our estimation problem becomes that of minimizing a cost function.

The large dimensionality of the unknown parameter X and the presence of local minima make it normally very difficult to find x_{opt}. For instance, if X is a 256×256 image with 256 gray-levels, the set Ω contains $256^{256\times256} = 2^{16,777,216}$ possible realizations, requiring a prohibitive amount of computation time to search for x_{opt}. Consequently, we are forced to settle for an approximation of the optimum solution.

1.1.3.1 Simulated Annealing

Simulated annealing (SA), which is also known as stochastic relaxation or Monte Carlo annealing, is an optimization technique that solves the combinatorial optimization problem by a partially random search of the configuration space Ω. It is based on the algorithm proposed by Metropolis *et al.* [35] to simulate the interactions between molecules in solids and their evolution to thermal equilibrium.

Metropolis Algorithm

Kirkpatrick *et al.* [36] and Černy [32] first recognized the connection between combinatorial optimization problems and statistical mechanics. The goal of combinatorial optimization is to minimize a function that depends on a large number of variables, whereas statistical mechanics analyzes systems consisting of a large number of atoms or molecules and aims at finding the lowest energy states.

For instance, to obtain the state of lowest energy of a substance, the substance could be melted and then gradually cooled down. The temperature must be lowered slowly to allow the substance to approach equilibrium and to avoid defects in the resulting crystals. Once the equilibrium has been reached, there will still be random changes of the state from one configuration to another. However, the probability that the substance is in a certain state x is then given by the Boltzmann-Gibbs distribution (1.8), whereby $U(x)$ is the energy of the configuration x. Notice that if the temperature is $T = 0$, the substance must be in a state of lowest energy.

To study these equilibrium properties for very large numbers of interacting atoms or molecules, Metropolis *et al.* proposed an iterative algorithm [35]. The annealing process is simulated by a Monte Carlo method [37] that generates a sequence of random samples so that the equilibrium state at a given temperature T is reached.

This algorithm can also be applied to our combinatorial optimization problem by replacing the energy with the cost function [32, 36]. The global minimum of the cost function then corresponds to the lowest energy ground state of the solid. Starting off from an arbitrary initial configuration $x^{(0)} \in \Omega$, a new candidate solution $x^{(n+1)}$ is generated in each iteration at random. The perturbation must be small so that $x^{(n+1)}$ is in the neighborhood of $x^{(n)}$. The new candidate is then accepted if it decreases the cost function. However, uphill moves that increase the cost function are also possible on a random basis to prevent the search getting trapped in a local minimum. The probability of accepting such a new candidate depends on the threshold $\exp(-\frac{\Delta Cost}{T})$, which is derived from the Boltzmann distribution. It is controlled by the temperature parameter T. Initially, the temperature T is very high so that nearly all uphill moves are accepted, but T is gradually lowered until the system reaches a steady-state and is frozen.

The *Metropolis algorithm* applied to the combinatorial optimization problem can be summarized as:

1. `Initialization:` $n = 0$, $T = T_{max}$ `(system is ''melted'');` `select an initial` $x^{(0)}$ `at random.`
2. `Generate new candidate` $x^{(n+1)}$ `at random by a small perturbation of` $x^{(n)}$.
3. `Compute` $\Delta Cost = Cost(x^{(n+1)}) - Cost(x^{(n)})$.
4. (a) $\Delta Cost \leq 0$: `accept` $x^{(n+1)}$.
   (b) $\Delta Cost > 0$: `draw a random number` P, `uniformly distributed between 0 and 1. If` $P < \exp(-\frac{\Delta Cost}{T})$ `then accept` $x^{(n+1)}$, `otherwise keep` $x^{(n)}$.
5. $n = n + 1$; `if` $n \leq I_{max}$ `then go to 2.`
6. `Equilibrium is approached sufficiently closely: reduce` T `according to an annealing schedule;` $n = 0$, $x^{(0)} = x^{(I_{max})}$; `if` $T > T_{min}$ `then go to 2.`
7. `System is frozen: STOP.`

The definition of "small" perturbation in step 2 depends on the particular optimization problem [32]. One possibility is to change the value at one site at a time, while leaving all other pixels unchanged. This is exactly the approach taken by the Gibbs sampler, which we will describe in the following.

Gibbs Sampler

The Gibbs sampler is a stochastic relaxation method introduced by Geman and Geman [20]. It is based on the idea of the Metropolis algorithm and was proposed to compute the MAP estimate in an image restoration problem, although this technique is not restricted to that type of application.

To obtain the MAP estimate (1.3), X is assumed to be a sample of an MRF so that $P(X)$ is a Gibbs distribution, whereas the conditional probability $P(O|X)$ is modeled by white Gaussian noise. The latter assumption has been successfully used in countless applications in image processing, because it often leads to solutions that can easily be implemented while giving satisfactory results. Both $P(X)$ and $P(O|X)$ are then exponential distributions and so will be their product. As a result, the posterior probability $P(X|O) \propto P(O|X)P(X)$ will be a Gibbs distribution as well. It is possible to extend the observation distribution $P(O|X)$ to more sophisticated models [20], but for reasons of computational efficiency it is important that the resulting posterior probability $P(X|O)$ is a Gibbs distribution.

In each iteration, the Gibbs sampler replaces one pixel (i,j) at a time. This change is random in accordance with the idea of the Metropolis algorithm, and is generated by sampling from a local conditional probability distribution. The new value for $X(i,j)$ is, however, not completely randomly chosen. Instead, the current values of the pixels in the neighborhood of (i,j) are taken into account. The more likely a value $X(i,j)$, given all available information, the more likely it will be selected.

To this end, the Gibbs sampler evaluates the local conditional probability distribution

$$P(X(i,j) \mid O,\ X(k,l),\ \text{all } (k,l) \neq (i,j))$$

for each possible value of $X(i,j)$. This is the probability of the value $X(i,j)$, given the observation O and the current values of all other pixels. It is easy to show that this probability only depends on the values of X and O in the neighborhood of (i,j) due to the Markovian property of $P(X|O)$. These local conditional probabilities are therefore easy to compute. Note that depending on the observation model, $P(O|X)$, this neighborhood might be larger than that of the prior distribution $P(X)$.

The likelihood of selecting a particular value for $X(i,j)$ is now proportional to its local conditional probability. To illustrate this, suppose $X(i,j)$ can take on four values, denoted by $X(i,j) \in \{0,1,2,3\}$. The drawing of a new value for $X(i,j)$ is then performed as follows. Firstly, compute $P(X(i,j) \mid O,\ X(k,l),\ \text{all } (k,l) \neq (i,j))$ for all possible values of $X(i,j)$. In our example, let these probabilities be 0.1, 0.5, 0.25, and 0.15 for $X(i,j) = 0,\ 1,\ 2,$ and 3, respectively. Then, a random number that is uniformly distributed between 0 and 1 is generated. If this random number falls into the range $[0 \ldots 0.1)$, then $X(i,j)$ will be assigned the new value 0. Accordingly, the ranges $[0.1 \ldots 0.6)$, $[0.6 \ldots 0.85)$, and $[0.85 \ldots 1)$ will lead to a new value of 1, 2, and 3, respectively. Thus, the interval lengths are equal to the conditional probabilities.

As mentioned above, one pixel is perturbed in each iteration. Pixels can be visited in any order, provided each pixel is visited infinitely often[2]. Since $P(X|O)$ is a Gibbs distribution, the conditional probability $P(X(i,j) \mid O,\ X(k,l),\ \text{all } (k,l) \neq (i,j))$ depends on a temperature parameter T. At the beginning, this temperature is high so that transitions will occur almost uniformly over the set of possible values for $X(i,j)$. As T is gradually lowered, it becomes more likely that values for $X(i,j)$ will be chosen which decrease the cost function.

The choice of the annealing schedule is enormously important. If the temperature T is decreased sufficiently slowly, the Gibbs sampler will be able to reach the global minimum. It was shown in [20] that if for every iteration n the temperature $T(n)$ satisfies

$$T(n) \geq \frac{T_{max}}{\log(1+n)} \tag{1.11}$$

with the constant T_{max}, then the solution $x^{(n)}$ after the nth iteration will converge to the global minimum as $n \to \infty$. Should there be multiple minima, $x^{(n)}$ will be uniformly distributed over those values of X that take on the global minimum. Notice that the constant T_{max} must be selected appropriately [20].

Unfortunately, the annealing schedule (1.11) is normally too slow for practical applications. Therefore, a faster schedule is often preferred to reduce the computational burden, although there is no longer any guarantee that a global minimum will be obtained. Furthermore, the solution will become dependent on the initial configuration $x^{(0)}$.

1.1.3.2 Deterministic Algorithms

The simulated annealing techniques are able to find the global minimum of the cost function, but a major drawback is their computational complexity.

[2]in practice, a suitably large number is sufficient

This often makes their application impossible in practical situations. Faster convergence can be accomplished by *deterministic algorithms* such as iterated conditional modes (ICM) [21] and highest confidence first (HCF) [16].

Iterated Conditional Modes (ICM)

As a computationally efficient alternative to the Gibbs sampler, Besag proposed the iterated conditional modes (ICM) algorithm, which belongs to the category of deterministic approximation methods. ICM, which is also known as the greedy algorithm, improves the estimate of X iteratively by updating one pixel at a time. Unlike the Gibbs sampler, only perturbations yielding a lower energy or higher probability of the configuration X are permitted. Hence, only downhill moves are allowed in contrast to simulated annealing. This makes ICM converge significantly faster, but at the cost of settling in a local minimum of the cost function.

Consider an image restoration problem where O denotes the degraded image and X the unknown original image to be estimated. Typically, X is assumed to be a sample of an MRF and therefore $P(X)$ is a Gibbs distribution. The degradation is modeled as zero-mean independent and identically distributed (i.i.d.) white Gaussian noise with variance σ^2 such that

$$P(O|X) = \prod_{\text{all } (i,j)} f(O(i,j)|X(i,j)) \tag{1.12}$$

with

$$f(O(i,j)|X(i,j)) = \frac{1}{\sqrt{2\pi\sigma^2}} \exp\left(-\frac{(O(i,j) - X(i,j))^2}{2\sigma^2}\right). \tag{1.13}$$

Similarly to the Gibbs sampler, the update of pixel (i,j) is based on the local conditional probability $P(X(i,j) \mid O,\ X(k,l),\ \text{all } (k,l) \neq (i,j))$. However, in ICM $X(i,j)$ is set to the value that maximizes this conditional probability. It is easy to show that due to the Markovian property of $P(X)$ and the whiteness of the noise in $P(O|X)$ the following relation holds

$$\begin{aligned} &P(X(i,j) \mid O,\ X(k,l),\ \text{all } (k,l) \neq (i,j)) \\ &\propto f(O(i,j)|X(i,j)) \cdot P(X(i,j) \mid X(k,l),\ (k,l) \in \mathcal{N}_{i,j}). \end{aligned} \tag{1.14}$$

Together with (1.8), (1.9), (1.12) and (1.13) we then arrive at

$$\begin{aligned} &P(X(i,j) \mid O,\ X(k,l),\ \text{all } (k,l) \neq (i,j)) \\ &\propto \exp\left(-\frac{(O(i,j) - X(i,j))^2}{2\sigma^2} - \frac{1}{T} \sum_{C \in C_{i,j}} V_C(X)\right). \end{aligned} \tag{1.15}$$

$C_{i,j}$ denotes the set of all cliques that contain the pixel (i,j). Thus, the local conditional probability only depends on $X(i,j)$, $O(i,j)$ and the neighbors of (i,j) in $\mathcal{N}_{i,j}$.

ICM can now be summarized as follows. Starting from an initial configuration, the estimate is iteratively improved by visiting and updating pixels in a raster scan order. For each pixel (i,j) in turn, $X(i,j)$ is replaced by the value that maximizes the conditional probability

$$P(X(i,j) \mid O,\ X(k,l),\ \text{all } (k,l) \neq (i,j)).$$

Hence, the value at (i,j) is replaced by the most likely $X(i,j)$, given all available information, which are the observation O and the current values of all other pixels. The algorithm then terminates after a prescribed number of iterations or when the estimated configuration X does not change anymore. The latter happens when a local minimum has been reached.

ICM can be regarded as a special case of the Gibbs sampler with constant temperature $T = 0$. Consequently, the cost is decreased by each replacement operation, and the algorithm converges much faster. However, ICM will terminate in a local minimum since no uphill moves are possible. The cost associated with the local minimum depends heavily on the initial estimate for X and might be far higher than that of the global minimum.

Apart from the initial estimate, the order in which pixels are visited has an effect on the result. The raster scan order that is commonly used has the undesirable property of propagating pixel values in the direction of the scan order, because the Gibbs distribution encourages adjacent pixels to have similar values.

Highest Confidence First (HCF)

Another deterministic numerical approximation method is highest confidence first (HCF) by Chou and Brown [16]. HCF is an iterative algorithm like ICM or the simulated annealing approaches, however, the number of visited pixels per iteration normally declines with each iteration. For each pixel in turn, HCF maximizes the conditional probability

$$P(X(i,j) \mid O,\ X(k,l),\ \text{all } (k,l) \neq (i,j))$$

in a similar way to ICM. In particular, no uphill moves are allowed, and consequently HCF will converge to a local minimum.

Nevertheless, HCF overcomes, at least partially, two of the problems associated with ICM - the order in which pixels are visited depends on the reliability of the available information, and no initial estimate is required.

To this end, the configuration space Ω is augmented by an additional label, the so-called *uncommitted* state.

Initially, all pixels are labeled as uncommitted. During the estimation process pixels will become committed, which means they will have a value assigned that is different from the *uncommitted* label. Once a pixel has committed itself to a label, it cannot go back to the uncommitted state, but it is allowed to change its label if required.

Rather than following a raster scan order, it would naturally be preferable to update first those pixels for which we are very confident about the change. HCF visits pixels in the order of confidence so that the most confident site will be updated first. Before defining confidence, consider the local conditional probability in (1.15). Obviously, this is a Gibbs distribution with the energy function

$$U_{i,j}(X(i,j)) = T\frac{(O(i,j) - X(i,j))^2}{2\sigma^2} + \sum_{C \in C_{i,j}} V_C(X), \tag{1.16}$$

where $C_{i,j}$ is the set of cliques that contain the pixel (i,j). Since unreliable pixels should not affect reliable pixels, the potential $V_C(X)$ is set to zero for all cliques C that contain one or more pixels that are still uncommitted. The resulting function $U_{i,j}(X(i,j))$ is referred to as the local energy at site (i,j). It is easy to see that a low local energy corresponds to a high likelihood of the value $X(i,j)$ and vice versa.

The confidence $c(i,j)$ of a committed site (i,j) is now defined as the difference between the current local energy and the minimum local energy. That is,

$$c(i,j) = \begin{cases} U_{i,j}(X(i,j)) - \min_l U_{i,j}(l), & \text{if } (i,j) \text{ committed, and} \\ \min_{l \neq k}\left(U_{i,j}(l) - \min_k U_{i,j}(k)\right), & \text{if } (i,j) \text{ uncommitted.} \end{cases} \tag{1.17}$$

Roughly speaking, a positive value of $c(i,j)$ indicates that a more stable (lower energy) estimate X will result if the value at (i,j) is changed from $X(i,j)$ to l. The larger $c(i,j)$, the more confident we are about the change. Further, notice that the confidence of uncommitted pixels is always positive.

HCF visits pixels in the order of decreasing confidence. The current value $X(i,j)$ of the visited site (i,j) is replaced by the value that maximizes the local conditional probability $P(X(i,j) \mid O,\ X(k,l),\ \text{all } (k,l) \neq (i,j))$, which is equivalent to minimizing the local energy $U_{i,j}(X(i,j))$. Immediately after the update of pixel (i,j), the confidence of the corresponding site will obviously be zero. However, if a neighbor of (i,j) gets updated, the confidence $c(i,j)$ might become positive again. This means that (i,j) would be visited again as soon as no other pixel with a higher confidence is left. The algorithm finally terminates when there are no pixels remaining with a positive value for the confidence $c(i,j)$.

For an efficient implementation of the HCF algorithm using a heap structure we refer to [16]. Generally, the results obtained by HCF are better than those of ICM, although both algorithms converge to local minima. In addition, HCF is more flexible than ICM, because it does not require an initial estimate. The price to be paid is a slight increase in computational complexity. Nevertheless, HCF is still much faster than the simulated annealing approaches.

1.2 Edge Detection

Often, segmentation techniques are classified into two categories [38]. In the first category, images are partitioned based on discontinuities or edges, whereas the second category groups pixels based on similarity. Only segmentation algorithms of the second category will be considered, because they promise to yield more useful results. Discontinuities detected by an edge operator seldom form connected contours. Consequently, an edge linking procedure must be employed to obtain a partition, which is tedious and often even more difficult than the actual task of segmentation. Indeed, most segmentation techniques nowadays are based on a similarity measure.

Nevertheless, a brief introduction to edge detection is given. Even though edge-linking will not be used to obtain the partitions, the information contained in gray-level or color discontinuities can be very useful for segmentation, as we will see later in Chapter 5.

Edges in an image are normally characterized by an anisotropic, abrupt change in luminance. Therefore, examining images by differentiating the luminance function appears to be the way to go. Let $I(x,y)$ be the luminance or gray-level of a discrete image at pixel (x,y). Since luminance is a discrete function, the simplest edge operators are obtained by replacing differentiation with discrete differences. For instance, the partial derivative $\frac{\partial I}{\partial x}$ would then become

$$\frac{\partial I}{\partial x} \to \frac{1}{2}(I(x+1,y) - I(x-1,y)). \tag{1.18}$$

Unfortunately, the success of this approach is limited, particularly in the presence of noise.

1.2.1 Gradient Operators: Sobel, Prewitt, Frei-Chen

The edge operator proposed by Sobel [39] is significantly more robust than the simple differencing in (1.18). To enable a proper differentiation of the luminance function at pixel (x_0, y_0), the discrete image $I(x, y)$ is replaced by an analytical function $\tilde{I}(x, y; x_0, y_0)$, which approximates $I(x, y)$ in the neighborhood of (x_0, y_0). That is, a linear function $\tilde{I}(x, y; x_0, y_0)$,

$$\tilde{I}(x, y; x_0, y_0) = a_0(x - x_0) + a_1(y - y_0) + a_2, \tag{1.19}$$

is fitted to the image $I(x, y)$ about pixel (x_0, y_0). Then, the partial derivatives at (x_0, y_0) are given by

$$\left.\frac{\partial I}{\partial x}\right|_{(x_0,y_0)} \approx \left.\frac{\partial \tilde{I}}{\partial x}\right|_{(x_0,y_0)} = a_0 \quad \text{and} \quad \left.\frac{\partial I}{\partial y}\right|_{(x_0,y_0)} \approx \left.\frac{\partial \tilde{I}}{\partial y}\right|_{(x_0,y_0)} = a_1. \tag{1.20}$$

Thus, the gradient $\nabla I(x_0, y_0) \approx (a_0, a_1)$ is obtained by finding the corresponding model parameters a_0 and a_1. These parameters are determined for each pixel (x_0, y_0) by minimizing

$$\Phi(a_0, a_1, a_2) = \sum_{x=x_0-1}^{x_0+1} \sum_{y=y_0-1}^{y_0+1} \Big(I(x, y) - \tilde{I}(x, y)\Big)^2 \cdot w(x - x_0, y - y_0) \tag{1.21}$$

with respect to a_0, a_1, and a_2. The function $\Phi(a_0, a_1, a_2)$ in (1.21) is the weighted quadratic error between the image $I(x, y)$ and the linear fit $\tilde{I}(x, y)$ in a 3×3 neighborhood centered at (x_0, y_0). The weights $w(x - x_0, y - y_0)$ take into account the different Euclidean distances of horizontal, vertical and diagonal neighbors. Sobel suggested the values

$$\begin{aligned} w(-1,0) = w(1,0) = w(0,-1) = w(0,1) = 2 \\ w(-1,-1) = w(-1,1) = w(1,-1) = w(1,1) = 1 \end{aligned} \tag{1.22}$$

for these weights; that is, the weight for diagonal neighbors is half of that for horizontally and vertically adjacent pixels. Notice that $w(0, 0)$ is not needed for the computation of a_0 and a_1.

The function $\Phi(a_0, a_1, a_2)$ is minimized by setting the derivatives $\frac{\partial \Phi}{\partial a_i}$ to zero for $i \in \{0, 1, 2\}$, leading to three equations in three unknowns. It is then easy to show that

$$a_0 = \frac{1}{8}\{\ I(x_0+1, y_0-1) - I(x_0-1, y_0-1) \\ + 2[I(x_0+1, y_0) - I(x_0-1, y_0)] \\ + I(x_0+1, y_0+1) - I(x_0-1, y_0+1)\ \} \tag{1.23}$$

and

$$a_1 = \frac{1}{8}\{\ I(x_0-1, y_0+1) - I(x_0-1, y_0-1) \\ + 2[I(x_0, y_0+1) - I(x_0, y_0-1)] \\ + I(x_0+1, y_0+1) - I(x_0+1, y_0-1)\ \}. \tag{1.24}$$

Hence, the parameters a_0 and a_1 are the result of a discrete convolution with the filters

$$h_0(k,l) = \frac{1}{8}\begin{bmatrix} -1 & -2 & -1 \\ 0 & 0 & 0 \\ 1 & 2 & 1 \end{bmatrix} \quad \text{and} \quad h_1(k,l) = \frac{1}{8}\begin{bmatrix} -1 & 0 & 1 \\ -2 & 0 & 2 \\ -1 & 0 & 1 \end{bmatrix}, \tag{1.25}$$

respectively. These filter masks are commonly known as the Sobel operator. Notice that the factors $\frac{1}{8}$ in (1.25) simply represent a scaling, and they are usually omitted.

By selecting different weights $w(\cdot,\cdot)$ in (1.22), other well-known gradient operators for edge detection are derived, such as the Prewitt operator [40] and the Frei-Chen operator [41].

1.2.2 Canny Operator

The gradient operators in Section 1.2.1 are probably the simplest and therefore fastest edge operators that are practical. However, the ingenious optimization approach by Canny has led to an edge operator that is widely considered to be the best edge detector [42].

Canny first defines three criteria that an ideal edge detector should meet. These are good detection, good localization, and only one response to a single edge. The first criterion requires the edge operator to have a low probability both for missing real edges and for false alarms. Good localization means that the detected edges should be as close as possible to the center of the true edge. The third and last criterion makes sure that a single edge does not result in multiple detected edges, particularly in the case of thick edges.

Edge detection is then formulated as a filter design problem. To this end, a mathematical form of encapsulating the above criteria is derived. Canny considers a one-dimensional edge of known cross-section with additive white Gaussian noise. This one-dimensional signal is convolved with a filter so that the center of the edge corresponds to a local maximum in the filter output. The objective is now to find a filter that yields the best performance with respect to the three criteria.

The optimal filters for different types of edges are derived using numerical optimization. Furthermore, it is shown that the impulse response of the optimal step edge operator can be approximated by the first derivative of a Gaussian function.

The mathematics behind the whole optimization process is rather tedious. However, the optimal edge detector turns out to have a surprisingly simple approximate implementation: edges are detected by smoothing the image with a Gaussian low-pass filter and identifying maxima in the gradient magnitude of the smoothed image. The low-pass filtering prior to calculating the gradients significantly contributes to a reduction in noise sensitivity of the Canny edge detector.

1.2.2.1 Implementation

Following the proposed approximation of the optimal edge detector, the Canny operator could be implemented as follows. Firstly, the input image $I(x, y)$ is smoothed by an isotropic Gaussian filter to reduce the effects of noise. The filter coefficients are given by

$$h(k,l) = Z^{-1} \exp\left(-\frac{k^2 + l^2}{2\sigma^2}\right), \tag{1.26}$$

where Z is a normalizing constant. For example, good results for different types of images can be obtained by setting the filter width to 6σ ($[-3\sigma \ldots 3\sigma]$) with $\sigma = 1$. This means that the filter support is given by $-3 \leq k, l \leq 3$. Notice that (1.26) is a separable filter and can therefore be efficiently implemented.

The next step is to calculate the gradient of the smoothed image $\tilde{I}(x, y)$. For that, the derivatives of $\tilde{I}(x, y)$ are calculated in horizontal, vertical, and the two diagonal directions. Since $\tilde{I}(x, y)$ is a discrete function, the derivatives are approximated by differences:

$$\begin{aligned}
\Delta\tilde{I}_{hor}(x,y) &= \{\tilde{I}(x,y+1)-\tilde{I}(x,y-1)\}/2 \\
\Delta\tilde{I}_{ver}(x,y) &= \{\tilde{I}(x+1,y)-\tilde{I}(x-1,y)\}/2 \\
\Delta\tilde{I}_{diag1}(x,y) &= \{\tilde{I}(x+1,y-1)-\tilde{I}(x-1,y+1)\}/(2\sqrt{2}) \\
\Delta\tilde{I}_{diag2}(x,y) &= \{\tilde{I}(x+1,y+1)-\tilde{I}(x-1,y-1)\}/(2\sqrt{2}).
\end{aligned} \tag{1.27}$$

The use of four derivatives instead of two (for example, the horizontal and vertical derivatives) leads to more robust results, because more edge orientations are examined. The gradient magnitude $|\nabla\tilde{I}(x,y)|$ is then defined as the maximum value of the four differences in (1.27), i.e.,

$$\begin{aligned}
|\nabla\tilde{I}(x,y)| \triangleq \max\{\ &|\Delta\tilde{I}_{hor}(x,y)|, |\Delta\tilde{I}_{ver}(x,y)|, \\
&|\Delta\tilde{I}_{diag1}(x,y)|, |\Delta\tilde{I}_{diag2}(x,y)|\ \}.
\end{aligned} \tag{1.28}$$

The gradient angle or direction, $\arg(\nabla\tilde{I}(x,y))$, is obtained in a conventional way from the horizontal and vertical derivatives $\Delta\tilde{I}_{hor}(x,y)$ and $\Delta\tilde{I}_{ver}(x,y)$ using the arctan function.

In many applications, a binary edge image is needed where each pixel is classified as edge or non-edge. Such an edge image is easily computed from the gradient image by thresholding the magnitude $|\nabla\tilde{I}(x,y)|$, as illustrated in Fig. 1.3. However, this often leads to undesired thick edges that must be removed (see Fig. 1.3 (c)).

To this end, an edge-thinning technique called *non-maximum suppression* can be applied. Each edge pixel (x,y) is tested to determine whether the gradient magnitude is a local maximum in the direction of the maximum difference as given by (1.28). If it is a local maximum, the pixel will be finally classified as edge; otherwise it is a non-edge pixel.

For example, suppose the vertical distance $\Delta\tilde{I}_{ver}(x,y)$[3] achieves the maximum value among the four distances in (1.27). Consequently, the gradient magnitude $|\nabla\tilde{I}(x,y)|$ would be set to $|\Delta\tilde{I}_{ver}(x,y)|$. Furthermore, the non-maximum suppression technique would have to compare the gradient magnitude of (x,y) with that of its two vertical neighbors. Thus, pixel (x,y) would be classified as an edge if and only if $|\nabla\tilde{I}(x,y)| > |\nabla\tilde{I}(x-1,y)|$ and $|\nabla\tilde{I}(x,y)| > |\nabla\tilde{I}(x+1,y)|$. The edge thinning effect of the non-maximum suppression method is clearly illustrated in Fig. 1.3 (d).

All in all, the Canny operator has several strengths. It is less sensitive to noise than other edge detectors [39, 40, 41, 43], and detected edge pixels tend to form connected edges rather than being isolated.

---

[3] Note that the x-coordinate corresponds to the row and the y-coordinate to the column in the image, respectively.

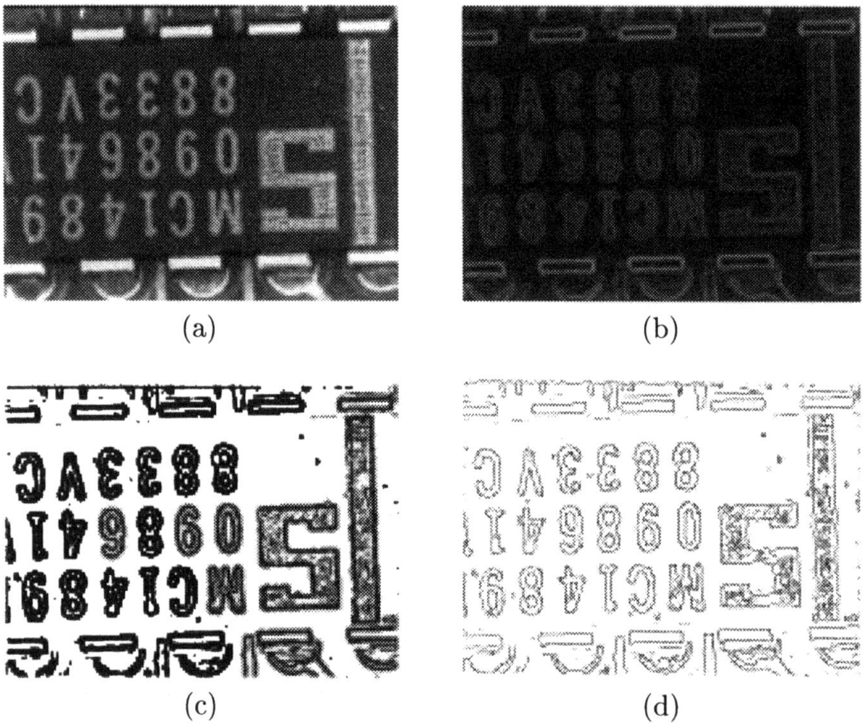

Figure 1.3: Canny edge detector [42]: (a) Original image *chip* and (b) corresponding gradient magnitude according to (1.28). (c) Binary edge image after thresholding the gradient magnitude in (b), and (d) final edge image obtained after non-maximum suppression.

1.3 Image Segmentation

Segmenting images or video sequences into regions that somehow go together is generally the first step in image analysis and computer vision, as well as for second-generation coding techniques. Unsupervised segmentation is certainly one of the most difficult tasks in image processing. The ongoing research in this field and the vast number of proposed approaches and algorithms, without offering a really satisfactory solution, are clear indicators of the difficulties.

The famous introduction by Haralick and Shapiro, which summarizes what a good image segmentation should be like [44], is a good starting point: "Regions of an image segmentation should be uniform and homogeneous with respect to some characteristic such as gray tone or texture. Region interiors should be simple and without many small holes. Adjacent regions of a segmentation should have significantly different values with respect to the characteristic on which they are uniform. Boundaries of each segment should be simple, not ragged, and must be spatially accurate."

Notice that the characteristic or similarity measure is a low-level feature such as color, intensity, or optical flow. Therefore, apart from very simple cases where the features directly correspond to objects, the resulting partitions do not have any semantical meaning attached to them. An interpretation of the scene must be obtained by a higher-level process, after the segmentation into primitive regions has been carried out.

A complete coverage of all the different image segmentation approaches would be far beyond the scope of this book. Some of the best known segmentation techniques, although not necessarily the best ones, are region growing [45, 46], thresholding [47, 48, 49], split-and-merge [50, 51, 52], and algorithms motivated by graph theory [53, 54]. There exist also introductory texts and papers on segmentation [38, 44, 55] that usually cover some of these simple methods. This book will concentrate on two approaches which have grown in popularity over the last few years; these are *morphological* and *Bayesian segmentation*. They both have in common that they are based on a sound theory.

Morphology refers to a branch of biology that is concerned with the form and structure of animals and plants. In image processing and computer vision, mathematical morphology denotes the study of topology and structure of objects from images. It is also known as a shape-oriented approach to image processing, in contrast to, for example, frequency-oriented approaches.

Mathematical morphology owes a lot of its popularity to the work by Serra [56], who developed much of the early foundation. The major strength of morphological segmentation is the elegant separation of the initialization step, the so-called marker extraction, from the decision step, where all pixels are labeled by the watershed algorithm. On the negative side is the lack of constraints to enforce spatial continuity on the segmentation.

Bayesian segmentation algorithms perform a maximum a posteriori (MAP) estimation of the unknown partition. For that purpose, segmentation label fields and images are assumed to be samples of two-dimensional random fields. Label fields are usually modeled as Markov random fields (MRFs). Although the use of MRFs to describe spatial interactions in physical systems can be traced back to the Ising model in the 1920s [33], it took until 1974 before MRFs became more practical [27]. Thanks to the Hammersley-

Clifford theorem, which states the duality of MRFs and Gibbs random fields, it became possible to specify MRFs by means of simple clique potential functions (see Section 1.1.2). With the increase in available computing power, the popularity of Bayesian segmentation techniques started growing rapidly in the 1980s.

A clear advantage of Bayesian segmentation methods over morphological techniques is the incorporation of spatial continuity constraints. On the other hand, the need for an initial estimate and the strong dependency of the resulting partitions on the infamous input parameter K, specifying the number of labels to be used, are some of its shortcomings.

1.3.1 Morphological Segmentation

Mathematical morphology is a shape-oriented approach to signal processing. In the context of image processing and computer vision, it provides useful tools for image simplification, segmentation and coding [57, 58, 59, 60, 61]. In particular, the watershed algorithm and simplification filters have become increasingly popular for segmentation and coding. Here, we are mainly concerned with the application of morphology to image and video sequence segmentation.

A typical morphological segmentation technique consists of three main steps: image simplification, marker extraction, and watershed algorithm [58, 61]. Firstly, the image is simplified by removing small dark and bright patches using a so-called morphological filter by reconstruction. The following marker extraction step then selects initial regions, for instance, by identifying large regions of constant gray-level. Based on these initial regions, the watershed algorithm labels pixels in a similar fashion to region growing techniques.

The separation of the feature or marker extraction step from the decision step, the watershed algorithm, is a major strength of morphological approaches.

1.3.1.1 Connected Operators

Before discussing filters by reconstruction, we must introduce a few definitions. To this end, we closely follow the notation in [58, 60, 62]. Mathematical morphology was originally applied to binary images and was only later extended to gray-level images. As a result, there are often separate definitions for the two cases. However, binary images can be viewed as a special case of images with two gray-levels. Therefore, we will here only consider gray-level operators.

As in Section 1.1.2, let $L = \{(x, y)|1 \leq x \leq M, 1 \leq y \leq N\}$ denote a finite rectangular lattice of $M \times N$ pixels so that the gray-level image $I(x, y)$ is defined on L. A partition $A = \{A_1, \ldots, A_m\}$ of L is then the set of disjoint connected components A_i such that the union of these components is equal to L; that is, $\cup_{i=1}^{m} A_i = L$.

Furthermore, a partition $A = \{A_1, \ldots, A_m\}$ is finer than another partition $B = \{B_1, \ldots, B_n\}$ if any pair of pixels belonging to the same component A_i also belongs to the same component B_j for some $j \in \{1 \ldots n\}$.

An important concept regarding filters by reconstruction is the partition of *flat zones* of image I. This is defined as the set of the largest connected components where the gray-level is constant. Some of these flat zones might consist of only one pixel. Thus, all pixels that belong to the same flat zone must have the same gray-level. Moreover, two flat zones which are neighbors of each other must have different gray-levels. It is easy to verify that the set of flat zones is indeed a partition of the image.

Finally, a connected operator Ψ for gray-level images I is an operator such that the partition of flat zones of I is finer than the partition of flat zones of $\Psi(I)$. In other words, connected operators process image I by merging flat zones of I [60].

1.3.1.2 Image Simplification Using "Filters by Reconstruction"

Some of the most powerful morphological tools are filters by reconstruction. They belong to the class of connected operators. An attractive property of these filters is that they simplify images without introducing blurring or changing contours like low-pass or median filters [58, 61], which are classical simplification tools. Morphological filters by reconstruction enable the user to control the amount of information that is kept, with the objective of making images easier to segment.

To start with, the two most basic operators, erosion and dilation, will be introduced. Let B denote a window or flat structuring element and let $B_{x,y}$ be the translation of B so that its origin is located at (x, y). Then, the *erosion* $\epsilon_B(I)$ of an image I by the structuring element B is defined as

$$\epsilon_B(I)(x, y) = \min_{(k,l) \in B_{x,y}} I(k, l). \tag{1.29}$$

Similarly, the *dilation* $\delta_B(I)$ of the image I by the structuring element B is given by

$$\delta_B(I)(x, y) = \max_{(k,l) \in B_{x,y}} I(k, l). \tag{1.30}$$

For example, consider a window B consisting of 3×3 pixels. Then, the erosion $\epsilon_B(I)$ replaces each pixel (x, y) with the minimum gray-level within the 3×3 neighborhood of (x, y). Because a lower value for $I(x, y)$ corresponds to a darker gray-level, the resulting image will look darker.

Using the erosion and dilation operators, two morphological filters can be defined. These are *morphological opening*, $\gamma_B(I)$,

$$\gamma_B(I) = \delta_B(\epsilon_B(I)), \tag{1.31}$$

and *morphological closing*, $\varphi_B(I)$,

$$\varphi_B(I) = \epsilon_B(\delta_B(I)), \tag{1.32}$$

The morphological opening operator $\gamma_B(I)$ applies an erosion $\epsilon_B(\cdot)$ followed by a dilation $\delta_B(\cdot)$. Erosion leads to darker images and dilation to brighter images. The combination of these two operators according to (1.31) has then the effect of simplifying the original image I by removing bright components that do not fit within the structuring element B. Similarly, morphological closing removes dark components.

To simplify images prior to the segmentation, one would have to apply both a morphological opening and closing, because both small dark and bright components should be removed. Depending on the order in which these operators are applied, the resulting filter is either called *morphological opening-closing* or *morphological closing-opening.* The disadvantage of these two filters is that they do not allow a perfect preservation of the contour information [58].

For that reason, so-called filters by reconstruction are preferred. Although similar in nature, they rely on different erosion and dilation operators, making their definitions slightly more complicated. The elementary *geodesic erosion* $\epsilon^{(1)}(I, R)$ of size one of the original image I with respect to the reference image R is defined as

$$\epsilon^{(1)}(I, R)(x, y) = \max\{\epsilon_B(I)(x, y), R(x, y)\}, \tag{1.33}$$

and the dual *geodesic dilation* $\delta^{(1)}(I, R)$ of I with respect to R is given by

$$\delta^{(1)}(I, R)(x, y) = \min\{\delta_B(I)(x, y), R(x, y)\}, \tag{1.34}$$

Thus, the geodesic dilation $\delta^{(1)}(I, R)$ dilates the image I using the classical dilation operator $\delta_B(I)$ of (1.30). As mentioned earlier, dilated gray values are greater or equal to the original values in I. However, geodesic dilation limits these to the corresponding gray values of R. The choice of the reference image R will be discussed shortly.

Geodesic erosions and dilations of arbitrary size are obtained by iterating the elementary versions $\epsilon^{(1)}(I,R)$ and $\delta^{(1)}(I,R)$ accordingly. In particular, the so-called *reconstruction by erosion*, $\varphi^{(rec)}(I,R)$, and the *reconstruction by dilation*, $\gamma^{(rec)}(I,R)$, are defined as

$$\begin{aligned}\varphi^{(rec)}(I,R) &= \epsilon^{(\infty)}(I,R) = \underbrace{\epsilon^{(1)} \circ \epsilon^{(1)} \circ \ldots \circ \epsilon^{(1)}}_{\infty \text{ times}}(I,R) \\ \gamma^{(rec)}(I,R) &= \delta^{(\infty)}(I,R) = \underbrace{\delta^{(1)} \circ \delta^{(1)} \circ \ldots \circ \delta^{(1)}}_{\infty \text{ times}}(I,R).\end{aligned} \tag{1.35}$$

Notice that $\varphi^{(rec)}(I,R)$ and $\gamma^{(rec)}(I,R)$ will reach stability after a certain number of iterations. Anyway, this is not important in practice, because Vincent [62] presented a very fast implementation of these reconstruction operators using FIFO queues so that no iterations are needed.

Finally, the two simplification filters, *morphological opening by reconstruction*,

$$\gamma^{(rec)}(\epsilon_B(I), I), \tag{1.36}$$

and *morphological closing by reconstruction*,

$$\varphi^{(rec)}(\delta_B(I), I), \tag{1.37}$$

are merely special cases of $\gamma^{(rec)}(I,R)$ and $\varphi^{(rec)}(I,R)$ in (1.35).

Like morphological opening in (1.31), morphological opening by reconstruction first applies the basic erosion operator $\epsilon_B(I)$ of (1.29) to eliminate bright components that do not fit within the structuring element B. However, instead of applying just a basic dilation afterwards, as in (1.31), the contours of components that have not been completely removed are restored by the reconstruction by dilation operator $\gamma^{(rec)}(\cdot,\cdot)$. The reconstruction is accomplished by choosing I as the reference image R, which guarantees that for each pixel the resulting gray-level will not be higher than that in the original image I[4].

The strength of the morphological opening (closing) by reconstruction filter is that it removes small bright (dark) components, while perfectly preserving other components and their contours. Obviously, the size of removed components depends on the structuring element B.

The simplification effect of *morphological opening-closing by reconstruction*[5] is illustrated in Fig. 1.4 for the image *palms*. In particular, notice that the intensity of the simplified image is more homogeneous and therefore

(a) (b)

Figure 1.4: (a) Original image *palms* and (b) output of morphological opening-closing by reconstruction with a structuring element B of size 7×7 pixels.

easier to segment.

Morphological opening-closing by reconstruction is one of the most widely used simplification tools, but there exist other morphological tools that serve this purpose, such as area opening-closing filters. For a more detailed treatment, we refer the reader to [60, 62].

1.3.1.3 Marker Extraction

After simplifying the image, the marker extraction step detects the presence of uniform areas. Each of these markers forms an initial seed for a region in the final segmentation. This step also decides implicitly how many regions there will be in the final partition. Notice that marker extraction is not concerned with the location of region boundaries. This will be accomplished by the watershed algorithm in the next step. Consequently, markers typically consist only of the interior of regions.

The marker extraction step often contains most of the know-how of the segmentation algorithm [57]. Both the simplification filters and the watershed algorithm are clearly specified, apart from the choice of some parameters, whereas the marker extraction process will depend on a particular application.

For instance, Fig. 1.4 demonstrated that morphological opening-closing

[4] Recall that the dilation operator has the effect of increasing gray values.

[5] morphological opening by reconstruction followed by a morphological closing by reconstruction

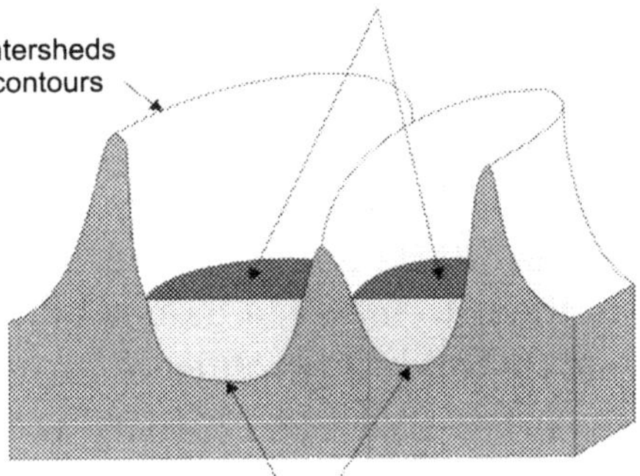

Figure 1.5: The watershed algorithm owes its name to the relief interpretation of the gradient image. Regions are represented by catchment basins, and the contours are given by the watersheds [57, 58].

by reconstruction leads to images with a more homogeneous luminance function. Therefore, markers could be extracted by identifying large regions of constant color or luminance in the simplified image. It is also possible to include partitions of previous frames of a video sequence into the marker extraction process, and some authors have suggested incorporating motion information [63, 64].

1.3.1.4 Watershed Algorithm

Undecided pixels are assigned a segmentation label in the decision step, the so called *watershed algorithm*, which is a technique similar to region-growing [57, 58]. The classical approach relies on the *morphological gradient* [57], although it was recently shown that this is not always the best choice [58, 61]. The morphological gradient $g(x, y)$ is defined as

$$g(x, y) = \delta_B(I)(x, y) - \epsilon_B(I)(x, y). \tag{1.38}$$

Notice that, according to (1.29) and (1.30), $g(x, y)$ is always greater or equal to zero. The gradient image can then be interpreted as a relief, as depicted in Fig. 1.5. Regions of the partition correspond to catchment basins and their contours are determined by the watershed lines.

Each marker obtained by the previous marker extraction step results in one region or basin. Because normally large flat zones are selected as markers, the morphological gradient in their interior will be zero. Consequently, these markers correspond to minima in the relief (see Fig. 1.5).

The watershed algorithm can now be viewed as a flooding procedure. Starting from the lowest altitude, the water gradually fills up the first catchment basin. When the water level of this basin reaches the altitude of another minimum, water also starts filling up that basin. As soon as water of two different basins is about to merge, a dam is built along the lines where the floods would merge to avoid the confluence.

Roughly speaking, pixels at lower altitudes are flooded first, and so are pixels that are closer to the water if they are on the same altitude. The flooding procedure terminates when the water level is higher that the maximum gradient value, and the region boundaries are given by the dams.

Efficient implementations of the watershed algorithm rely on clever scanning. Like the reconstruction operators for simplification (1.35), they make use of hierarchical FIFO queues [58].

All in all, morphological segmentation techniques are computationally efficient, and there is no need to specify in advance the number of objects as with some Bayesian approaches. This is automatically accomplished by the marker or feature extraction step. However, by its very nature, the watershed algorithm suffers from the problems associated with other simple region-growing techniques. For instance, it only takes one path of slowly changing gray-levels from one region to a neighboring one to cause these regions to merge [44].

1.3.2 Bayesian Segmentation

Arguably the most widely used approach to image segmentation is the Bayesian framework. The objective of such algorithms is to maximize the posterior probability of the unknown segmentation label field X, given the observed image or video sequence O [16, 17, 18]. Bayesian inference has also been applied to image understanding and scene interpretation by incorporating task specific knowledge [65].

From equation (1.2) we know that two probability distributions must be specified: the conditional probability $P(O|X)$ and the prior likelihood $P(X)$. To determine the latter distribution, X is usually assumed to be a Markov random field. Bayesian segmentation techniques then differ in the observation model $P(O|X)$ and the choice of the energy function $U(X)$ for the Gibbs distribution $P(X)$ (see (1.8)). There are also variations regarding the numerical optimization method employed.

The basics of Bayesian inference were already introduced in Section 1.1. Therefore, let us here consider an example that highlights different aspects of Bayesian segmentation. To this end, we will describe the well-known algorithm proposed by Pappas [17], because it is representative of the Bayesian approach.

1.3.2.1 Pappas' Method [17]

Let O be the observed gray-scale image and $O(i,j)$ the intensity of the pixel at location (i,j). The unknown segmentation of the image is denoted by X. Each pixel (i,j) is assigned a label $m \in \{0,\dots,K-1\}$ so that $X(i,j)=m$ means (i,j) belongs to region m. Notice that K, which is usually specified as an input parameter, is not the number of regions in the resulting partition. Normally, there will be far more regions than K, hence different regions are allowed to share the same label m as long as these regions are not neighbors of each other.

The aim is to find the MAP estimate of X. Thus, we want to find the most likely segmentation X, given the gray-scale image O. According to Bayes' theorem (1.2), the two probability distributions $P(X)$ and $P(O|X)$ must be defined.

The prior likelihood $P(X)$ describes the prior expectation on X. Intuition tells us that two neighboring pixels are more likely to belong to the same region than to different regions. Such interactions are local in nature, which suggests that X is ideally modeled by an MRF. Due to the Hammersley-Clifford theorem [27], $P(X)$ must then be a Gibbs distribution (1.8). Furthermore, $P(X)$ is completely specified by defining the energy function $U(X)$ in (1.9).

Pappas proposes an energy function $U(X=x)$ with non-zero contributions coming only from two-point cliques. The clique potential $V_C(x)$ associated with such pairs of horizontally, vertically, or diagonally adjacent pixels is given by

$$V_C(x) = \begin{cases} -\beta, & \text{if } x(i,j) = x(k,l) \text{ and } (i,j),\ (k,l) \in C \\ +\beta, & \text{if } x(i,j) \neq x(k,l) \text{ and } (i,j),\ (k,l) \in C. \end{cases} \tag{1.39}$$

Recall that a low potential or energy corresponds to a high probability and vice versa. By choosing a positive value for β, two neighboring pixels (i,j) and (k,l) are assigned a higher probability if they belong to the same region. Moreover, increasing β increases the strength of these correlations, resulting in larger regions and smoother boundaries.

To derive the conditional distribution $P(O|X)$, Pappas considers the gray-scale image O as a collection of regions with uniform or slowly varying luminance. The only sharp transitions in gray-level occur at region boundaries. More precisely, the intensity of region m is modeled as a constant signal μ_m plus additive, zero-mean white Gaussian noise with variance σ^2. The value of μ_m is computed by taking the average gray-level of all pixels that belong to region m in the current estimate of the segmentation field[6]. It follows then that

$$P(O = o|X = x) = \prod_{(i,j)} \frac{1}{\sqrt{2\pi\sigma^2}} \exp\left(-\frac{(o(i,j) - \mu_{x(i,j)})^2}{2\sigma^2}\right), \quad (1.40)$$

so that the posterior probability to be maximized, $P(X|O) \propto P(O|X)P(X)$, has the form

$$P(X = x|O = o) \propto \exp\left(-\frac{1}{T}\sum_{\text{all cliques } C} V_C(x) - \sum_{(i,j)} \frac{(o(i,j) - \mu_{x(i,j)})^2}{2\sigma^2}\right). \quad (1.41)$$

The constants $\frac{1}{Z}$ and $\frac{1}{\sqrt{2\pi\sigma^2}}$ have been omitted because they do not depend on X. The resulting probability distribution (1.41) is also a Gibbs distribution, and its energy function consists of one-point and two-point clique potentials.

In Section 1.1.3, it was outlined that finding the global maximum of (1.41) is computationally prohibitive for practical applications. Pappas approximated the optimal solution using ICM [21], which maximizes

$$P(X(i,j)|O, X(k,l), \text{ all } (k,l) \neq (i,j))$$

for each pixel (i,j) in turn. That is, it maximizes the probability of $X(i,j)$ in the light of all available information.

ICM can also be viewed as maximizing (1.41), for each pixel (i,j) in turn, with respect to $X(i,j)$ only. Due to the Markovian property of (1.41), only a few terms depend on $X(i,j)$, and we obtain

$$P(X(i,j)|O, X(k,l), \text{ all } (k,l) \neq (i,j)) \propto \exp\left(-\frac{1}{T}\sum_{C \in C_{i,j}} V_C(x) - \frac{(o(i,j) - \mu_{x(i,j)})^2}{2\sigma^2}\right), \quad (1.42)$$

[6]Pappas actually proposed a $\mu_m^{(i,j)}$ that also depends on the pixel (i,j). To this end, the average luminance is taken of all pixels that belong to region m within a window centered at (i,j) [17].

where $C_{i,j}$ is the set of two-point cliques that contain (i, j). This set usually consists of eight cliques, unless (i, j) is at an image boundary.

Finally, maximizing (1.42) is obviously equivalent to minimizing its negative logarithm. Moreover, it is easy to see that the parameters T, β, and σ^2 are interdependent. Therefore, we can set $T = 1$ and $2\sigma^2 = 1$ to simplify the expression. This results in the following cost or objective function to be minimized with respect to $X(i, j)$:

$$Cost(X(i,j)) = \sum_{C \in C_{i,j}} V_C(x) + \left(o(i,j) - \mu_{x(i,j)}\right)^2. \tag{1.43}$$

The parameter β, which is needed to evaluate $V_C(x)$, is expected as an input parameter to the segmentation algorithm.

The cost function (1.43) consists of a spatial continuity term and a close-to-data term. The spatial continuity term, derived from the Gibbs distribution, encourages adjacent pixels to have the same segmentation label. In fact, a partition consisting of one region only would yield the minimum cost. On the other hand, such a segmentation would not describe the observation O very well. The close-to-data term prefers a segmentation where (i, j) is assigned to the region that is closest with respect to the gray-level $o(i, j)$. The spatial continuity and the close-to-data terms complement each other and comprise a trade-off which is controlled by the input parameter β.

As shown in Section 1.1.3, ICM requires an initial estimate of X. This is necessary in order to evaluate $V_C(x)$ and to calculate initial estimates of μ_m for all regions m. To obtain an initial estimate, Pappas applies the K-means algorithm [66], which is a special case of (1.43) with $\beta = 0$. Based on the output of K-means, ICM can then iteratively approximate the optimal solution X by minimizing $Cost(X(i, j))$ for each pixel (i, j) in turn. Obviously, this update selects a value for $X(i, j)$ that minimizes the cost under the constraint of fixing the remaining values in X.

After each iteration, the μ_m's are updated according to the current partition so that the μ_m's become gradually more meaningful. Finally, ICM terminates when a local minimum is reached or after a prescribed number of iterations.

The necessity of an initial estimate and the strong dependence on the input parameter K, denoting the number of labels to be used, are two of the major drawbacks of Bayesian segmentation compared to morphological approaches. The latter automatically select, in an elegant manner, initial regions in their marker extraction step. To avoid these weaknesses, a different Bayesian approach is described in [67]. The initialization step is separated from the actual labeling process, as previously proposed for morphological segmentation. This segmentation algorithm can therefore be seen as a combination of the advantages of Bayesian and morphological techniques.

1.3.2.2 Multi-resolution Segmentation

Bayesian estimation is particularly well suited to multi-resolution segmentation [18, 68]. The key idea is to segment images first at a coarse resolution, and then to proceed to finer resolutions to refine the partitions. Finally, at the finest resolution, which is the original image itself, individual pixels are assigned a segmentation label.

At each resolution, the MAP estimate of the segmentation is computed using a conventional Bayesian segmentation technique. The resulting partitions then serve as an initial estimate for the segmentation at the next finer level, whereby an upsampling of the partitions is required.

Clearly, multi-resolution segmentation requires a multi-resolution representation of images, such as the Laplacian or Gaussian pyramid [69]. For instance, the Gaussian pyramid starts with the original image I_0 at the highest resolution. By filtering I_0 using a Gaussian low-pass filter and downscaling the filtered image by a factor two, an image I_1 is obtained with both decreased resolution and number of pixels. If this process is repeated, we get a sequence of images $I_2, I_3, \ldots,$ of progressively decreasing resolution and sample size. Each image I_n then corresponds to a level in a quad tree so that a pixel at one resolution corresponds to four pixels at the next finer resolution.

There are several benefits of multi-resolution segmentation. The computational load is often reduced, because labels can propagate quickly across images at coarse resolutions due to the smaller size of images. Furthermore, the segmentation algorithm becomes more robust. Coarse resolution images do not contain details, which means that in the beginning the segmentation is guided by dominant features of the image. The partitions will adapt to details only at finer resolutions. Multi-resolution approaches have proven to be particularly useful for segmentation of texture and high resolution images, where the information is spread over large areas [18, 68].

1.4 Motion

So far only still image segmentation has been considered in this chapter. However, recently there has been a growing interest in video sequence segmentation, mainly due to the development of MPEG-4 [11, 12, 70, 71, 72], which is set to become the new video coding standard for multimedia communication.

Physical objects are often characterized by a coherent motion that is different from that of the background. This makes motion a very useful feature for video sequence segmentation. It can complement other features such as color, intensity, or edges that are commonly used for the segmentation of still images (see Section 1.3). In fact, some motion segmentation algorithms are based solely on motion.

One of the earliest systems to segment scenes into regions based on motion was described in [73]. The motion of objects is determined by identifying the position of spatial gray scale discontinuities or edges in successive frames. The resulting system is very simple and can only handle rectangular shaped objects undergoing translation.

1.4.1 Real Motion and Apparent Motion

The rather vague term motion shall be defined first. Let $I(\mathbf{x};t)$ denote the intensity or luminance of the image with $\mathbf{x} = (x, y)$ being the spatial coordinates and t the temporal variable. In most practical cases, $\mathbf{x}$ will specify a discrete pixel location and t the discrete frame number. The projection onto the image plane of the true 3-D motion of objects in the scene will be referred to as *real motion*. The only available observation, on the other hand, is the time-varying intensity $I(\mathbf{x};t)$. The variations of these brightness patterns are perceived as *apparent motion*.

Apparent motion can be characterized by a *correspondence vector field* or by an *optical flow field*. The correspondence vector $\mathbf{d}(\mathbf{x}) = (p(x,y), q(x,y))$ describes the displacement of pixel $\mathbf{x}$ between t and $t + \Delta t$ resulting from changes of $I(\mathbf{x};t)$, whereas the optical flow $\mathbf{u}(\mathbf{x}) = (u(x,y), v(x,y))$ refers to a velocity of the point $(\mathbf{x};t)$ induced by variations of the brightness pattern $I(\mathbf{x};t)$:

$$\mathbf{u}(\mathbf{x}) = (u(x,y), v(x,y)) = (\frac{dx}{dt}, \frac{dy}{dt}). \tag{1.44}$$

For a sufficiently small Δt, the velocity can be approximated as being constant during that time interval. It follows that $\mathbf{d}(\mathbf{x}) = \mathbf{u}(\mathbf{x}) \cdot \Delta t$, which means that the correspondence vector is proportional to the optical flow. If Δt is set to unity, optical flow and correspondence vectors can even be used interchangeably.

It has been shown that real motion and apparent motion are in general different [74, 75]. Consider, for instance, a static scene with time-varying illumination. The real motion is obviously zero because no 3-D motion is present, while the change in intensity induces optical flow and therefore apparent motion. Furthermore, moving objects must contain sufficient texture to generate optical flow. A circle of uniform luminance rotating about its center, for example, does not produce any optical flow.

To segment a scene into independent moving objects we need to know the real motion, but only apparent motion can be observed. As a result, it is normally more or less implicitly assumed that real and apparent motion are the same, although it has been shown that they are in many cases different.

Another important issue in motion estimation is noise sensitivity. From the definition in (1.44) it can be seen that apparent motion is highly sensitive to noise, which can cause large discrepancies with respect to the real motion.

1.4.2 The Optical Flow Constraint (OFC)

Motion estimation algorithms rely on the fundamental idea that the luminance of a point P on a moving object remains constant along P's motion trajectory. This can be written as

$$I(\mathbf{x};t) = I(\mathbf{x}+\Delta\mathbf{x};t+\Delta t) \tag{1.45}$$

where the projection $\mathbf{x}$ of P is a function of the time t. The right-hand side of (1.45) can be approximated by a first-order Taylor series about $(\mathbf{x};t)$ as

$$I(\mathbf{x}+\Delta\mathbf{x};t+\Delta t) = I(\mathbf{x};t) + \Delta x\frac{\partial I}{\partial x} + \Delta y\frac{\partial I}{\partial y} + \Delta t\frac{\partial I}{\partial t}. \tag{1.46}$$

By substituting (1.45) into (1.46), dividing both sides of (1.46) by Δt and taking the limit as Δt approaches zero, we obtain the well-known optical flow constraint (OFC)

$$\frac{\partial x}{\partial t}\frac{\partial I}{\partial x} + \frac{\partial y}{\partial t}\frac{\partial I}{\partial y} + \frac{\partial I}{\partial t} = \mathbf{u}^T(\mathbf{x})\cdot\nabla I(\mathbf{x}) + I_t(\mathbf{x}) = 0, \tag{1.47}$$

with $\nabla I(\mathbf{x})$ denoting the spatial gradient at $\mathbf{x}$, $I_t(\mathbf{x})$ the partial derivative with respect to time, and $\mathbf{u}(\mathbf{x})$ the optical flow (1.44).

For each site $\mathbf{x}$, $\nabla I(\mathbf{x})$ and $I_t(\mathbf{x})$ can be computed by approximating the derivatives by differences taken in a small neighborhood of $\mathbf{x}$. The OFC (1.47) then defines a linear constraint for the two unknowns $u(x,y)$ and $v(x,y)$. Any point $\mathbf{u}(\mathbf{x})$ on this constraint line, which is depicted in Fig. 1.6, satisfies the OFC. Note that this constraint is local in the sense that only information from a small neighborhood of $\mathbf{x}$ is considered.

One equation is of course not enough to solve for two unknowns. In fact, it is easy to show that only the normal flow vector in the direction

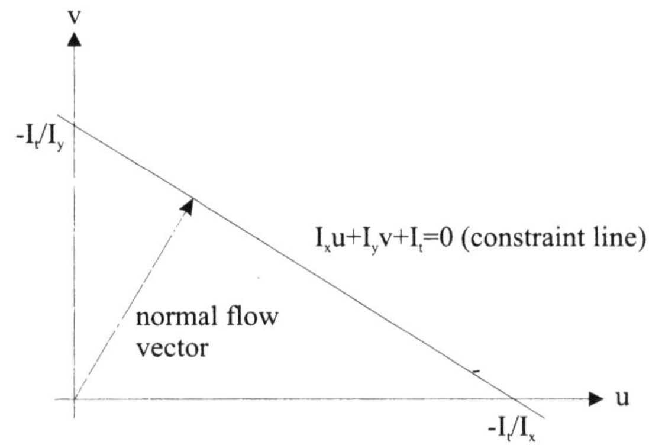

Figure 1.6: Optical flow constraint line.

of the local image gradient can be derived from the OFC [75]. This is also known as the aperture problem of motion estimation and is illustrated in Fig. 1.7. The true motion cannot be computed by considering just a small neighborhood. Instead, only the motion normal to the object contour is observable. Corners and regions with sufficient texture, however, are not affected by the aperture problem.

Solving for the optical flow field using the OFC (1.47) is, in the absence of additional constraints, a classical ill-posed problem [76]. In fact, there are infinitely many motion fields consistent with the observed $I(\mathbf{x};t)$. To overcome the aperture problem, additional information from a larger neighborhood is required. This can be incorporated by imposing smoothness constraints on the optical flow field to achieve continuity or by deriving models for the projection of object surfaces onto the image plane. These two approaches are also referred to as non-parametric and parametric representations, respectively, of the motion field. Block-matching, for instance, achieves smoothness by keeping the correspondence vector constant over a whole block.

1.4.3 Non-parametric Motion Field Representation

Non-parametric algorithms estimate a dense motion field so that each pixel is assigned a correspondence or flow vector [23, 24, 77, 75, 78, 79, 80, 81, 82, 83]. The aperture problem is tackled by incorporating a smoothing constraint that enforces neighbor pixels to have similar motion vectors. Block matching and variants thereof are among the most popular non-parametric

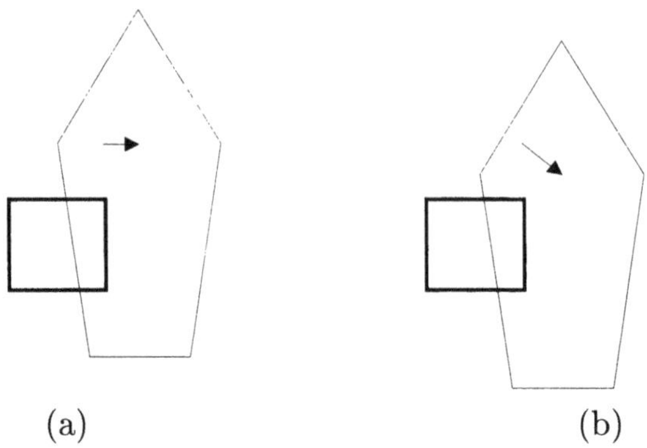

Figure 1.7: Illustration of the aperture problem. By considering only the local window it is not possible to distinguish between the two different motions in (a) and (b). Only the component normal to the object contour is uniquely defined.

approaches due to their simplicity.

A drawback of non-parametric algorithms is the blurring of motion edges introduced by the smoothness constraint. This can pose a problem for segmentation techniques that are based solely on the estimated motion field. If the motion boundaries are blurred, then an exact boundary location cannot be expected. On the other hand, the rather generic assumption of smoothness makes non-parametric methods applicable for a broad range of situations and applications.

Non-parametric dense field representations are, however, not directly suitable for segmentation. Apart from the simple case of pure translation, an object moving in 3-D space generates a spatially varying 2-D motion field even within the same object. Hence, it would be difficult to group pixels based on the similarity of their flow vectors. For that reason, parametric models are commonly used in segmentation algorithms. However, dense field estimation is often the first step in calculating the required model parameters. A detailed description of non-parametric motion estimation techniques will be given in Section 1.5.

1.4.4 Parametric Motion Field Representation

Parametric models derive the additional constraint required to solve the aperture problem by modeling the projection onto the image plane of surfaces moving in the 3-D space. Consequently, they rely on a segmentation of the frame into independently moving regions representing these surfaces. The motion of each region is described by a set of a few parameters, making it very compact in contrast to the non-parametric dense field description. These parameters are sufficient to synthesize or reconstruct the motion vector of any pixel in the image. If $\mathbf{u}(\mathbf{x})$ is the flow vector $(u(x,y), v(x,y))$ for pixel $\mathbf{x} = (x,y)$, then the model defines a mapping

$$\mathbf{u}(\mathbf{x}) = \mathbf{u}(\mathbf{x}; \mathbf{m}_p) \tag{1.48}$$

with $\mathbf{m}_p$ being the vector containing the model parameters of the region that $\mathbf{x}$ belongs to.

Another advantage of parametric representations is that they are less sensitive to noise because many pixels contribute to the estimation of a few parameters. Furthermore, there is no blurring of motion boundaries as long as they coincide with region boundaries. The necessity of a segmentation and some possibly restrictive assumptions on the scene and motion are among the drawbacks of parametric representations.

Note that the requirements on the segmentation here are not the same as for VOP extraction. Pixels are grouped into regions that obey the same rather simple motion model. As a result, one VOP would normally be described by several surfaces and their parameters.

In the following, some commonly used parametric models will be examined. By (X, Y, Z) and (X', Y', Z') we denote the 3-D coordinates of a point on an object in frames k and $k+1$, respectively. The corresponding coordinates in the image plane are (x, y) and (x', y').

The displacement from frame k to $k+1$ of a point on the surface of an object undergoing translation, rotation, and linear deformation is then given by [84]:

$$\begin{bmatrix} X' \\ Y' \\ Z' \end{bmatrix} = \underbrace{\begin{bmatrix} s_{11} & s_{12} & s_{13} \\ s_{21} & s_{22} & s_{23} \\ s_{31} & s_{32} & s_{33} \end{bmatrix}}_{S} \cdot \begin{bmatrix} X \\ Y \\ Z \end{bmatrix} + \underbrace{\begin{bmatrix} t_1 \\ t_2 \\ t_3 \end{bmatrix}}_{T} \tag{1.49}$$

T is a 3-D translation vector, while S is often defined as a 3×3 rotation matrix R that can be described using Eulerian angles of rotation about the three coordinate axes. The model (1.49) can also include scaling by choosing $S = DR$ with the scaling matrix D or deformable motion by setting $S = (D + R)$ where D is an arbitrary deformation matrix [84].

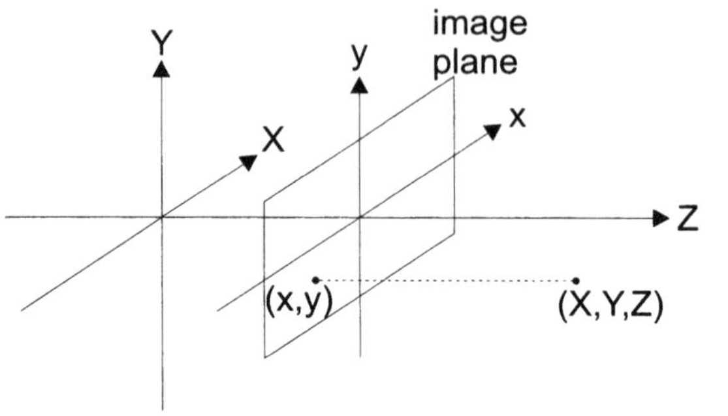

Figure 1.8: Projection of pixel (X, Y, Z) onto image plane (x, y) under orthographic (parallel) projection.

For motion estimation, real-world objects are often approximated by piecewise planar 3-D surfaces. This, at least locally, is a reasonable assumption. The points on such a planar patch in frame k satisfy

$$aX + bY + cZ = 1. \tag{1.50}$$

Together with (1.49) we then obtain the so-called affine motion model under orthographic projection and the so-called eight-parameter model under perspective projection.

As can be seen from Fig. 1.8, the 3-D and image plane coordinates are related under the *orthographic (parallel) projection* by

$$(x, y) = (X, Y) \quad \text{and} \quad (x', y') = (X', Y'). \tag{1.51}$$

This projection is computationally efficient and a good approximation if the distance between the objects and the camera is large compared to the depth of the objects.

By combining (1.49), (1.50) and (1.51) we obtain

$$\begin{aligned} x' &= a_1 x + a_2 y + a_3 \\ y' &= a_4 x + a_5 y + a_6 \end{aligned} \tag{1.52}$$

with $a_1 = (s_{11} - s_{13}\frac{a}{c})$, $a_2 = (s_{12} - s_{13}\frac{b}{c})$, $a_3 = (t_1 + s_{13}\frac{1}{c})$, $a_4 = (s_{21} - s_{23}\frac{a}{c})$, $a_5 = (s_{22} - s_{23}\frac{b}{c})$, and $a_6 = (t_2 + s_{23}\frac{1}{c})$. Equation (1.52) is the well-known *affine motion model.*

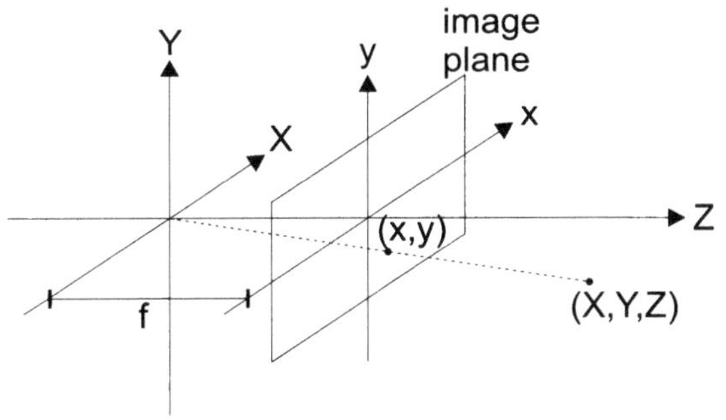

Figure 1.9: Projection of pixel (X, Y, Z) onto image plane (x, y) under perspective (central) projection.

In the case of the more realistic *perspective (central) projection* it can be seen from Fig. 1.9 that

$$(x, y) = (f\frac{X}{Z}, f\frac{Y}{Z}) \quad \text{and} \quad (x', y') = (f\frac{X'}{Z'}, f\frac{Y'}{Z'}). \tag{1.53}$$

Together with (1.49) and (1.50) this results in the *eight-parameter model*

$$\begin{aligned} x' &= \frac{a_1x + a_2y + a_3}{a_7x + a_8y + 1} \\ y' &= \frac{a_4x + a_5y + a_6}{a_7x + a_8y + 1} \end{aligned} \tag{1.54}$$

where $a_1 = \frac{s_{11}+at_1}{s_{33}+ct_3}$, $a_2 = \frac{s_{12}+bt_1}{s_{33}+ct_3}$, $a_3 = f\frac{s_{13}+ct_1}{s_{33}+ct_3}$, $a_4 = \frac{s_{21}+at_2}{s_{33}+ct_3}$, $a_5 = \frac{s_{22}+bt_2}{s_{33}+ct_3}$, $a_6 = f\frac{s_{23}+ct_2}{s_{33}+ct_3}$, $a_7 = \frac{1}{f}\frac{s_{31}+at_3}{s_{33}+ct_3}$, and $a_8 = \frac{1}{f}\frac{s_{32}+bt_3}{s_{33}+ct_3}$. The parameters $a_1, \ldots, a_8$ are also known as the eight pure parameters [85].

The parallel projection (1.51) of a parabolic surface

$$Z = aX^2 + bXY + cY^2 + dX + eY + g \tag{1.55}$$

moving according to (1.49) leads to the *twelve-parameter quadratic model*

$$\begin{aligned} x' &= a_1x^2 + a_2y^2 + a_3xy + a_4x + a_5y + a_6 \\ y' &= a_7x^2 + a_8y^2 + a_9xy + a_{10}x + a_{11}y + a_{12} \end{aligned} \tag{1.56}$$

with $a_1 = s_{13}a$, $a_2 = s_{13}c$, $a_3 = s_{13}b$, $a_4 = (s_{11} + s_{13}d)$, $a_5 = (s_{12} + s_{13}e)$, $a_6 = (t_1 + s_{13}g)$, $a_7 = s_{23}a$, $a_8 = s_{23}c$, $a_9 = s_{23}b$, $a_{10} = (s_{21} + s_{23}d)$, $a_{11} = (s_{22} + s_{23}e)$, and $a_{12} = (t_2 + s_{23}g)$.

Independent of what model is used, each region is described by one set of parameters that must be estimated. This could theoretically be done by identifying corresponding point pairs in the two image frames. The eight-parameter model (1.54), for instance, requires at least four independent point pairs to solve for the parameters. Unfortunately, to find such pairs without supervision is not an easy task. As a result, the parameters are usually obtained either by fitting the model in the least-squares sense to a dense motion field obtained by a non-parametric method or directly from the signal $I(\mathbf{x};t)$ and gradient information. We will examine both approaches later in Section 1.6.

Parametric model-based motion estimation and segmentation algorithms are indeed very popular. In model-based coding schemes, regions typically represent areas of similar image characteristics such as color or intensity and are therefore relatively small. The assumption of the 3-D motion (1.49) and locally planar surfaces (1.50) are normally valid approximations for such regions. In the case of layered scene descriptions like in MPEG-4, however, all these requirements are not well met. Thus, describing whole physical objects with possibly strongly non-rigid motion by one set of model parameters cannot be justified. Instead, one VOP must be represented by several smaller regions or patches.

1.4.5 The Occlusion Problem

Besides the aperture problem and the fact that only apparent motion can be observed, motion estimation also suffers from the so-called occlusion problem, which is demonstrated in Fig. 1.10. A moving object naturally uncovers and covers background. Obviously, no correspondence vectors exist for the uncovered background and background to be covered. Most motion estimation techniques neither identify these so-called occlusion regions nor treat them specially. Instead, they are simply accepted as regions of high compensation error. For segmentation, however, occlusion regions cannot be neglected because this would have a negative effect on the accuracy of the motion boundary location.

All the difficulties affecting motion estimation mentioned above suggest that the resulting motion field has to be carefully interpreted. Apparent motion alone is not well-suited for segmentation because an accurate motion field is required. Thus, it seems to be inevitable that additional information such as color or intensity must be included to accurately and reliably detect boundaries of moving objects.

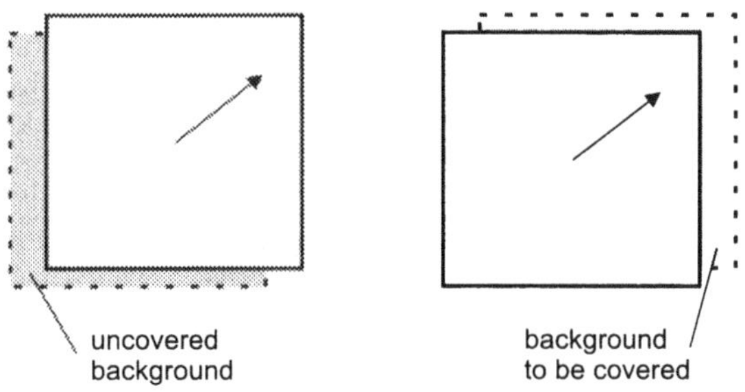

Figure 1.10: Illustration of the occlusion problem. No correspondence can be established for pixels in occlusion areas, i.e., in (a) uncovered background and (b) background to be covered.

1.5 Motion Estimation

Virtually all motion estimation algorithms in video communication have been developed for coding purposes with different objectives from those of motion segmentation. They aim at minimizing the prediction error after motion-compensation so that only a comparatively small residue must be encoded. By removing the high temporal redundancy present in video sequences, high compression ratios can be achieved.

Recovering the true motion of objects with high motion boundary accuracy, which is crucial for segmentation, plays only a minor role in coding as long as the prediction error is low. Schunck [77] commented on this issue by stating "... Image compression has not forced the development of image flow estimation algorithms that handle discontinuities because image compression does not require perfect estimation of the motion and does not require the detection of motion boundaries. Any discrepancy between frames caused by inaccurate estimation of the motion is transmitted as a correction. ..." Motion segmentation, on the other hand, depends very much on the accuracy of the estimated motion field.

Classical approaches to motion estimation belong to the group of non-parametric techniques, because their only interest is in computing the motion field. Consequently, we will focus here on these algorithms. Parametric motion estimation techniques involve some kind of segmentation and they will be discussed in Section 1.6. Note that motion estimation itself has been a very active research area and numerous techniques have been published so that even describing only the most important of these algorithms would be far beyond the scope of this book. For a more detailed treatment of motion estimation we recommend [84, 86, 87] as a starting point.

All motion estimation methods rely on the principle of intensity conservation; that is, they more or less implicitly assume that the luminance of pixels does not change along their motion trajectories. Depending on the approach they take, motion estimation techniques can be classified as gradient-based [77, 75], block-based [78, 79, 80], pixel-recursive [81, 82, 83], or Bayesian [23, 24] methods.

1.5.1 Gradient-based Methods

Gradient-based methods directly utilize the OFC (1.47) and incorporate an additional constraint to tackle the aperture problem [77, 75]. The latter is normally designed to achieve continuity of the estimated flow field by forcing neighboring pixels to have similar flow vectors.

The classical algorithm by Horn and Schunck [75] seeks an optical flow field that minimizes the deviation from the OFC (1.47) with minimum pixel-to-pixel variations of flow vectors. The total error to be minimized is given by

$$E^2 = \sum_{\mathbf{x}} \left(\alpha^2 E_c^2(\mathbf{x}) + E_b^2(\mathbf{x})\right) \tag{1.57}$$

where the first term $E_c^2(\mathbf{x}) = \|\nabla u(x,y)\|^2 + \|\nabla v(x,y)\|^2$ penalizes departure from smoothness in the flow field, the second term $E_b^2(\mathbf{x}) = (\mathbf{u}^T(\mathbf{x}) \cdot \nabla I(\mathbf{x}) + I_t(\mathbf{x}))^2$ measures the deviation from the OFC (1.47), and the weighting factor α^2 controls the strength of smoothing. By increasing the value of α a smoother flow field will be obtained.

An iterative solution based on the Gauss-Seidel method [88] was derived. Let the flow vector at pixel $\mathbf{x}$ after the n-th iteration be denoted by $(u^{(n)}, v^{(n)})$ and the corresponding local average at $\mathbf{x}$ taken in a 3×3 spatial neighborhood by $(\bar{u}^{(n)}, \bar{v}^{(n)})$. The iteration is then given by

$$\begin{aligned} u^{(n+1)} &= \bar{u}^{(n)} - I_x \frac{I_x \bar{u}^{(n)} + I_y \bar{v}^{(n)} + I_t}{\alpha^2 + I_x^2 + I_y^2} \\ v^{(n+1)} &= \bar{v}^{(n)} - I_y \frac{I_x \bar{u}^{(n)} + I_y \bar{v}^{(n)} + I_t}{\alpha^2 + I_x^2 + I_y^2}. \end{aligned} \tag{1.58}$$

While the flow cannot be directly estimated in uniform areas where the gradient ∇I is zero, the motion information from the region boundaries

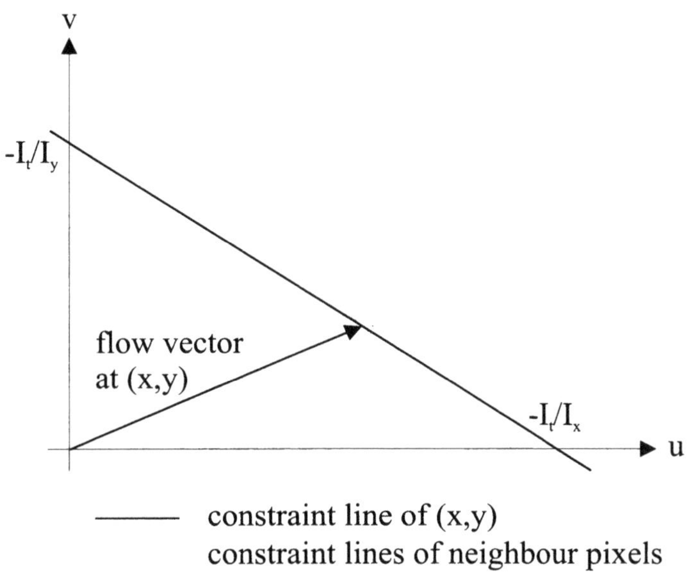

Figure 1.11: The constraint line of $\mathbf{x}$ is intersected with the constraint lines of neighboring pixels. The cluster of intersections indicates the correct flow vector for $\mathbf{x}$.

will propagate inwards to these pixels due to the average term $(\bar{u}^{(n)}, \bar{v}^{(n)})$. Therefore, the number of iterations should be larger than the maximum distance across the largest region that must be filled in. Note that the smoothing term E_c^2 in (1.57) is not capable of handling motion field discontinuities, which means that motion boundaries will be blurred.

It was shown in Section 1.4.2 that the OFC (1.47) defines a constraint line for the two unknowns $u(x, y)$ and $v(x, y)$ at pixel $\mathbf{x} = (x, y)$. Since any point $\mathbf{u}(\mathbf{x})$ on that line satisfies the OFC, additional information is necessary to obtain a unique solution. Schunck developed an elegant constraint line clustering algorithm [77] that solves this aperture problem. He examines the intersections of the constraint line at $\mathbf{x}$ with the constraint lines of the neighborhood pixels as depicted in Fig. 1.11.

For a $n \times n$ neighborhood one obtains $(n^2 - 1)$ intersections unless some constraint lines are parallel to that of $\mathbf{x}$. Pixels that are part of the same moving object as $\mathbf{x}$ have similar flow vectors and the corresponding intersections should form a tight cluster on the constraint line indicating the

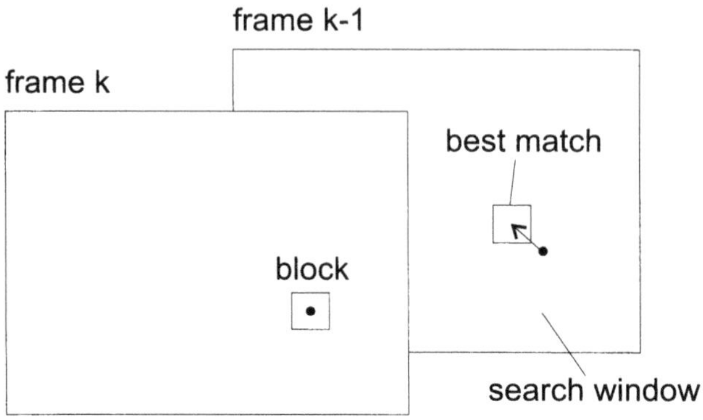

Figure 1.12: For each block, the best match in the previous frame is computed by examining a search window centered at the block. This is referred to as backward motion estimation. Note that the center of the search window corresponds to a zero displacement.

position of the true flow vector $\mathbf{u}(\mathbf{x})$. The intersections of other pixels in the neighborhood are spread along the constraint line. The center of the shortest interval on the constraint line of $\mathbf{x}$ containing half of the intersection points is selected as the estimate for $\mathbf{u}(\mathbf{x})$. Note that the required cluster analysis of intersections is a one-dimensional process along the flow constraint line of $\mathbf{x}$.

As long as a majority of intersections form a tight cluster, outliers will not influence the result. This means that near motion boundaries a few pixels with different motion will not affect the estimation of $\mathbf{u}(\mathbf{x})$. Consequently, there is relatively little blurring of motion boundaries.

1.5.2 Block-based Techniques

Block-matching and variants thereof are among the most popular techniques due to their computational simplicity [78, 79, 80]. They subdivide the current frame into blocks of normally equal size and compute for each block the best match in the next or previous frame (see Fig. 1.12). All pixels of a block are assumed to undergo the same translation and are assigned the same correspondence vector. The various block-matching algorithms differ in the block sizes, the search window in which to look for the best match, the search strategy, and the matching criterion.

Mean Absolute Difference (MAD) is the most widely used matching criterion because of its low computational cost and ease of VLSI implementation. For a block B of size $M \times N$, the MAD is given by

$$MAD(p, q) = \frac{1}{MN} \sum_{(x,y)\in B} |I(x, y; k) - I(x + p, y + q; k - 1)|, \tag{1.59}$$

where (p, q) is the displacement of the block B between frame k and $k - 1$. The performance of MAD deteriorates compared to that of the Mean Squared Difference (MSD), which uses the squared difference instead of the absolute difference in (1.59), when the search window becomes larger in faster moving sequences.

The Pixel Difference Classification (PDC) was proposed in [80]. Its performance lies somewhere between that of MAD and MSD, however, at lower computational cost. The PDC classifies each pixel in the block either as matching or mismatching. If the absolute difference $|I(x, y; k) - I(x + p, y + q; k - 1)|$ is smaller than a threshold T, the pixel (x, y) is labeled as matching, and otherwise as mismatching. The largest number of matching pixels then identifies the best match.

The search window restricts the maximum displacement d_{max} allowed in either direction to limit the computation time. Unfortunately, a full search of just the search window is often too costly. A good searching strategy that is a compromise between speed and quality is the 2-D logarithmic search [79]. It can be thought of as a hierarchical search where first a rough estimate is found that is subsequently refined.

Generally, the computational load for block-matching increases dramatically with the maximum allowed displacement in either direction. For that reason it is advantageous to compute large displacements at lower image resolution. In a hierarchical image representation, large displacements can be computed at lower resolution in order to reduce the risk of wrong matches, while the estimates are refined at higher resolutions.

Bierling [78] observed the importance of the selection of the block size. Large blocks might contain more than one motion and cannot accurately locate motion boundaries, whereas small blocks often result in mismatches because the presence of very similar patterns or blocks becomes more likely for smaller blocks. As a result, Bierling proposed a hierarchical block-matching algorithm with variable block size. Firstly, a large block size is used to find the major component of the displacement. This rough estimate, which is very robust due to the large block size, serves as an initial value for lower levels of the hierarchy where the motion field is refined using smaller block sizes. The search window is also reduced at lower levels to avoid mismatches for the smaller blocks. At the lowest level, relatively small blocks are employed to estimate the local displacement within a small search window.

A weakness of block-matching algorithms is their inability to cope with rotations, zooming, and deformations as well as the limited accuracy along motion boundaries due to their blocky nature. There exist extensions to deformable blocks that can handle these types of motion better, but this results in increased complexity. Computational efficiency, on the other hand, is one of the major strengths that have made block-based techniques so popular.

1.5.3 Pixel-recursive Algorithms

Netravali and Robbins proposed in [81] a pixel-recursive motion estimation technique. It is based on a prediction-update principle and revises the motion estimate iteratively at each pixel in turn until the estimates converge. Let $\mathbf{d}(\mathbf{x})$ be the correspondence vector at pixel $\mathbf{x}$ and $\mathbf{d}^{(i)}(\mathbf{x})$ the estimated correspondence vector after the ith iteration. Then, the update is carried out according to

$$\mathbf{d}^{(i)}(\mathbf{x}) = \mathbf{d}^{(i-1)}(\mathbf{x}) + \epsilon \cdot \mathbf{u}^{(i-1)}(\mathbf{x}), \tag{1.60}$$

where $\mathbf{d}^{(i-1)}(\mathbf{x})$ is the current estimate and $\epsilon \cdot \mathbf{u}^{(i-1)}(\mathbf{x})$ is the update term.

With predictive coding of television signals in mind, the algorithm aims at minimizing the resulting prediction error. This error, after motion-compensation or reconstruction from the estimated motion field, can be expressed by the so-called *displaced frame difference (DFD)*. The DFD for pixel $\mathbf{x}$ with displacement $\mathbf{d}$ between frame $n-1$ and n is given by

$$DFD(\mathbf{x};\mathbf{d}) = I(\mathbf{x};n) - I(\mathbf{x}-\mathbf{d};n-1). \tag{1.61}$$

Likewise, the DFD for $\mathbf{x}$ after the ith iteration is $DFD(\mathbf{x};\mathbf{d}^{(i)}) = I(\mathbf{x};n) - I(\mathbf{x}-\mathbf{d}^{(i)};n-1)$.

By minimizing $DFD^2(\mathbf{x};\mathbf{d})$ for each pixel in turn with respect to $\mathbf{d}(\mathbf{x})$, the resulting prediction error will be minimized. This can be achieved using a recursive numerical optimization method such as steepest-descent [88], which updates the current estimate in the direction of the local gradient. This leads to the following iterations

$$\begin{aligned}\mathbf{d}^{(i)}(\mathbf{x}) &= \mathbf{d}^{(i-1)}(\mathbf{x}) - \alpha \cdot \nabla_{\mathbf{d}} \left(DFD^2(\mathbf{x};\mathbf{d}^{(i-1)}) \right) \\ &= \mathbf{d}^{(i-1)}(\mathbf{x}) - 2\alpha \cdot DFD(\mathbf{x};\mathbf{d}^{(i-1)}) \nabla_{\mathbf{d}} DFD(\mathbf{x};\mathbf{d}^{(i-1)})\end{aligned} \tag{1.62}$$

It can be shown that this is essentially the same as minimizing the departure from the OFC (1.47) [84].

The gradient of the DFD with respect to $\mathbf{d}$ can be expressed using (1.61) as

$$\nabla_{\mathbf{d}} DFD(\mathbf{x}; \mathbf{d}^{(i-1)}) = +\nabla_{\mathbf{x}} I(\mathbf{x} - \mathbf{d}^{(i-1)}; n-1). \qquad (1.63)$$

By combining (1.62), (1.63), and setting $\epsilon = 2\alpha$ we obtain the following iteration to update the motion estimate at $\mathbf{x}$

$$\mathbf{d}^{(i)}(\mathbf{x}) = \mathbf{d}^{(i-1)}(\mathbf{x}) - \epsilon \cdot DFD(\mathbf{x}, \mathbf{d}^{(i-1)}) \nabla_{\mathbf{x}} I(\mathbf{x} - \mathbf{d}^{(i-1)}; n-1). \qquad (1.64)$$

Both the DFD and the image gradient $\nabla_{\mathbf{x}} I$ on the right-hand side of (1.64) can easily be computed since the estimate $\mathbf{d}^{(i-1)}(\mathbf{x})$ is known.

By comparing (1.64) with (1.60), the update term can clearly be identified. It is proportional to the motion-compensated prediction error DFD. Further, note that the estimate $\mathbf{d}^{(i)}(\mathbf{x})$ is only corrected in the direction of the image gradient, which is a consequence of the aperture problem.

The parameter ϵ is critical for the speed of convergence and stability of the iterations. A small value means that the estimate will converge slowly in fine steps, leading to a small prediction error, while a large value of ϵ allows quick adjustment to rapid changes in motion at the price of reduced accuracy. Netravali and Robbins suggested a value of $\frac{1}{1024}$ for ϵ and they clipped the update term to a maximum of $\pm\frac{1}{16}$ pixels per iteration. Thus, an update of a few pixels requires already a large number of iterations. Walker and Rao proposed an adaptive ϵ that becomes smaller near edges and larger in uniform areas [82].

1.5.4 Bayesian Approaches

As it was shown in Section 1.1, the Bayesian framework provides an elegant formalism for estimation problems. Consequently, several researchers have investigated into formulating motion estimation as a probabilistic estimation problem [23, 24, 25, 26, 89]. Some of these techniques are based on parametric models and involve segmentation. They will be described later in Section 1.6. Here we are interested in the estimation of dense motion fields.

Konrad and Dubois recognized that motion estimation, which is an ill-posed problem without further assumptions, can be regularized using a Bayesian estimation approach [23]. To this end, two probability mass functions must be defined: the observation model and the prior model (see Section 1.1). As usual, let $I(\mathbf{x}; n)$ be the gray-level of pixel $\mathbf{x}$ in frame n and

$\mathbf{d}(\mathbf{x})$ the displacement of $\mathbf{x}$ between frame n and frame $n-1$. Further, let I_n denote the whole frame n and D_n the correspondence vector field between frame n and frame $n-1$. The most likely motion field D_n given the frames I_n and I_{n-1} is obtained according to Bayes' rule by maximizing

$$P(D_n|I_n, I_{n-1}) \propto P(I_n|D_n, I_{n-1})P(D_n|I_{n-1}). \tag{1.65}$$

The displacement field D_n, which is assumed to be independent of the observation I_{n-1} (i.e., $P(D_n|I_{n-1}) = P(D_n)$), is modeled by a Markov random field (MRF) and therefore $P(D_n)$ is Gibbs distributed [27]. The corresponding potential function is chosen as

$$V_C(\mathbf{d}(\mathbf{x}_i), \mathbf{d}(\mathbf{x}_j)) = \|\mathbf{d}(\mathbf{x}_i) - \mathbf{d}(\mathbf{x}_j)\|^2, \tag{1.66}$$

where $\mathbf{x}_i$ and $\mathbf{x}_j$ are neighboring pixels. Since low values for the potential mean high probability, this prior model enforces smoothness on the estimated motion field. The conditional probability $P(I_n|D_n, I_{n-1})$, on the other hand, models the DFD of each pixel by zero-mean white Gaussian noise with variance σ^2.

Then, the motion field is estimated by minimizing the objective function

$$\begin{aligned} f(D_n) = & \sum_{\text{all cliques } C = \{x_i, x_j\}} \|\mathbf{d}(\mathbf{x}_i) - \mathbf{d}(\mathbf{x}_j)\|^2 \\ & + \frac{1}{2\sigma^2} \sum_{\mathbf{x}} (I(\mathbf{x}; n) - I(\mathbf{x} - \mathbf{d}(\mathbf{x}); n-1))^2 \end{aligned} \tag{1.67}$$

with respect to D_n using a Gibbs sampler [20]. The first term achieves continuity of the motion field and the second term enforces intensity conservation along motion trajectories. A major drawback of this technique is the enormous computational load, especially due to the use of a simulated annealing method for optimization.

The motion estimation algorithm by Zhang and Hanauer contains two auxiliary MRFs to avoid blurring of motion boundaries and to accommodate occlusion regions [24]. The sites of the line field are placed between neighboring pixels; that is, each pixel has one line field site above, below, to its left, and to its right. The line field is binary and defines whether there is a motion field discontinuity between the corresponding pixels or not. The second auxiliary field is a binary segmentation field specifying for which pixels a motion vector is defined. This allows excluding occlusion areas when searching for correspondence vectors.

The optimization is performed using the mean field theory. This reduces the computational load compared to simulated annealing techniques, however, the two additional auxiliary fields which must be estimated along with the motion field lead to a dramatic increase in the number of unknowns.

1.6 Motion Segmentation

Video sequence segmentation algorithms in the field of video communication and coding can be classified based upon their motivation into two main groups: motion segmentation and video object plane extraction. The latter aims at enabling content-based coding with MPEG-4 by decomposing scenes into semantically meaningful objects.

Most motion segmentation techniques are inspired by the so-called second generation coding methods [1, 2, 90] with the main goal of achieving high compression ratios. The major innovation of second generation methods is the use of better and more sophisticated source models by taking into account the characteristics of the human visual system. Motion segmentation algorithms attempt to partition the frame into regions of similar intensity, color, and/or motion characteristics. The contour, texture, and motion of each region can then be efficiently encoded. For instance, the gray-level within a region is relatively uniform, leading to high coding gains, and the motion of each region is described in a very compact way by one set of parameters of a parametric motion model (see Section 1.4.4).

The partitions resulting from motion segmentation consist of entities that correspond more to physical objects compared to the pixels and blocks in first generation coding schemes. They are, however, still different from the content-based representation in MPEG-4. Video object planes are normally larger than these regions and are not necessarily characterized by similar intensity, color, or motion. Thus, motion segmentation techniques usually obtain a finer partition than VOP extraction algorithms. This is depicted in Fig. 1.13 using the hierarchical object representation model by Zhong and Chang [91]. At the bottom are primitive regions that are consistent over space and time with respect to motion, color, or luminance. Motion segmentation algorithms typically partition frames into such primitive regions according to their motion and possibly luminance. VOP segmentation aims at extracting meaningful objects, which can be found at the next higher level. These objects normally consist of several primitive regions. Note that it is very difficult, if not impossible, to find a feature that allows direct segmentation of these higher-level objects. Some prior knowledge or user input might be necessary to extract objects from generic video sequences. At the

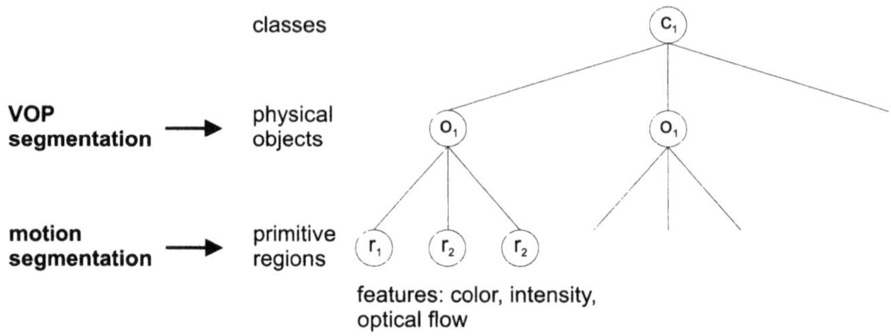

Figure 1.13: Hierarchical object representation model [91]. Motion segmentation algorithms segment frames into primitive regions of homogeneous color, intensity, or motion. VOP segmentation techniques, on the other hand, try to extract higher-level objects that typically consist of several primitive regions.

highest level we have the scene which comprises several objects.

As we will see later, many VOP segmentation techniques appear to be more ad-hoc approaches compared to motion segmentation algorithms, which can be nicely formulated in a Bayesian framework or using mathematical morphology. This only highlights the difficulty of formulating high-level semantic concepts in an algorithm. In the following, a comprehensive review of motion segmentation algorithms will be given. VOP extraction techniques will be described later in Section 5.1.

There exist many ways of classifying motion segmentation algorithms. For instance, they could be described by the approach they take such as morphological segmentation or Bayesian estimation. Here the various techniques will be distinguished based on the information that they exploit for the segmentation. This leads to the following four groups: 3-D segmentation, segmentation based on motion information only, spatio-temporal segmentation, and joint motion estimation and segmentation.

1.6.1 3-D Segmentation

The proposals in [58, 19, 61] consider video sequences to be three-dimensional signals. They extend conventional 2-D methods by adding a third dimension for time, although the time axis does not play the same role as the two spatial axes. In that sense, they are actually not true motion segmentation techniques.

The Bayesian framework provides an elegant formalism and is among the most popular approaches to motion segmentation. The key idea is to find the MAP estimate of the segmentation S for some given observation O, i.e., to maximize $P(S|O) \propto P(O|S)P(S)$. Techniques that make use of Bayesian inference are more plausible than some rather ad-hoc methods. They can also easily incorporate mechanisms to achieve spatial and temporal continuity. On the negative side, Bayesian approaches suffer from higher computational complexity and many algorithms need the number of objects or regions in the scene as an input parameter.

Hinds and Pappas [19] extended the 2-D adaptive clustering algorithm of [17], which was described in Section 1.3.2, to video sequences. They find the MAP estimate of the unknown segmentation S given the 3-D volume O of image frames that form the video sequence. According to Bayes' theorem two probability functions must be defined: the prior probability $P(S)$ modeling the segmentation label field and the conditional probability $P(O|S)$ describing how well the observed video signal fits the segmentation. For the prior model, the label field S is assumed to be a sample of a 3-D Markov random field (MRF), whereby the energy function of the corresponding Gibbs distribution $P(S)$ comprises two components to achieve spatial and temporal continuity of labels. The temporal potential function encourages pixels to have the same label in consecutive frames. However, this does not reflect the temporal connectivity required for moving objects. If $\mathbf{d}$ is the displacement of pixel $\mathbf{x}$ between two frames due to motion, then $\mathbf{x} + \mathbf{d}$ should have the same label as $\mathbf{x}$ and not the same site $\mathbf{x}$. Finally, in order to obtain $P(O|S)$ the difference between a pixel's gray value and the mean gray-level of the region it belongs to is modeled by zero-mean white Gaussian noise.

Morphological tools such as the watershed algorithm and simplification filters have been widely used both for segmentation and coding. Salembier and Pardàs [58] proposed a segmentation algorithm for 3-D video signals that has the typical structure of morphological approaches, as described in Section 1.3.1. In a first step, the image is simplified by a morphological "opening-closing by partial reconstruction" filter to remove small dark and bright patches. The size of these patches depends on the structuring element used. The color or intensity of the resulting simplified images is relatively homogeneous. The following marker extraction step detects the presence of homogeneous 3-D areas by identifying large regions or volumes of constant intensity. Each extracted marker is then the seed for a region in the final segmentation. Undecided pixels are assigned a label in the decision step by a 3-D version of the watershed algorithm. A quality estimation is performed as the last step to determine which regions require re-segmentation.

The technique by Salembier *et al.* in [61] is very similar, but the segmentation is performed on a frame-by-frame basis. Temporal continuity and linking of the segmentation is achieved through an additional projection step that warps the previous partition onto the current frame. This projection is also computed by the watershed algorithm using the previous partition as markers.

The regions obtained by 3-D segmentation algorithms are obviously homogeneous with respect to intensity as this is the only information used, but it is not assured that these regions can be efficiently described in terms of motion. Temporal linkage of the partition is automatically accomplished in the case of the 3-D segmentation [58, 19] or can be achieved in a frame-based scheme by projecting the partition of the previous frame onto the current frame [61]. The fundamental flaw of 3-D video segmentation algorithms is the way temporal continuity of the segmentation is enforced. A pixel $\mathbf{x}$ is expected to have the same segmentation label in frame n as it had in the previous frame $n-1$. While this might be reasonable for stationary areas, it certainly does not hold for moving objects where the continuity should be enforced along the motion trajectory of $\mathbf{x}$. Thus, motion information is not only useful as a cue for segmentation, it also enables a better way of establishing temporal continuity of the label field.

1.6.2 Segmentation Based on Motion Information Only

Many researchers have reported segmentation techniques that partition the scene based solely on motion information [6, 7, 92, 93, 94]. A classical approach among these is the segmentation of an estimated dense motion field [92, 93, 94]. Notice that simply applying one of the segmentation methods of Section 1.3 directly to the flow field does not produce useful results, because apart from the case of pure translation, a moving object generates a spatially varying flow field. Consequently, parametric motion field representations are used, and pixels are grouped together according to how well they are described by a common motion model.

In his early work, Adiv [92] proposed a hierarchically structured three-stage algorithm. The flow field is first segmented using the Hough transform [95, 96] into connected components such that the motion of each component can be modeled by the six-parameter affine transformation (1.52). Each flow vector votes for those points in the six-dimensional parameter space for which the associated transformation is consistent with the flow vector. Points in the parameter space that receive many votes indicate the motion of large areas in the flow field. Adjacent components are then merged in the second stage into segments if they obey the same eight-parameter quadratic flow model. This model describes the perspective projection of the 3-D velocity of a planar patch undergoing translation, rotation, and linear deformation. It is based on the same assumptions as the eight-parameter model (1.54) except that it describes a flow field instead of a displacement field. In the last stage, neighboring segments that are consistent with the same 3-D motion (1.49) are combined, resulting in the final segmentation. This technique has no mechanism incorporated to achieve linkage and temporal continuity of the partition.

The Bayesian technique by Murray and Buxton [93] uses an estimated flow field as observation O. As it is common, the label field S is assumed to be a sample of a Markov random field, whereby the energy function of the corresponding Gibbs distribution comprises three components. These are a spatial smoothness term, a temporal continuity term, and a line field as in [20] to allow for motion discontinuities. To define the observation probability $P(O|S)$, the parameters of a quadratic flow model [92] are calculated for each region by linear regression. The mismatch between this synthesized flow and the flow field given in O is modeled by zero-mean white Gaussian noise. The resulting probability function $P(O|S)P(S)$ is maximized by simulated annealing with the partition of the previous frame as the initial estimate. Major drawbacks of this proposal are its computational complexity and that the number of objects likely to be found has to be specified. In addition, as for the 3-D segmentation techniques described above, temporal continuity is enforced for pixels at the same spatial location in successive frames and not along motion trajectories.

A similar approach was taken by Bouthemy and François [94]. The energy function of their MRF consists only of a spatial smoothness term. The observation O contains the temporal and spatial gradients of the intensity function, which are related to the optical flow by the OFC (1.47). For each region, the affine motion parameters (1.52) are computed in the least-squares sense and $P(O|S)$ models the deviation of this synthesized flow from the optical flow constraint (1.47) by zero-mean white Gaussian noise. The optimization is performed by ICM (see Section 1.1.3), which is faster than simulated annealing but is likely to get trapped in a local minimum. To achieve temporal continuity, the segmentation result of the previous frame is used as the initial estimate for the current frame. The algorithm then alternates between updating the segmentation labels S, estimating the affine motion parameters, and updating the number of regions in the scene.

The object-oriented analysis-synthesis coding algorithms proposed by

Hötter and Thoma [6] and Musmann *et al.* [7] aim at a segmentation where the motion of each region can be described by one set of motion parameters. They do not explicitly estimate a motion field. Instead, the required parameters are obtained directly from the spatio-temporal image intensity function $I(\mathbf{x}; n)$ and its gradient. The segmentation is hierarchically structured and is initialized by dividing the current frame into changed and unchanged areas, whereby each connected changed region is interpreted as one object. After estimating the motion parameters for each object, the frame is reconstructed by motion-compensation and compared with the original frame. Objects with high prediction error are further subdivided into smaller objects and analyzed in subsequent levels of the hierarchy. The algorithm sequentially refines the segmentation and motion estimation until all changed regions are accurately compensated. An eight-parameter model (1.54) is employed to describe the motion, and the parameters are obtained directly from the frame difference and spatial gradients. A Taylor series expansion of the luminance function $I(\mathbf{x}; n)$ about $(\mathbf{x}; n)$ allows expressing the *frame difference (FD)* at pixel $\mathbf{x}$,

$$FD(\mathbf{x}) = I(\mathbf{x}; n) - I(\mathbf{x}; n-1), \tag{1.68}$$

in terms of spatial intensity gradients and the unknown parameters. Both the frame difference (1.68) and the gradients are easy to compute, with the latter being approximated by discrete differences. Each pixel of an object contributes one equation, although noisy observation points are identified by means of a simple statistical test and are excluded. The resulting overdetermined system of linear equations is then solved for the model parameters by linear regression.

None of the techniques in [6, 7, 92, 93, 94] makes use of intensity, color, or spatial edges. They provide only motion information for the segmentation decision, which means that they inevitably suffer from the problems associated with motion estimation described in Section 1.4 and 1.5. This will certainly limit the accuracy of object boundaries.

1.6.3 Spatio-Temporal Segmentation

Many researchers have reported that motion boundaries usually coincide with intensity boundaries [8, 9, 63, 64, 97, 98]. Gray-level information is indeed very helpful, especially along motion boundaries, and should complement the information conveyed by the motion field to avoid the occlusion problem.

Diehl described an object-oriented analysis-synthesis coding algorithm in [8] that is very similar to [6, 7]. He uses the twelve-parameter quadratic motion model (1.56) describing a parabolic surface under parallel projection instead of the eight-parameter model (1.54) in [6, 7]. The parameters are estimated by minimizing the mean squared prediction error (MSE) between the original and the motion-compensated frame using a modified Newton algorithm [88]. To improve the accuracy of object boundaries, the resulting segmentation is refined by combining it with a spatial segmentation. To this end, a spatial partition is derived from a computed intensity edge image by closing the contours or edges. Contour-closing is, however, a non-trivial task and it is not specified how it is performed.

Bayesian approaches were taken in [9, 97]. Chang *et al.* [97] include intensity information and an estimated displacement vector field into the observation O. The energy function of the MRF describing the label field $P(S)$ consists of a spatial continuity term and a motion-compensated temporal term. The latter enforces temporal continuity of segmentation labels along motion trajectories in contrast to 3-D segmentation techniques [58, 19, 61] or [93], which consider the same spatial location in successive frames. To model the conditional probability $P(O|S)$, two methods of generating a synthesized displacement field for each region are suggested: the eight-parameter quadratic model in [92] and the mean displacement vector of the region calculated from the field given in O. For $P(O|S)$, it is then assumed that the absolute difference between the observed displacement and the synthesized displacement, as well as the deviation of a pixel's gray-level from the mean gray-level of the region it belongs to, obey zero-mean Gaussian distributions. More weight can be put on the motion data in cases where it is reliable, i.e., for small values of the DFD, and more weight on the gray-level information in areas with unreliable motion data by controlling the variances of these two Gaussian distributions. The optimization is then performed by ICM.

The technique by Konrad and Dang [9] aims at a rate-efficient segmentation of video sequences. Firstly, an overly fine initial partition is derived from a spatial still image segmentation algorithm. For each of these regions, the affine motion parameters (1.52) are computed. The region fusion stage merges these regions by minimizing an objective function that is inspired by MRF models. This function consists of three terms in order to minimize the intensity residual or DFD, to achieve spatial and temporal continuity of the segmentation, and to reduce the amount of data to be encoded by keeping the number of regions to a minimum. Note that this merging process works with regions as entities and not pixels. The improved quality of motion estimates after merging is then exploited to readjust the boundary pixels.

Dufaux *et al.* also start from a spatial segmentation [98]. The video sequence is first simplified by a morphological opening-closing by reconstruction, followed by a spatial segmentation using the K-means algorithm [66]. For each region obtained, one set of affine motion parameters (1.52) is calculated. Regions with high prediction error are then further split, while regions with similar motion are merged. A shortcoming of this technique is the lack of a criterion to achieve temporal continuity of the segmentation, although the use of a tracking algorithm based on a Kalman filter is suggested to establish temporal linking.

A morphological video segmentation algorithm was proposed by Choi *et al.* [63, 64]. In a first step, so-called joint markers are extracted by detecting areas that are not only homogeneous in luminance but also in motion. For that, the frames are simplified by a morphological opening-closing by reconstruction and large regions of constant intensity are identified. The affine motion parameters (1.52) are then calculated for each of these intensity markers by linear regression from an estimated dense flow field. Intensity markers for which the affine model is not accurate enough are split into smaller markers that are homogeneous with respect to motion. As a result, multiple joint markers might be obtained from a single intensity marker. The watershed algorithm, which performs the actual segmentation, also uses a joint similarity measure that incorporates luminance and motion. In a last stage, the segmentation is simplified by merging regions with similar affine motion. A drawback of this technique is the lack of temporal correspondence to enforce continuity in time.

1.6.4 Joint Motion Estimation and Segmentation

It is well-known that motion estimation and segmentation are interdependent [6, 7, 8, 25, 26, 89, 99]. Motion estimation requires the knowledge of motion boundaries where the smoothing constraint must be switched off, while segmentation needs the estimated motion field to identify motion boundaries. Joint motion estimation and segmentation algorithms have been proposed to break this cycle. Most of them alternate between motion estimation and segmentation until the result converges. Here only those techniques are considered that recalculate the dense motion field in each iteration. The methods in [6, 7, 8], which have been described above, only update the model parameters of every region. The actual motion estimation is performed prior to the segmentation and remains unchanged during these iterations.

The class of joint motion estimation and segmentation algorithms is clearly dominated by Bayesian approaches [25, 26, 89, 99, 100]. The motion field is now no longer part of the observation O and has to be estimated along with the segmentation.

The proposal by Heitz and Bouthemy [100] uses the temporal derivatives of the intensity function and spatial intensity edges detected by the Canny operator [42] as observation O. It jointly estimates a dense flow field and a line field indicating motion discontinuities. The sites of the line field are placed between the pixels of the motion field. A statistical test identifies pixels in occlusion areas for which no correspondence exists. For the remaining pixels $\mathbf{x}$, the deviation of the flow $\mathbf{u}(\mathbf{x})$ from the OFC (1.47) is assumed to be zero-mean Gaussian distributed. Motion discontinuities specified by the line field are enforced to coincide with the observed spatial edges. Both the dense flow field and the line field are modeled by MRFs to achieve continuity of the motion field, whereby the smoothness constraint is suspended across motion discontinuities. ICM is then used to perform the MAP estimation. The technique in [100] is not a true segmentation algorithm because it only computes a line field of motion discontinuities that generally do not form closed contours. A proper segmentation yielding connected regions with closed contours is obtained by [25, 26, 89, 99].

Chang *et al.* [26] use both a parametric and a dense correspondence field representation of the motion. The parameters of the eight-parameter model (1.54) are obtained for each region in the least-squares sense from the dense field. The objective function to be minimized resulting from the MAP criterion consists of three terms, each derived from an MRF. The first term measures how good the prediction is and is minimized when both the synthesized and dense motion field minimize the DFD. The second term is minimized if the dense motion field is smooth and the parametric representation is consistent with the dense field. However, smoothness is only enforced for pixels having the same segmentation label; that is, the smoothness constraint is suspended across region boundaries. The third and last term is a standard spatial continuity term to enforce a smooth label field. Since the number of unknowns is three times higher when the motion field has to be estimated as well, the computational complexity is significantly larger. Chang *et al.* decomposed the objective function into two terms and alternate between estimating the motion field and the segmentation labels using HCF and ICM (see Section 1.1.3), respectively. A shortcoming of this algorithm is the lack of a constraint to ensure temporal continuity of the partition. Furthermore, neither color nor luminance is exploited to locate region boundaries. Intensity information is only considered to minimize the prediction error DFD.

The technique proposed by Stiller in [89] and extended in [25] is similar, but no parametric motion field representation is necessary. The main objective is dense motion field estimation and the segmentation is merely used to accommodate motion boundaries. In [89], the objective function consists of two terms derived from the observation and prior model. The DFD generated by the dense motion field is modeled by a zero-mean generalized Gaussian distribution whose parameters can vary between different regions. Note that non-zero values for the DFD can be interpreted as being caused by an additive noise term that prevents intensity conservation along the motion trajectories. The prior model is described by an MRF to ensure segmentwise smoothness of the motion field and spatial continuity of the segmentation. In [25], the DFD is also assumed to obey a zero-mean generalized Gaussian distribution, however, occluded regions are detected and no correspondence is required for them. The MRF modeling the motion field and segmentation is made up of four terms enforcing spatial and temporal continuity of the segmentation, segmentwise spatial smoothness of the motion field and temporal continuity of motion vectors along motion trajectories. Although a deterministic relaxation technique similar to ICM is used to obtain the MAP estimate, the computational burden of this algorithm is enormous.

The algorithms [25, 26, 89] are targeted at a smooth motion and label field where the region boundaries coincide with motion boundaries. However, they do not guarantee that these regions are also coherent with respect to luminance. Intensity information is only employed to minimize the prediction error. Han *et al.* [99], on the other hand, start with a simple region-growing method to obtain a spatial partition. This partition is not reestimated during the following iterations. It merely serves as a guide for the motion segmentation. The posterior probability of the motion and label field, given two consecutive frames, consists of three terms as in [26]. The first term aims at a small prediction error by minimizing the DFD. The second and third terms impose smoothness on the motion and label fields. Spatial continuity of the flow field within the same region is accomplished, as well as temporal continuity of the motion and label fields along the motion trajectories. Smoothness of the label field is only enforced if two neighboring pixels belong to the same region in the partition obtained by the region-growing algorithm. The resulting algorithm alternates between updating the motion field and segmentation using ICM.

None of the motion segmentation techniques in this chapter achieves a partition into semantically meaningful objects, as required for the content-based functionalities in MPEG-4. Regions obtained by the segmentation methods described here are typically homogeneous with respect to motion and color or intensity, and they could be used by some second-generation coding techniques. However, segmentation algorithms that specifically target the extraction of physical objects to support the new functionalities provided by MPEG-4 will be described later in Section 5.1.

References

[1] M. Kunt, A. Ikonomopoulos, and M. Kocher, "Second-generation image-coding techniques," *Proceedings of the IEEE*, vol. 73, no. 4, pp. 549–574, Apr. 1985.

[2] M. Kunt, M. Bénard, and R. Leonardi, "Recent results in high-compression image coding," *IEEE Trans. Circuits and Systems*, vol. CAS-34, no. 11, pp. 1306–1336, Nov. 1987.

[3] G.K. Wallace, "The JPEG still picture compression standard," *Communications of the ACM*, vol. 34, no. 4, pp. 30–44, Apr. 1991.

[4] W.B. Pennebaker and J.L. Mitchell, *JPEG - Still Image Data Compression Standard*, Van Nostrand Reinhold, New York, NY, 1993.

[5] K.R. Rao and P. Yip, *Discrete Cosine Transform - Algorithms, Advantages, Applications*, Academic Press, Boston, MA, 1990.

[6] M. Hötter and R. Thoma, "Image segmentation based on object oriented mapping parameter estimation," *Signal Processing*, vol. 15, no. 3, pp. 315–334, Oct. 1988.

[7] H.G. Musmann, M. Hötter, and J. Ostermann, "Object-oriented analysis-synthesis coding of moving images," *Signal Processing: Image Communication*, vol. 1, no. 2, pp. 117–138, Oct. 1989.

[8] N. Diehl, "Object-oriented motion estimation and segmentation in image sequences," *Signal Processing: Image Communication*, vol. 3, no. 1, pp. 23–56, Feb. 1991.

[9] J. Konrad and V.N. Dang, "Coding-oriented video segmentation inspired by MRF models," in *IEEE Int. Conf. on Image Processing, ICIP'96, Lausanne, Switzerland*, Sept. 1996, vol. 1, pp. 909–912.

[10] C. Stiller, "Object-oriented video coding employing dense motion fields," in *IEEE Int. Conf. on Acoustics, Speech, and Signal Processing, ICASSP'94, Adelaide, Australia*, Apr. 1994, vol. V, pp. 273–276.

[11] MPEG Video Group, "MPEG-4 video verification model version 11.0," in *ISO/IEC JTC1/SC29/WG11 MPEG98/N2172, Tokyo, Japan*, Mar. 1998.

[12] T. Sikora, "The MPEG-4 video standard verification model," *IEEE Trans. Circuits Syst. for Video Technol.*, vol. 7, no. 1, pp. 19–31, Feb. 1997.

[13] J. Pearl, *Probabilistic Reasoning in Intelligent Systems: Networks of Plausible Inference*, Morgan Kaufmann Publishers, San Mateo, CA, 1988.

[14] C.P. Robert, *The Bayesian Choice - A Decision-Theoretic Motivation*, Springer-Verlag, New York, NY, 1994.

[15] J. Pearl, "On evidential reasoning in a hierarchy of hypotheses," *Artificial Intelligence*, vol. 28, pp. 9–15, 1986.

[16] P.B. Chou and C.M. Brown, "The theory and practice of Bayesian image labeling," *Int. Journal of Computer Vision*, vol. 4, pp. 185–210, 1990.

[17] T.N. Pappas, "An adaptive clustering algorithm for image segmentation," *IEEE Trans. Signal Processing*, vol. 40, no. 4, pp. 901–914, Apr. 1992.

[18] C. Bouman and B. Liu, "Multiple resolution segmentation of textured images," *IEEE Trans. Pattern Analysis and Machine Intelligence*, vol. 13, no. 2, pp. 99–113, Feb. 1991.

[19] R.O. Hinds and T.N. Pappas, "An adaptive clustering algorithm for segmentation of video sequences," in *IEEE Int. Conf. on Acoustics, Speech, and Signal Processing, ICASSP'95, Detroit, MI, USA*, May 1995, vol. 4, pp. 2427–2430.

[20] S. Geman and D. Geman, "Stochastic relaxation, Gibbs distributions, and the Bayesian restoration of images," *IEEE Trans. Pattern Analysis and Machine Intelligence*, vol. PAMI-6, no. 6, pp. 721–741, Nov. 1984.

[21] J. Besag, "On the statistical analysis of dirty pictures," *Journal Royal Statist. Soc. B*, vol. 48, no. 3, pp. 259–279, 1986.

[22] F.C. Jeng and J.W. Woods, "Compound Gauss-Markov random fields for image estimation," *IEEE Trans. Signal Processing*, vol. 39, no. 3, pp. 683–697, Mar. 1991.

[23] J. Konrad and E. Dubois, "Estimation of image motion fields: Bayesian formulation and stochastic solution," in *IEEE Int. Conf. on Acoustics, Speech, and Signal Processing, ICASSP'88, New York, NY, USA*, Apr. 1988, vol. 2, pp. 1072–1075.

[24] J. Zhang and G.G. Hanauer, "The application of mean field theory to image motion estimation," *IEEE Trans. Image Processing*, vol. 4, no. 1, pp. 19–32, Jan. 1995.

[25] C. Stiller, "Object-based estimation of dense motion fields," *IEEE Trans. Image Processing*, vol. 6, no. 2, pp. 234–250, Feb. 1997.

[26] M.M. Chang, M.I. Sezan, and A.M. Tekalp, "An algorithm for simultaneous motion estimation and scene segmentation," in *IEEE Int. Conf. on Acoustics, Speech, and Signal Processing, ICASSP'94, Adelaide, Australia*, Apr. 1994, vol. V, pp. 221–224.

[27] J. Besag, "Spatial interaction and the statistical analysis of lattice systems," *Journal Royal Statist. Soc. B*, vol. 36, no. 2, pp. 192–236, 1974.

[28] R. Kindermann and J.L. Snell, *Markov Random Fields and their Applications*, American Mathematical Society, Providence, RI, 1980.

[29] H. Derin and P.A. Kelly, "Discrete-index Markov-type random processes," *Proceedings of the IEEE*, vol. 77, no. 10, pp. 1485–1510, Oct. 1989.

[30] H. Derin and H. Elliott, "Modeling and segmentation of noisy and textured images using Gibbs random fields," *IEEE Trans. Pattern Analysis and Machine Intelligence*, vol. PAMI-9, no. 1, pp. 39–55, Jan. 1987.

[31] Z. Fan and F.S. Cohen, "Textured image segmentation as a multiple hypothesis test," *IEEE Trans. Circuits and Systems*, vol. 35, no. 6, pp. 691–702, June 1988.

[32] V. Černy, "Thermodynamical approach to the traveling salesman problem: An efficient simulation algorithm," *Journal of Optimization Theory and Applications*, vol. 45, no. 1, pp. 41–51, Jan. 1985.

[33] E. Ising, "Beitrag zur Theorie des Ferromagnetismus," *Zeitschrift Physik*, vol. 31, pp. 253–258, 1925.

[34] P.J.M. van Laarhoven and E.H.L. Aarts, *Simulated Annealing: Theory and Applications*, Kluwer Academic Publishers, Dordrecht, The Netherlands, 1987.

[35] N. Metropolis, A.W. Rosenbluth, M.N. Rosenbluth, A.H. Teller, and E. Teller, "Equations of state calculations by fast computing machines," *Journal of Chemical Physics*, vol. 21, no. 6, pp. 1087–1092, June 1953.

[36] S. Kirkpatrick, C.D. Gelatt Jr., and M.P. Vecchi, "Optimization by simulated annealing," *Science*, vol. 220, no. 4598, pp. 671–680, May 1983.

[37] G.S. Fishman, *Monte Carlo - Concepts, Algorithms, and Applications*, Springer-Verlag, New York, NY, 1996.

[38] R.C. Gonzalez and R.E. Woods, *Digital Image Processing*, Addison-Wesley, Reading, MA, 1993.

[39] L.S. Davis, "A survey of edge detection techniques," *Computer Graphics and Image Processing*, vol. 4, pp. 248–270, 1975.

[40] B.S. Lipkin and A. Rosenfeld, *Picture Processing and Psychopictorics*, Academic Press, New York, NY, 1970.

[41] W. Frei and C.C. Chen, "Fast boundary detection: A generalization and a new algorithm," *IEEE Trans. Computers*, vol. C-26, no. 10, pp. 988–998, Oct. 1977.

[42] J. Canny, "A computational approach to edge detection," *IEEE Trans. Pattern Analysis and Machine Intelligence*, vol. PAMI-8, no. 6, pp. 679–698, Nov. 1986.

[43] D. Marr and E. Hildreth, "Theory of edge detection," *Proc. Royal Soc. London, Series B*, vol. 207, pp. 187–217, 1980.

[44] R.M. Haralick and L.G. Shapiro, "Image segmentation techniques," *Computer Vision, Graphics, and Image Processing*, vol. 29, pp. 100–132, 1985.

[45] C.R. Brice and C.L. Fennema, "Scene analysis using regions," *Artificial Intelligence*, vol. 1, pp. 205–226, 1970.

[46] T. Asano and N. Yokoya, "Image segmentation schema for low-level computer vision," *Pattern Recogn.*, vol. 14, pp. 267–273, 1981.

[47] J.S. Weszka, "A survey of threshold selection techniques," *Computer Graphics and Image Processing*, vol. 7, no. 2, pp. 259–265, Apr. 1978.

[48] P.K. Sahoo, S. Soltani, and A.K.C. Wong, "A survey of thresholding techniques," *Computer Vision, Graphics, and Image Processing*, vol. 41, pp. 233–260, 1988.

[49] D.M. Tsai and Y.H. Chen, "A fast histogram-clustering approach for multi-level thresholding," *Pattern Recognition Letters*, vol. 13, no. 4, pp. 245–252, Apr. 1992.

[50] S.L. Horowitz and T. Pavlidis, "Picture segmentation by a tree traversal algorithm," *Journal of the Association for Computing Machinery*, vol. 23, no. 2, pp. 368–388, Apr. 1976.

[51] Y. Fukada, "Spatial clustering procedures for region analysis," *Pattern Recogn.*, vol. 12, pp. 395–403, 1980.

[52] P.C. Chen and T. Pavlidis, "Image segmentation as an estimation problem," *Computer Graphics and Image Processing*, vol. 12, no. 2, pp. 153–172, Feb. 1980.

[53] O.J. Morris, M.J. Lee, and A.G. Constantinides, "Graph theory for image analysis: An approach based on the shortest spanning tree," *IEE Proceedings, Pt. F*, vol. 133, no. 2, pp. 146–152, Apr. 1986.

[54] Z. Wu and R. Leahy, "An optimal graph theoretic approach to data clustering: Theory and its applications to image segmentation," *IEEE Trans. Pattern Analysis and Machine Intelligence*, vol. 15, no. 11, pp. 1101–1113, Nov. 1993.

[55] W.K. Pratt, *Digital Image Processing*, John Wiley & Sons, New York, NY, 1991.

[56] J. Serra, *Image Analysis and Mathematical Morphology*, Academic Press, London, UK, 1982.

[57] F. Meyer and S. Beucher, "Morphological segmentation," *Journal of Visual Communication and Image Representation*, vol. 1, no. 1, pp. 21–46, Sept. 1990.

[58] P. Salembier and M. Pardàs, "Hierarchical morphological segmentation for image sequence coding," *IEEE Trans. Image Processing*, vol. 3, no. 5, pp. 639–651, Sept. 1994.

[59] P. Salembier, L. Torres, F. Meyer, and C. Gu, "Region-based video coding using mathematical morphology," *Proceedings of the IEEE*, vol. 83, no. 6, pp. 843–857, June 1995.

[60] P. Salembier and J. Serra, "Flat zones filtering, connected operators, and filters by reconstruction," *IEEE Trans. Image Processing*, vol. 4, no. 8, pp. 1153–1160, Aug. 1995.

[61] P. Salembier, P. Brigger, J.R. Casas, and M. Pardàs, "Morphological operators for image and video compression," *IEEE Trans. Image Processing*, vol. 5, no. 6, pp. 881–898, June 1996.

[62] L. Vincent, "Morphological grayscale reconstruction in image analysis: Applications and efficient algorithms," *IEEE Trans. Image Processing*, vol. 2, no. 2, pp. 176–201, Apr. 1993.

[63] J.G. Choi, S.W. Lee, and S.D. Kim, "Video segmentation based on spatial and temporal information," in *IEEE Int. Conf. on Acoustics, Speech, and Signal Processing, ICASSP'97, Munich, Germany*, Apr. 1997, vol. 4, pp. 2661–2664.

[64] J.G. Choi, S.W. Lee, and S.D. Kim, "Spatio-temporal video segmentation using a joint similarity measure," *IEEE Trans. Circuits Syst. for Video Technol.*, vol. 7, no. 2, pp. 279–286, Apr. 1997.

[65] I.Y. Kim and H.S. Yang, "An integration scheme for image segmentation and labeling based on Markov random field model," *IEEE Trans. Pattern Analysis and Machine Intelligence*, vol. 18, no. 1, pp. 69–73, Jan. 1996.

[66] J.S. Lim, *Two-Dimensional Signal and Image Processing*, Prentice-Hall, Englewood Cliffs, NJ, 1990.

[67] T. Meier, K.N. Ngan, and G. Crebbin, "A robust Markovian segmentation based on highest confidence first (HCF)," in *IEEE Int. Conf. on Image Processing, ICIP'97, Santa Barbara, CA, USA*, Oct. 1997, vol. I, pp. 216–219.

[68] M.L. Comer and E.J. Delp, "Multiresolution image segmentation," in *IEEE Int. Conf. on Acoustics, Speech, and Signal Processing, ICASSP'95, Detroit, MI, USA*, May 1995, vol. IV, pp. 2415–2418.

[69] P.J. Burt and E.H. Adelson, "The Laplacian pyramid as a compact image code," *IEEE Trans. Comm.*, vol. COM-31, no. 4, pp. 532–540, Apr. 1983.

[70] F. Pereira, "MPEG-4: A new challenge for the representation of audio-visual information," in *Int. Picture Coding Symposium, PCS'96, Melbourne, Australia*, Mar. 1996, vol. 1, pp. 7–16.

[71] T. Ebrahimi, "MPEG-4 video verification model: A video encoding/decoding algorithm based on content representation," *Signal Processing: Image Communication*, vol. 9, pp. 367–384, 1997.

[72] L. Chiariglione, "MPEG and multimedia communications," *IEEE Trans. Circuits Syst. for Video Technol.*, vol. 7, no. 1, pp. 5–18, Feb. 1997.

[73] J.L. Potter, "Velocity as a cue to segmentation," *IEEE Trans. Systems, Man, and Cybernetics*, pp. 390–394, May 1975.

[74] A. Verri and T. Poggio, "Motion field and optical flow: Qualitative properties," *IEEE Trans. Pattern Analysis and Machine Intelligence*, vol. 11, no. 5, pp. 490–498, May 1989.

[75] B.K.P. Horn and B.G. Schunck, "Determining optical flow," *Artificial Intelligence*, vol. 17, pp. 185–203, 1981.

[76] M. Bertero, T.A. Poggio, and V. Torre, "Ill-posed problems in early vision," *Proceedings of the IEEE*, vol. 76, no. 8, pp. 869–889, Aug. 1988.

[77] B.G. Schunck, "Image flow segmentation and estimation by constraint line clustering," *IEEE Trans. Pattern Analysis and Machine Intelligence*, vol. 11, no. 10, pp. 1010–1027, Oct. 1989.

[78] M. Bierling, "Displacement estimation by hierarchical blockmatching," in *SPIE Visual Communications and Image Processing, VCIP'88, Cambridge, MA, USA*, Nov. 1988, vol. 1001, pp. 942–951.

[79] J.R. Jain and A.K. Jain, "Displacement measurement and its application in interframe image coding," *IEEE Trans. Comm.*, vol. COM-29, no. 12, pp. 1799–1808, Dec. 1981.

[80] H. Gharavi and M. Mills, "Blockmatching motion estimation algorithms - new results," *IEEE Trans. Circuits and Systems*, vol. 37, no. 5, pp. 649–651, May 1990.

[81] A.N. Netravali and J.D. Robbins, "Motion compensated television coding: Part I," *Bell Syst. Tech. J.*, vol. 58, pp. 631–670, Mar. 1979.

[82] D.R. Walker and K.R. Rao, "Improved pel-recursive motion compensation," *IEEE Trans. Comm.*, vol. COM-32, no. 10, pp. 1128–1134, Oct. 1984.

[83] J.N. Driessen, L. Böröczky, and J. Biemond, "Pel-recursive motion field estimation from image sequences," *Journal of Visual Communication and Image Representation*, vol. 2, no. 3, pp. 259–280, Sept. 1991.

[84] A.M. Tekalp, *Digital Video Processing*, Prentice-Hall, Upper Saddle River, NJ, 1995.

[85] R.Y. Tsai and T.S. Huang, "Estimating three-dimensional motion parameters of a rigid planar patch," *IEEE Trans. Acoustics, Speech, and Signal Processing*, vol. ASSP-29, no. 6, pp. 1147–1152, Dec. 1981.

[86] G. Tziritas and C. Labit, *Motion Analysis for Image Sequence Coding*, Elsevier, Amsterdam, The Netherlands, 1994.

[87] A. Singh, *Optic Flow Computation*, IEEE Computer Society Press, Los Alamitos, CA, 1991.

[88] W.A. Smith, *Elementary Numerical Analysis*, Harper & Row, New York, NY, 1979.

[89] C. Stiller, "A statistical image model for motion estimation," in *IEEE Int. Conf. on Acoustics, Speech, and Signal Processing, ICASSP'93, Minneapolis, MN, USA*, Apr. 1993, vol. V, pp. 193–196.

[90] L. Torres and M. Kunt, *Video Coding - The Second Generation Approach*, Kluwer Academic Publishers, Dordrecht, The Netherlands, 1996.

[91] D. Zhong and S.F. Chang, "Video object model and segmentation for content-based video indexing," in *IEEE Int. Symposium on Circuits and Systems, ISCAS'97, Hong Kong*, June 1997, vol. 2, pp. 1492–1495.

[92] G. Adiv, "Determining three-dimensional motion and structure from optical flow generated by several moving objects," *IEEE Trans. Pattern Analysis and Machine Intelligence*, vol. PAMI-7, no. 4, pp. 384–401, July 1985.

[93] D.W. Murray and B.F. Buxton, "Scene segmentation from visual motion using global optimization," *IEEE Trans. Pattern Analysis and Machine Intelligence*, vol. PAMI-9, no. 2, pp. 220–228, Mar. 1987.

[94] P. Bouthemy and E. François, "Motion segmentation and qualitative dynamic scene analysis from an image sequence," *Int. Journal of Computer Vision*, vol. 10, no. 2, pp. 157–182, 1993.

[95] R.O. Duda and P.E. Hart, "Use of the Hough transformation to detect lines and curves in pictures," *Communications of the ACM*, vol. 15, no. 1, pp. 11–15, Jan. 1972.

[96] R.O. Duda and P.E. Hart, *Pattern Classification and Scene Analysis*, John Wiley & Sons, New York, NY, 1973.

[97] M.M. Chang, A.M. Tekalp, and M.I. Sezan, "Motion-field segmentation using an adaptive MAP criterion," in *IEEE Int. Conf. on Acoustics, Speech, and Signal Processing, ICASSP'93, Minneapolis, MN, USA*, Apr. 1993, vol. V, pp. 33–36.

[98] F. Dufaux, F. Moscheni, and A. Lippman, "Spatio-temporal segmentation based on motion and static segmentation," in *IEEE Int. Conf. on Image Processing, ICIP'95, Washington, DC, USA*, Oct. 1995, vol. 1, pp. 306–309.

[99] S.C. Han, L. Böröczky, and J.W. Woods, "Joint motion estimation / segmentation for object-based video coding," in *Eurasip EUSIPCO'96, Trieste, Italy*, Sept. 1996, number ME.3.

[100] F. Heitz and P. Bouthemy, "Motion estimation and segmentation using a global Bayesian approach," in *IEEE Int. Conf. on Acoustics, Speech, and Signal Processing, ICASSP'90, Albuquerque, NM, USA*, Apr. 1990, vol. 4, pp. 2305–2308.

Chapter 2

Face Segmentation

2.1 Face Segmentation Problem

The task of finding a person's face in a picture seems to be effortless for humans to perform. However it is far from simple for machine of current technology to do the same. In fact, the development of such machine or system has been widely and actively studied in the field of image understanding for the past few decades with applications such as machine vision and face recognition in mind. Moreover, in recent years, the research activities in this area have intensified as a result of its applications being extended towards video representation and coding purposes, and also of the increasing interests in multimedia.

The main objective of this research is to design a system that can find a person's face from a given image data. This problem is commonly referred to as face location, face extraction or face segmentation. Regardless of which terminology, they all share the same objective. However, note that the problem usually deals with finding the position and contour of a person's face since its location is unknown, but given the knowledge of its existence. If not, then there is also a need to discriminate between "images containing faces" and "images not containing faces". This is known as face detection. Nevertheless, this chapter focuses on face segmentation.

Although the research on face segmentation has been pursued at a feverish pace, there are still many problems yet to be fully and convincingly solved as the level of difficulty of the problem depends highly on the complexity level of the image content and its application. Many existing methods only work well on simple images with benign background and frontal view of the person's face. To cope with more complicated images and conditions, many more assumptions will have to be made.

The content of the input video typically consists of a head-and-shoulders image of a person and a background scene. The video data can either be a still image or a sequence of images, as well as in either gray-level or other color space formats. The common factors that contribute to the complexity of the image content include:

- unknown size and position of the person's face;
- variations in pose due to tilting and turning of the person's head, e.g. not having a frontal view;
- occlusions, e.g. faces that are partially hidden by other objects;
- variations in lighting condition as well as level of contrast;
- level of uniformity, structure and texture of the background scene, e.g. having a cluttered and non-uniform background.

In the case of video sequence input, there are additional factors to consider such as:

- whether the background is stationary or moving;
- and also whether there is any camera movement, such as panning, zooming and vibration caused by external means, e.g. in the case of car or hand-held videophones. With camera movement, the sequence can be considered as having an apparent foreground and background motion in addition to the actual moving foreground object.

The complexity level of the input video data will vary depending on the type of applications. Consequently, by knowing what the face segmentation algorithm will be used for, appropriate assumptions can be made to reduce the complexity of the problem.

Note that the studies of face segmentation in the past have focused on images taken in highly constrained environments. Nowadays, however, researchers are shifting their focuses towards less controlled or natural environments whereby images are taken with little or no constraint on the size and orientation of the faces, and with consideration of more complex background scene environments.

2.2 Various Approaches

Undoubtedly, there are various approaches to the face segmentation problem. These approaches usually employ shape analysis, motion analysis,

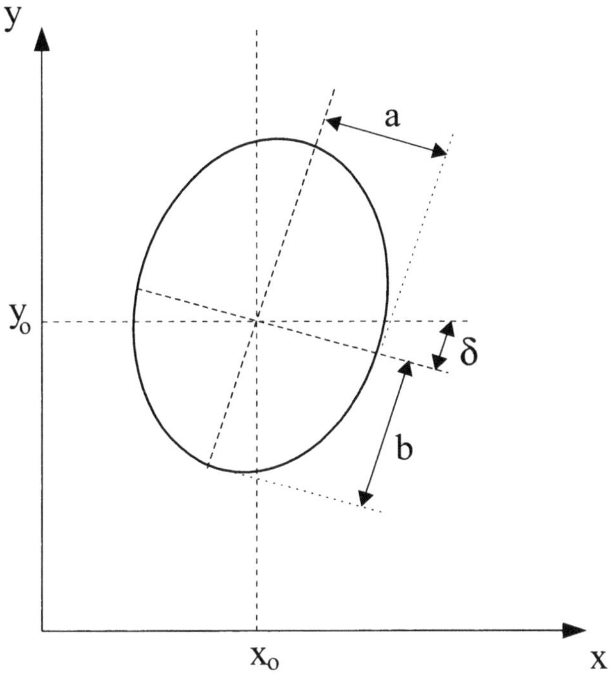

Figure 2.1: An elliptical face location model.

statistical analysis, or color analysis, or more often a combination of them. A discussion of each of these analyses is presented below.

2.2.1 Shape Analysis

One of the common methods used in the shape analysis approach is the ellipse fitting method. It is a common observation that the appearance of a human face resembles an oval shape, and hence an ellipse is employed to approximate the shape of the face. The use of this method can be found in recent papers such as those published by Eleftheriadis and Jacquin [1, 2, 3], Shimada [4], Nefian *et al.* [5], and Sobottka and Pitas [6, 7, 8].

The ellipse fitting process is applied after the possible outline of the person's head has been extracted by methods that are based on a variety of characteristics of the image, such as edge, texture, color or motion. A person's silhouette or a connected skin-color region or a moving foreground object can all lead to possible head outline.

An elliptical face location model is shown in Fig. 2.1, whereby an ellipse is defined by its center (x_o, y_o), its orientation δ and the length a and b of its minor and major axis. The objective of ellipse fitting is therefore to find x_o, y_o, δ, a and b parameters. Depending on the model accuracy, this method can be computationally intensive. For example, computation complexity can be reduced if assumption of zero head tilting (i.e., $\delta = 0$) is made; in such case, model accuracy has been compromised.

2.2.2 Motion Analysis

The use of motion information will require the input data to be a video sequence instead of just a single still image. This approach involves the interframe operator. The simplest and also the most popular of its kind is the frame difference operator. This operator is used to detect changed area due to object movement by subtracting two successive image frames. Hence it can partition a moving person from a stationary background. Generally, for motion analysis to work, the input images have to be restricted to only those with stationary backgrounds, moreover, there may also be a need to distinguish the person's face from other moving foreground objects. In addition, this method is very sensitive to noise and it cannot produce useful results consistently. Consequently, the interframe operator is typically used to complement other approaches in the pre-processing or post-processing domain. In some face segmentation methodologies, movement of the face is an essential feature for the initial face localization process because the appearance of the face is unknown. A simple frame difference between two successive images offers rapid pinpointing of interesting parts of the image to other processing modules. For instance, frame difference operator is used to obtain the silhouette of a person before the ellipse fitting method is applied [1, 4]. An approach that used frame difference operator to obtain movement information and then combined with color and shape information can be found in [9] and [10]. Another multi-modal system that used shape, color and motion information but with a slightly more sophisticated motion analysis that helps suppress noise can be found in [11].

2.2.3 Statistical Analysis

The statistical analysis approach offers sound theoretical based techniques such as higher order statistics [12, 13], statistical feature detectors [14] and maximum likelihood detection [15]. These techniques, however, are computationally intensive and rely on many assumptions for it to operate in a practical application. Furthermore, accurate and reliable results are difficult to achieve in this approach.

2.2.4 Color Analysis

In recent years, a new approach that uses color information has been introduced to the face segmentation problem. This approach is superior to the others in many ways. For example, unlike ellipse fitting, color analysis is robust against variable size and orientation of the person's face. It can also cope with variable lighting condition as well as high level of structure and texture of the background scene. In addition, color analysis requires only a single image, and therefore background and camera motions do not pose a problem.

The study of color information has gained increasing attention since its introduction to the face segmentation problem. Some recent publications that have reported this study include those by Li and Forchheimer [16], Hunke and Waibel [9], Matsuhashi *et al.* [17], Chen *et al.* [18], Sobottka and Pitas [6], Saxe and Foulds [19], Kjeldsen and Kender [20], Chai and Ngan [21], Cornall and Pang [22], and Zhang *et al.* [23]. They have all shown, in one way or another, that color is a powerful descriptor that has practical use in the extraction of face location. Although the use of color information and its potential to become a useful tool in face segmentation problem have been much talked about some years ago, a robust universal model of human skin color has only been realized recently.

The color information is typically used for region rather than edge segmentation. This region segmentation can be classified into two general approaches as illustrated in Fig. 2.2. One approach is to employ color as a feature for partitioning an image into a set of homogeneous regions. For instance, the color component of the image can be used in the region growing technique as demonstrated in [24], or as a basis for a simple thresholding technique as shown in [23]. The other approach, however, makes use of color as a feature for identifying a specific object in an image. In this case, the skin color can be used to identify the human face. This is feasible because human faces have a special color distribution that differs significantly (although not entirely) from those of the background objects. Hence this approach requires a color map that models the skin color distribution characteristics.

The skin-color map can be derived from two approaches, one approach is to pre-define or manually obtain the map that suits an individual [16] while the other approach is to design a reference map for all people [21, 25, 22, 7]. The modeling of human skin color is closely looked at in Section 2.4.

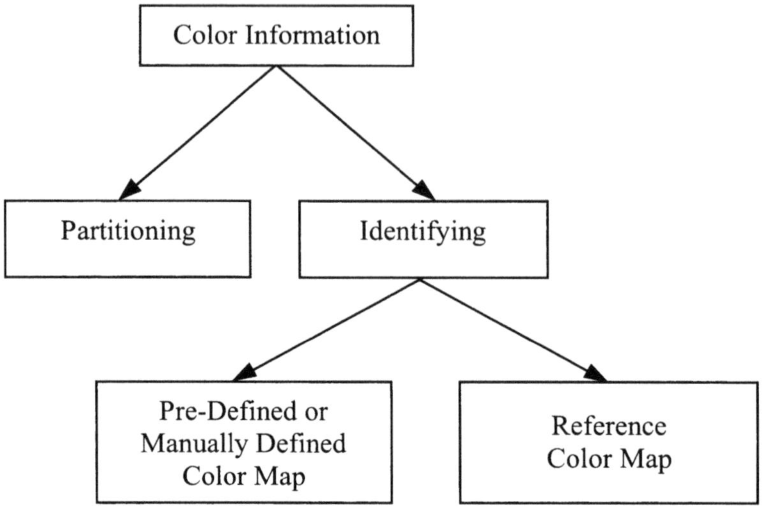

Figure 2.2: The use of color information for region segmentation.

2.3 Applications

Face segmentation holds an important key to future advances in human-to-human and human-to-machine communications. The significance of this problem can be illustrated by its vast applications.

The segmentation of facial region provides a content-based representation of the image where it can be exploited for numerous purposes such as image/video coding, manipulation, enhancement, indexing, modeling, pattern recognition, object tracking and human interface study. In fact, the information of face position can be applied to a myriad of systems that deal with human face video contents, and some of the major applications are discussed below.

2.3.1 Coding Area of Interest with Better Quality

The knowledge of the speaker's face position can be used to improve the subjective quality of the encoded videophone sequence by coding the facial image region that is of interest to viewers at higher quality. It is, however, achieved at the expense of reducing the objective quality of the less important background scene. This method is commonly referred to as foreground/background [26] or knowledge-based [27] or model-assisted [1]

Figure 2.3: *Carphone* image with the area of interest (i.e., facial region) encoded at higher quality than the background area using a foreground/background coding technique described in [30].

coding technique. This technique allows the facial area to be coded with high fidelity and hence produces images with better-rendered facial features. The use of face segmentation information in video coding has proven to be a very popular topic in recent time. This technique has been integrated and studied on coders such as wavelet [28, 29], 3D subband-based [1, 2], H.261 [3, 30, 31] and H.263 [26, 32] videoconferencing coders.

Fig. 2.3 illustrates an encoded image obtained from using the method described in [30]. The facial region, which is the area of interest, of this so-called *Carphone* image was encoded at a higher quality than the background scene. Notice that the background scene contains high level of distortion while the facial area is clear and sharp. This approach essentially produces a spatially variable quality encoded image. By taking account of the psychovisual consideration, the removal of the objectionable blocking artifacts from the area of the picture that is of importance to viewers has provided a significantly better subjective viewing quality.

2.3.2 Content-based Representation and MPEG-4

Face segmentation is a useful tool to facilitate MPEG-4 [33] content-based functionality. It provides content-based representation of the image, which can subsequently be used for coding, editing or other interactivity purposes. For example, the extracted facial region can be defined as a video object (VO) while the remaining background image region can be defined as another VO [34]. Depending upon its content, each VO can be encoded using different types of coder and coding parameters.

2.3.3 3D Human Face Model Fitting

The delimitation of the person's face is the fundamental requirement of 3D human face model fitting used in model-based coding, computer animation and morphing. Interested readers of model-based coding are referred to Chapter 4. Work related to adaptation of generic 3D face model to the actual face can be found in [24], [35] and [36]. Fig. 2.4 shows the *Miss America* image and the 3D wire frame model fitted onto her face.

2.3.4 Image Enhancement

Face segmentation information can be used in a post-processing task for enhancing images, such as automatic adjustment of tint in the facial region. Satyanarayana and Dalal [37] proposed an intelligent color enhancement module that automatically adjusts the color saturation on a field-by-field basis for television pictures, as these pictures are not always at their best color saturation settings. In their approach, incoming pictures are first classified into facial tone and non-facial tone categories so that any oversaturated or undersaturated pictures in both facial and non-facial tone categories can be detected and corrected.

2.3.5 Face Recognition, Classification and Identification

Finding the person's face is the first important step in the human face recognition, classification and identification systems. Readers who are interested in face recognition may find references [38], [39], [40] and [41] useful.

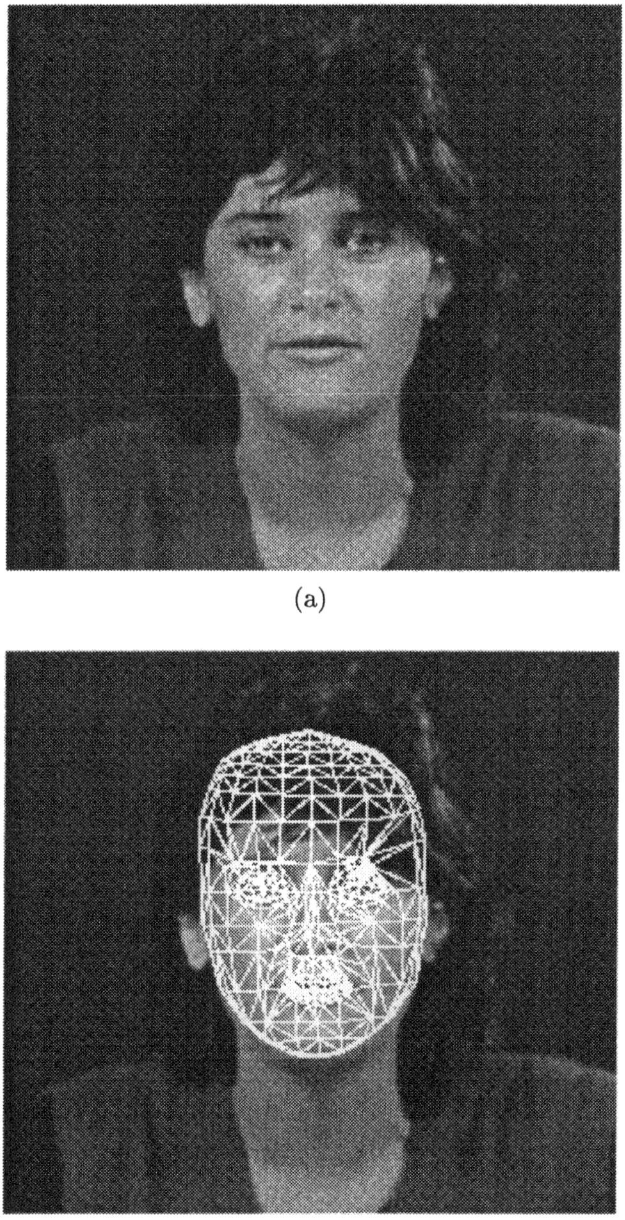

(a)

(b)

Figure 2.4: (a) A still image from the *Miss America* video sequence that shows a neutral (i.e., no expression exerted on the face), upright face in front of a plain background, and (b) the 3D wire frame model fitted onto the face.

2.3.6 Face Tracking

Face location can be used to design a video camera system that tracks a person's face in a room. It can be used as part of an intelligent vision system or simply in video surveillance. For example, Hunke and Waibel [9] proposed a face tracker that keeps a person's face located at all times in an arbitrary environment and maintains a centered position and relatively constant size of the face within the image by manipulating the orientation and zoom of the camera. Similarly, Collobert *et al.* [10] described a face localization and tracking technique that has application in automatic image framing. In the framework of an individual audiovisual communication terminal, automatic framing allows a person to move freely around the room while still being continuously framed by the camera. McKenna and Gong [42] dealt with the task of tracking faces in complex and low image quality scenes arise from surveillance applications. In addition, face tracker can be used to provide user location as input to a beam steering system. An application so-called adaptive beamforming uses a microphone array to efficiently pick up the speech produced by a speaker, who is free to move and free from attached microphone, while reducing competing acoustic signals from other sources.

2.3.7 Facial Expression Study

Besides face segmentation and tracking, the extraction of facial features is also a prerequisite for lip reading and facial expression estimation in human interface study. Wu *et al.* [43] presented a method that works hierarchically. It first locates the position of human face then the position of facial features, after that it approximates their contours and then extracts the facial feature points. An earlier work on facial feature extraction and facial expression tracking can be found in [44]. Recent works on lip movement analysis and synthesis can be found in [45] and [46].

2.3.8 Multimedia Database Indexing

In recent years, we have seen increased activities in digitizing and integrating many media such as broadcasting, publishing, movies and communications into the so-called multimedia environment. As a consequence, there is a need to structure a video database for indexing and search. In terms of video data with human face content, face indexing can be used to classify the television news articles or video documents into the proper categories such as politics, economics, culture, amusements, sports and so on [47]. Conversely, face indexing can also be used to retrieve the associated articles

Figure 2.5: *Foreman* image with a white contour highlighting the facial region.

or documents.

2.4 Modeling of Human Skin Color

As mentioned previously, the color information can be used as a feature for identifying a person's face in an image. This approach is feasible because human faces have indeed a special color distribution that differs significantly, although not entirely, from those of the background objects. Here, the design of a color map that models the skin color distribution characteristics is discussed.

The skin-color map can be derived in two ways on account that not all faces have identical color feature. One approach is to pre-define or manually obtain the map such that it suits only an individual color feature. For example, the skin color feature of the subject in a standard head-and-shoulders test image called *Foreman* is to be obtained. Although this is a color image in YCrCb format, its gray-scale version is shown in Fig. 2.5. The figure also shows a white contour highlighting the facial region. The histograms of the color information (i.e., Cr and Cb values) bounded within this contour are obtained as shown in Fig. 2.6. The diagrams show that the chrominance values in the facial region are narrowly distributed, which implies that the skin color is fairly uniform. Therefore this individual color feature can simply be defined by the presence of Cr values within, say, 136 and 156, and Cb values within 110 and 123. Using these ranges of values, the subject's face in another frame of *Foreman* and also in a completely different scene (a standard test image called *Carphone*) are located, as can be seen in Figs. 2.7 and 2.8 respectively. This approach was suggested in a very general manner by Li and Forchheimer in [16].

In another approach, the skin-color map can be designed by adopting histograming technique on a given set of training data and subsequently used as a reference for any human face. Such method was successfully adopted by Chai and Ngan [21, 34], Sobottka and Pitas [7], and Cornall and Pang [22].

Among the two approaches, the first is likely to produce better segmentation result in terms of reliability and accuracy by virtue of using a precise map. However, it is realized at the expense of having a face segmentation process that is either too restrictive because it uses a pre-defined map, or requires human interaction to manually define the necessary map. Therefore, the second approach is more practical and appealing as it attempts to cater for all personal color features in an automatic manner, albeit less precise. This, however, raises a very important issue regarding the coverage of all human races with one reference map. In addition, the general use of skin-color model for region segmentation prompts two other questions, namely, which color space to use, and how to distinguish other parts of the body and background objects with skin color appearance from the actual facial region.

2.4.1 Color Space

An image can be presented in a number of different color space models [48, 49], such as:

- *RGB*: This stands for the three primary colors: red, green and blue. It is a hardware-oriented model and well known for its color monitor display purpose.

- *HSV*: An abbreviation of *Hue-Saturation-Value*. *Hue* is a color attribute that describes a pure color, while *saturation* defines the relative purity or the amount of white light mixed with a *hue*, and *value* refers to the brightness of the image. This model is commonly used for image analysis.

- *YCrCb*: This is yet another hardware-oriented model. However, unlike the RGB space, here the luminance is separated from the chrominance data. The Y value represents the luminance (or brightness) component

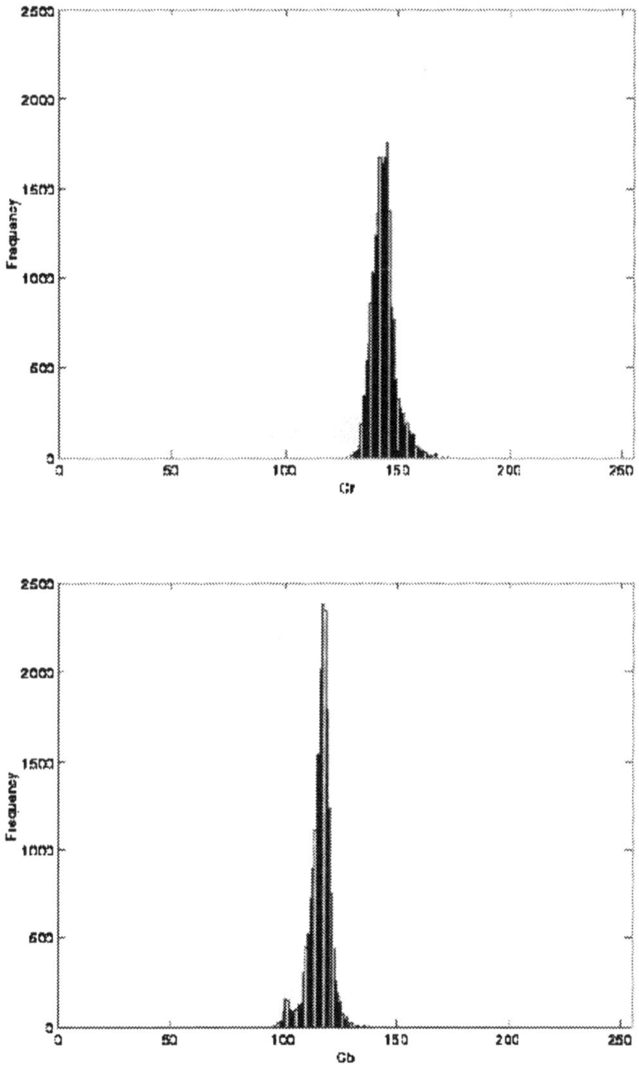

Figure 2.6: The histograms of Cr and Cb components in the facial region.

Figure 2.7: *Foreman* image and the result of color segmentation using his own skin-color map.

while the Cr and Cb values, also known as the color difference signals, represent the chrominance component of the image.

These are some of the color space models available in image processing. Therefore it is important to choose the appropriate color space for modeling human skin color. The factors that need to be considered are *application* and *effectiveness*. The intended purpose of the face segmentation will usually determine which color space to use, at the same time, it is essential that an effective and robust skin-color model can be derived from the given color space. For instance, Chai and Ngan [25] proposed the use of the YCrCb color space, and the reason is twofold. First, an effective use of the chrominance information for modeling human skin color can be achieved in this color space. Second, this format is typically used in video coding, and there-

Figure 2.8: *Carphone* image and the result of color segmentation using the same pre-defined skin-color map as the one used in Fig. 2.7.

fore the use of the same, instead of another, for segmentation will avoid the extra computation required in conversion. On the other hand, both Sobottka and Pitas [7], and Saxe and Foulds [19] have opted for the HSV color space as it is compatible to the human color perception, and the *hue* and *saturation* components have also been reported to be sufficient discriminating color information for modeling skin color. However, this color space is not suitable for video coding. Hunke and Waibel [9], and Graf *et al.* [11] used a normalized RGB color space. The normalization was employed to minimize the dependence on the luminance values.

On this note, it is interesting to point out that unlike the YCrCb and HSV color spaces whereby the brightness component is decoupled from the color information of the image, the RGB color space is not. Therefore,

Graf *et al.* have suggested pre-processing calibration in order to cope with unknown lighting condition. From this point of view, the skin-color model derived from the RGB color space will be inferior to those obtained from the YCrCb or HSV color spaces. Based on the same reasoning, Chai and Ngan [50] hypothesized that a skin-color model can remain effective regardless of the variation of skin color (e.g. black, white or yellow) if the derivation of the model is independent of the brightness information of the image. Further discussions are provided later.

2.4.2 Limitations of Color Segmentation

A simple region segmentation based on the skin-color map can provide accurate and reliable results if there is a good contrast between skin color and those of the background objects. However, if the color characteristics of the background are similar to that of the skin, then pinpointing the exact face location is more difficult as there will be more falsely detected background regions with skin color appearance. Note that in the context of face segmentation, other parts of the body are also considered as background objects. There are a number of methods to discriminate between the face and the background objects, and they include the use of other cues such as motion and shape.

Provided the temporal information is available and *a priori* knowledge of a stationary background and no camera motion, simple motion analysis can be incorporated into the face localization system to identify non-moving skin-color regions as background objects. Alternatively, shape analysis involving ellipse-fitting can also be employed to identify the facial region from among the detected skin-color regions. An ellipse is used to approximate a human face as it resembles an oval shape. Alternatively, a set of regularization processes can be used, which are based on the spatial distribution and the corresponding luminance values of the detected skin-color pixels. This approach overcomes the restriction of motion analysis and avoids the extensive computation of the ellipse-fitting method.

In addition to poor color contrast, there are other limitations of color segmentation when input image is taken in some particular lighting conditions. The color process will encounter some difficulties when input image has either:

1. a 'bright spot' on the subject's face due to reflection of intense lighting, or

2. a dark shadow on the face as a result of the use of strong directional lighting that has partially blackened the facial region, or

3. captured with the use of color filters.

Note that these types of images (particularly in case 1 and 2) are posing great technical challenges not only to the color segmentation approach but also to a wide range of other face segmentation approaches, especially those approaches that utilize edge image, intensity image or facial feature points extraction.

However, it has been found that the color analysis approach is immune to moderate illumination changes and shading resulting from slightly unbalance light source, as these conditions do not alter the chrominance characteristics of the skin-color model.

2.5 Skin Color Map Approach

Here, a practical solution to the face segmentation problem is presented, which was proposed by Chai and Ngan [21, 25, 50]. Their method can automatically segment out the person's face from a given image that consists of a head-and-shoulders view of the person and a complex background scene. It involves a fast, reliable and effective algorithm that exploits the spatial distribution characteristics of human skin color. A robust universal skin-color map is derived and used on the chrominance component of the input image to detect pixels with skin color appearance. Then, based on the spatial distribution of the detected skin-color pixels and their corresponding luminance values, the algorithm employs a set of novel regularization processes to reinforce regions of skin-color pixels that are more likely to belong to the facial regions and eliminate those that are not. The performance of this face segmentation algorithm is illustrated by some simulation results carried out on various head-and-shoulders test images.

2.5.1 Face Segmentation Algorithm

This approach is automatic in the sense that it uses an unsupervised segmentation algorithm, and hence no manual adjustment of any design parameter is needed in order to suit any particular input image. Moreover, the algorithm can be implemented in real-time and its underlying assumptions are minimal. In fact, the only principal assumption is that the person's face must be present in the given image since the face is to be located and not detected. Thus, the input information required by the algorithm is a single color image that consists of a head-and-shoulders view of the person and a background scene, and the facial region can be as small as only a 32×32

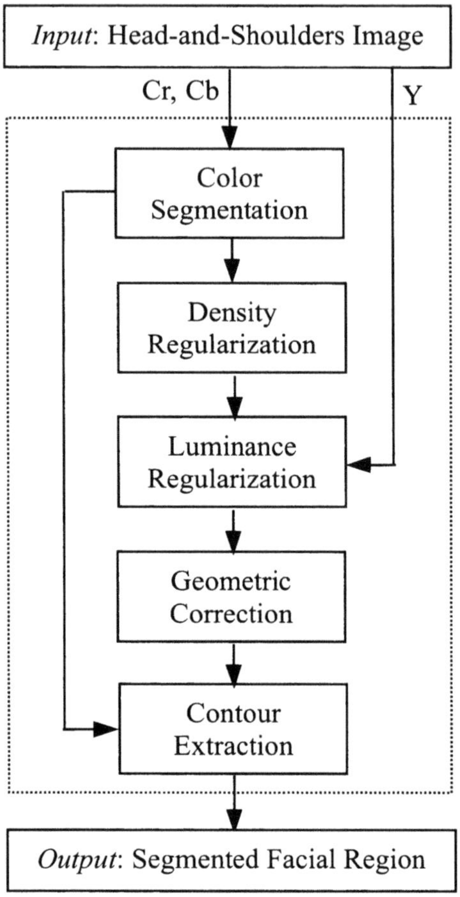

Figure 2.9: Block diagram of the automatic face segmentation algorithm.

pixels window (or 1%) of a CIF-size (352 × 288) input image. The format of the input image is to follow the YCrCb color space, based on the reason given previously. The spatial sampling frequency ratio of Y, Cr and Cb is 4:1:1. So, for a CIF-size image, Y has 288 lines and 352 pixels per line while both Cr and Cb have 144 lines and 176 pixels per line each.

The algorithm consists of five operating stages, as outlined in Fig. 2.9. It begins by employing a low-level process like color segmentation in the first stage, and then it uses higher-level operations that involve some heuristic knowledge about the local connectivity of the skin-color pixels in the later stages. Thus each stage makes full use of the result yielded by its preceding

Figure 2.10: The input image of *Miss America*.

stage in order to refine the output result. Consequently, all the stages must be carried out progressively according to the given sequence.

A detail description of each stage is presented below. For illustration purposes, a studio-based head-and-shoulders image called *Miss America* is used to present the intermediate results obtained from each stage of the algorithm. This input image is shown in Fig. 2.10.

2.5.2 Stage One - Color Segmentation

The first stage of the algorithm involves the use of color information in a fast, low-level region segmentation process. The aim is to classify pixels of the input image into skin-color and non-skin-color. To do so, a skin-color reference map in YCrCb color space has been devised.

The skin-color region can be identified by the presence of a certain set of chrominance (i.e., Cr and Cb) values that is narrowly and consistently distributed in the YCrCb color space. The location of these chrominance values has been found and can be illustrated using the CIE chromaticity diagram as shown in Fig. 2.11. Let R_{Cr} and R_{Cb} denote the respective ranges of Cr and Cb values that correspond to skin color, which subsequently define our skin-color reference map. The ranges that have been found to be the most suitable for all the input images are $R_{Cr} = [133, 173]$ and $R_{Cb} = [77, 127]$. This map has been experimentally proven to be very robust against different types of skin color. The conjecture is that the different skin color that we perceive from video image cannot be differentiated from the chrominance information of that image region. So, a map that is derived from Cr and

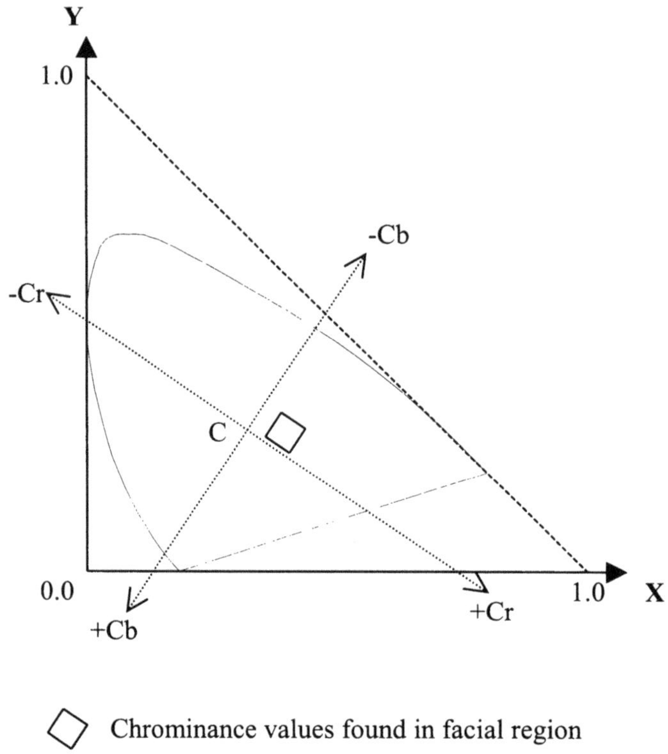

Figure 2.11: Skin-color region in CIE chromaticity diagram.

Cb chrominance values will remain effective regardless of skin color variation (see Section 2.5.7 for the experimental results). Moreover, the intuitive justification for the manifestation of similar Cr and Cb distributions of skin color of all human races is that the apparent difference in skin color that viewers perceive is mainly due to the darkness or fairness of the skin; these features are characterized by the difference in the brightness of the color, and the brightness of the color is governed by Y value but not Cr and Cb values.

With this skin-color reference map, the color segmentation can now begin. Since only the color information is to be utilized, the segmentation requires only the chrominance component of the input image. Consider an input image of $M \times N$ pixels and therefore the dimension of Cr and Cb is $M/2 \times N/2$. The output of the color segmentation, and hence stage one of

Figure 2.12: Bitmap produced by stage one.

the algorithm, is a bitmap of $M/2 \times N/2$ size, described as

$$O_1(x,y) = \begin{cases} 1, & \text{if } [\mathrm{Cr}(x,y) \in R_{Cr}] \bigcap [\mathrm{Cb}(x,y) \in R_{Cb}] \\ 0, & \text{otherwise} \end{cases} \tag{2.1}$$

where $x = 0, \ldots, M/2-1$ and $y = 0, \ldots, N/2-1$. The output pixel at point (x, y) is classified as skin-color and set to 1 if both the Cr and Cb values at that point fall inside their respective ranges, R_{Cr} and R_{Cb}. Otherwise, the pixel is classified as non-skin-color and set to 0. To illustrate this, color segmentation is performed on the input image of *Miss America*, and the bitmap produced can be seen in Fig. 2.12. The output value of 1 is shown in black while the value of 0 is shown in white (this convention will be used throughout this chapter).

Among all the stages, this first stage is the most vital one. Based on the model of the human skin color, the color segmentation has to remove as many pixels as possible that are unlikely to belong to the facial region while catering for a wide variety of skin color. However, if it falsely removes too many pixels that belong to the facial region, then the error will propagate down the remaining stages of the algorithm, and consequently causes a failure to the entire algorithm. Hence this has to be taken into account when designing a skin-color reference map.

Nonetheless, the result of color segmentation is the detection of pixels in facial area and may also include other areas where the chrominance values coincide with those of the skin color (as is the case in Fig. 2.12). Hence the successive operating stages of the algorithm are used to remove these unwanted areas.

2.5.3 Stage Two - Density Regularization

This stage considers the bitmap produced by the previous stage to contain the facial region that is corrupted by noise. The noise may appear as small holes on the facial region due to undetected facial features such as eyes and mouth, or it may also appear as objects with skin-color appearance in the background scene. Therefore this stage performs simple morphological operations [51] such as *dilation* to fill in any small hole in the facial area and *erosion* to remove any small object in the background area. The intention is not necessarily to remove entirely, but to reduce the amount and size of the noise.

To distinguish between these two areas, regions of the bitmap that have higher probability of being the facial region need to be identified. The probability measure used here is derived from the observation that the facial color is very uniform, and therefore the skin-color pixels belonging to the facial region will appear in a large cluster, while the skin-color pixels belonging to the background may appear as large clusters or small isolated objects. Thus, the density distribution of the skin-color pixels detected in stage one is studied. An $M/8 \times N/8$ array of density values called density map, $D(x, y)$, is computed as

$$D(x, y) = \sum_{i=0}^{3} \sum_{j=0}^{3} O_1(4x + i, 4y + j) \tag{2.2}$$

where $x = 0, \ldots, M/8 - 1$ and $y = 0, \ldots, N/8 - 1$. It first partitions the output bitmap of stage one, $O_1(x, y)$, into non-overlapping groups of 4×4 pixels, then it counts the number of skin-color pixels within each group and assigns this value to the corresponding point of the density map.

According to the density value, each point is classified into three types, namely zero ($D = 0$), intermediate ($0 < D < 16$) and full ($D = 16$). A group of points with zero-density value will represent a non-facial region, while a group of full-density points will signify a cluster of skin-color pixels and a high probability of belonging to a facial region. Any point of intermediate-density value will indicate the presence of noise. The density map of *Miss America* with the three density classifications is depicted in Fig. 2.13. The point of zero density is shown in white, intermediate density in gray and full density in black.

Once the density map is derived, the process termed as density regularization can then begin. This involves the following three steps:

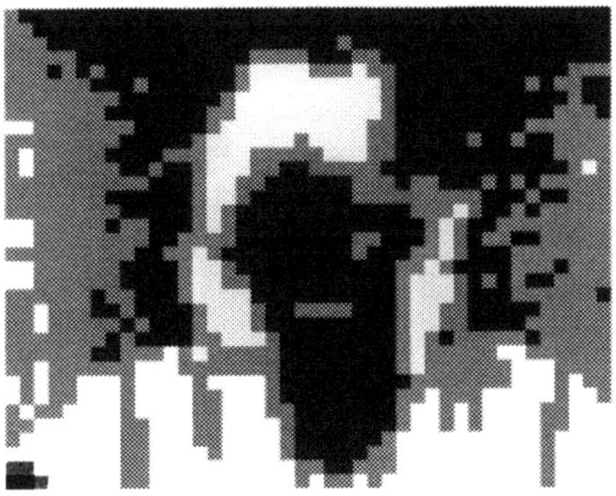

Figure 2.13: The density map after classification.

1. Discard all points at the edge of the density map, i.e., set

$$D\Big(0, y\Big) = D\Big(\frac{M}{8} - 1, y\Big) = D\Big(x, 0\Big) = D\Big(x, \frac{N}{8} - 1\Big) = 0 \qquad (2.3)$$

   for all $x = 0, \ldots, M/8 - 1$ and $y = 0, \ldots, N/8 - 1$.

2. Erode[1] any full-density point (i.e., set to 0) if it is surrounded by less than 5 other full-density points in its local 3×3 neighborhood.

3. Dilate[1] any point of either zero or intermediate density (i.e., set to 16) if there are more than 2 full-density points in its local 3×3 neighborhood.

After this process, the density map is converted to the output bitmap of stage two as

$$O_2(x, y) = \begin{cases} 1, & \text{if } D(x, y) = 16 \\ 0, & \text{otherwise} \end{cases} \qquad (2.4)$$

for all $x = 0, \ldots, M/8 - 1$ and $y = 0, \ldots, N/8 - 1$.

The result of stage two for the *Miss America* image is displayed in Fig. 2.14. Note that this bitmap is now four times lower in spatial resolution than that of the output bitmap in stage one, and eight times lower than the original input image.

[1]Readers are referred to Section 1.3.1 or reference [52] for the basic working knowledge of *erosion* and *dilation* operations.

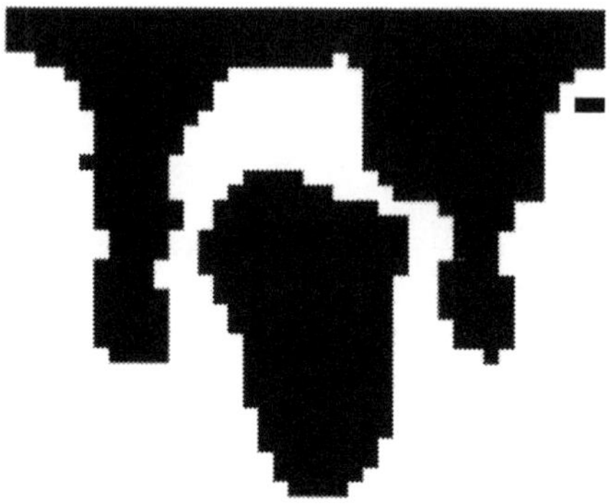

Figure 2.14: Bitmap produced by stage two.

2.5.4 Stage Three - Luminance Regularization

In a typical videophone image, the brightness is non-uniform throughout the facial region, while the background region tends to have a more even distribution of brightness. Hence based on this characteristic, background region that was previously detected due to its skin color appearance can be further eliminated.

The analysis employed in this stage involves the spatial distribution characteristics of the luminance values since they define the brightness of the image. Standard deviation is used as the statistical measure of the distribution. Note that the size of the previously obtained bitmap $O_2(x, y)$ is $M/8 \times N/8$, and hence each point corresponds to a group of 8×8 luminance values, denoted by W, in the original input image. For every skin-color pixels in $O_2(x, y)$, the standard deviation, denoted as $\sigma(x, y)$, of its corresponding group of luminance values can be calculated using

$$\sigma(x, y) = \sqrt{\mathrm{E}[W^2] - (\mathrm{E}[W])^2}. \tag{2.5}$$

Fig. 2.15 depicts the standard deviation values calculated for the *Miss America* image.

If the standard deviation is below a value of 2 then the corresponding 8×8 pixels region is considered as too uniform, and therefore, unlikely to be part of the facial region. As a result, the output bitmap of stage three, $O_3(x, y)$, is derived as

$$O_3(x, y) = \begin{cases} 1, & \text{if } O_2(x, y) = 1 \text{ and } \sigma(x, y) \geq 2 \\ 0, & \text{otherwise} \end{cases} \tag{2.6}$$

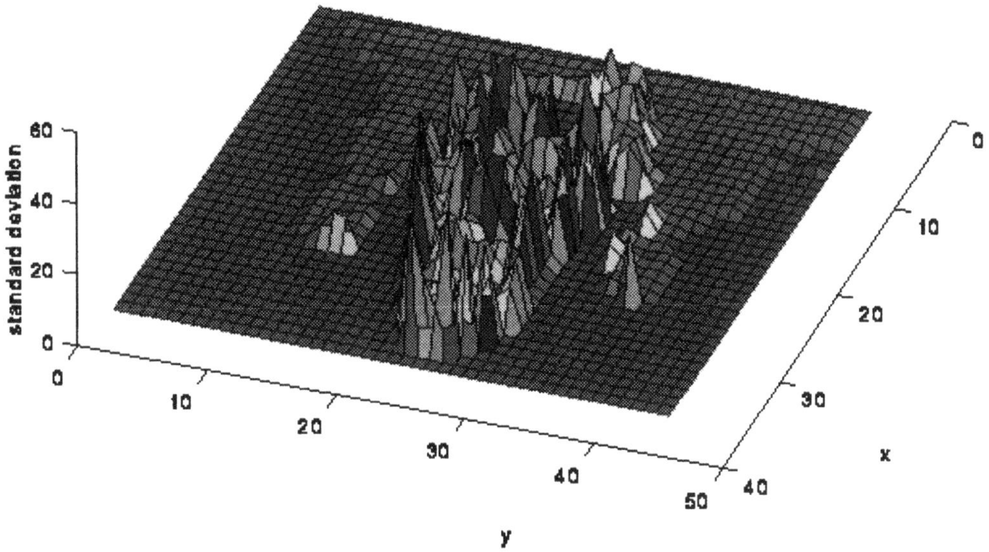

Figure 2.15: Standard deviation values of the detected pixels in $O_2(x, y)$.

for all $x = 0, \ldots, M/8 - 1$ and $y = 0, \ldots, N/8 - 1$. The output bitmap of this stage for the *Miss America* image is presented in Fig. 2.16. The figure shows that a significant portion of the unwanted background region was eliminated at this stage.

2.5.5 Stage Four - Geometric Correction

A horizontal and vertical scanning process is performed to identify the presence of any odd structure in the previously obtained bitmap, $O_3(x, y)$, and subsequently remove it. This is to ensure that a correct geometric shape of the facial region is obtained. However, prior to the scanning process, the face segmentation algorithm attempts to further remove any more noise by using a similar technique as initially introduced in stage two. Therefore, a pixel in $O_3(x, y)$ with the value of 1 will remain as detected pixel if there are more than 3 other pixels, in its local 3×3 neighborhood, with the same value. At the same time, a pixel in $O_3(x, y)$ with the value of 0 will be reconverted to the value of 1 (i.e., as a potential pixel of the facial region) if

Figure 2.16: Bitmap produced by stage three.

it is surrounded by more than 5 pixels, in its local 3×3 neighborhood, with the value of 1. These simple procedures will ensure that noise appearing on the facial region are filled in and that isolated noise objects on the background are removed. Then, it commences the horizontal scanning process on the "filtered" bitmap. Its searches for any short continuous run of pixels that are assigned with the value of 1. For a CIF-size image, the threshold for a group of connected pixels to belong to the facial region is 4. Therefore, any group of less than 4 horizontally connected pixels with the value of 1 will be eliminated and assigned to 0. Similar process is then performed in the vertical direction. The rationale behind this method is that, based on our observation, any such short horizontal or vertical run of pixels with the value of 1 is unlikely to be part of a reasonable size and well detected facial region. As a result, the output bitmap of this stage should contain the facial region with minimal or no noise, as demonstrated in Fig. 2.17.

2.5.6 Stage Five - Contour Extraction

In this final stage, the $M/8 \times N/8$ output bitmap of stage four is converted back to the dimension of $M/2 \times N/2$. To achieve the increase in spatial resolution, it utilizes the edge information that is already made available by the color segmentation in stage one. Therefore all the boundary points in the previous bitmap will be mapped into the corresponding group of 4×4 pixels with the value of each pixel as defined in the output bitmap of stage one. The representative output bitmap of this final stage of the algorithm is shown in Fig. 2.18.

Figure 2.17: Bitmap produced by stage four.

Figure 2.18: Bitmap produced by stage five.

2.5.7 Experimental Results

The experimental results of this face segmentation methodology is organized into two parts. The first part presents the testing of the skin-color reference map, whereas the second part shows the results of the face segmentation algorithm that makes use of the skin-color reference map.

2.5.7.1 Skin-Color Reference Map Results

The skin-color reference map is intended to work on a wide range of skin color including people of European, Asian and African decent. Therefore, to show that it works on subject with skin color other than white (i.e., as it is the case with *Miss America* image), the same map is used to perform the color segmentation process on subjects with black and yellow skin color. The results obtained were very good, as can be seen in Figs. 2.19 and 2.20. The skin-color pixels were correctly identified in both input images with only a small amount of noise appearing, as expected, in the facial regions and the background scene, which can be removed by the remaining stages of the algorithm.

Further testing of the skin-color map was carried out using 30 samples of images. Skin colors were classified into 3 classes: white, yellow and black. 10 samples, each of which contained the facial region of different subject and captured in different lighting condition, were taken from each class to form the test set. Three normalized histograms for each sample in the separate Y, Cr and Cb components is constructed. The normalization process for the histograms was used to account for the variation of facial region size in each sample. The average results from the 10 samples of each class were taken. These average normalized histogram results for class of white, yellow and black are presented in Figs. 2.21, 2.22 and 2.23 respectively.

Since all samples were taken from different and unknown lighting conditions, the histograms of Y component for all three classes cannot be used to verify whether the variations of luminance values in these image samples were caused by the different skin color or by the different lighting condition. However the use of such samples illustrated that the variation in illumination does not seem to affect the skin color distribution in the Cr and Cb components. On the other hand, the histograms of Cr and Cb components for all three classes clearly showed that the chrominance values are indeed narrowly distributed, and more importantly, the distributions are consistent across different classes. This demonstrated that an effective skin-color reference map could be achieved based on the Cr and Cb components of the input image.

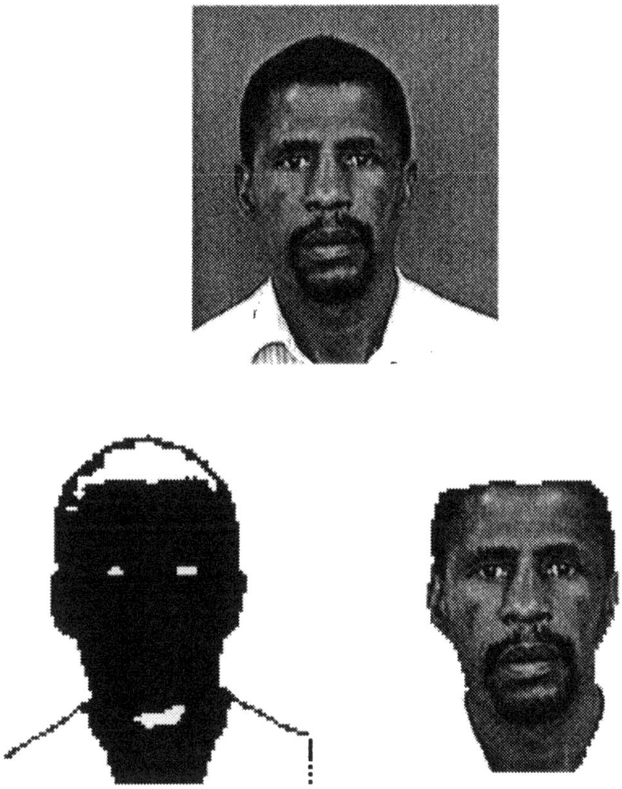

Figure 2.19: The results produced by the color segmentation process in stage one and the final output of the face segmentation algorithm, which was performed on subject with black skin color.

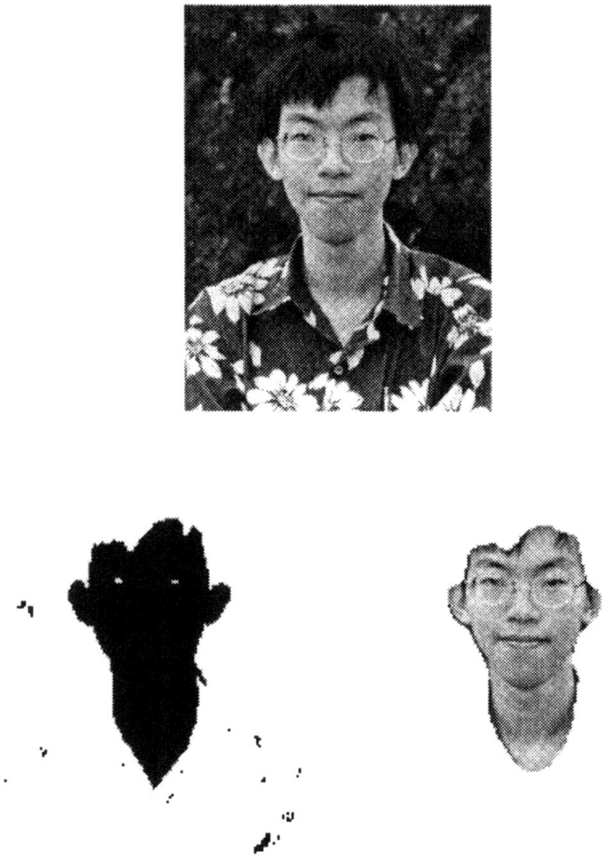

Figure 2.20: The results produced by the color segmentation process in stage one and the final output of the face segmentation algorithm, which was performed on subject with yellow skin color.

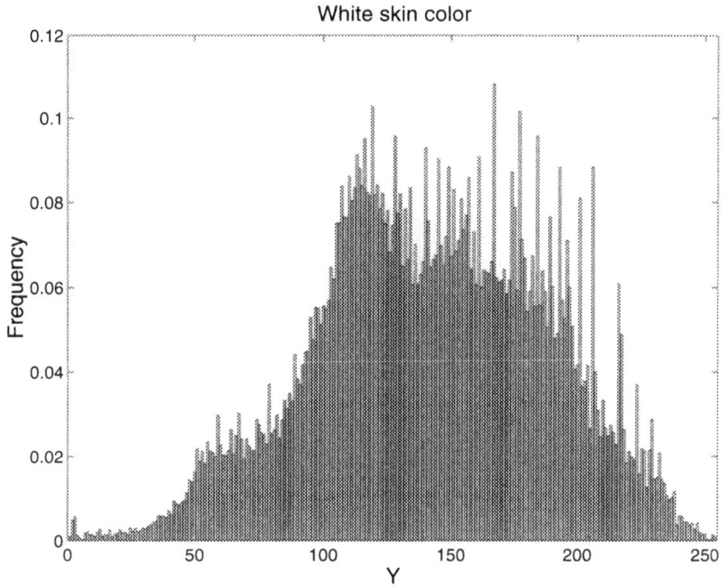

Figure 2.21: The histograms of Y, Cr and Cb values for white skin color.

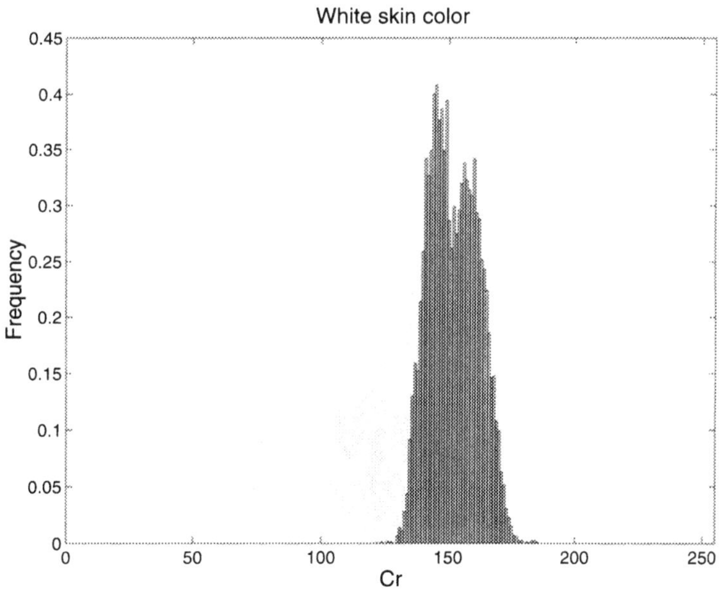

Figure 2.21: *Cont.*

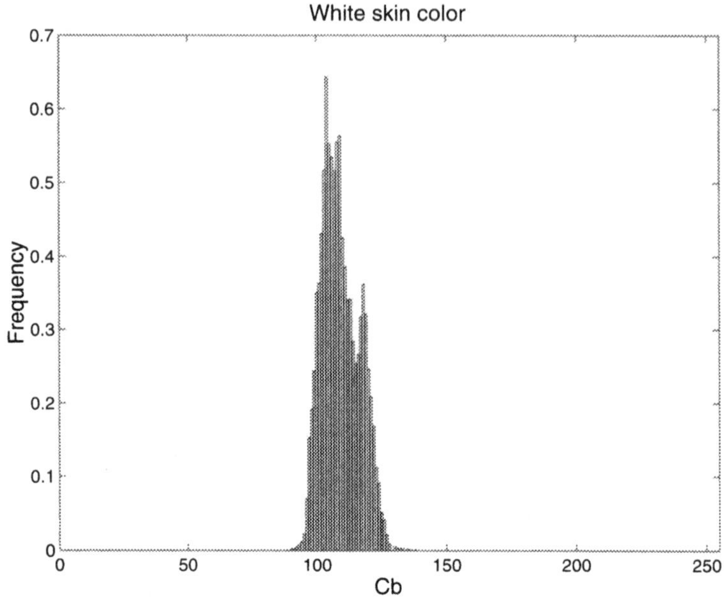

Figure 2.21: *Cont.*

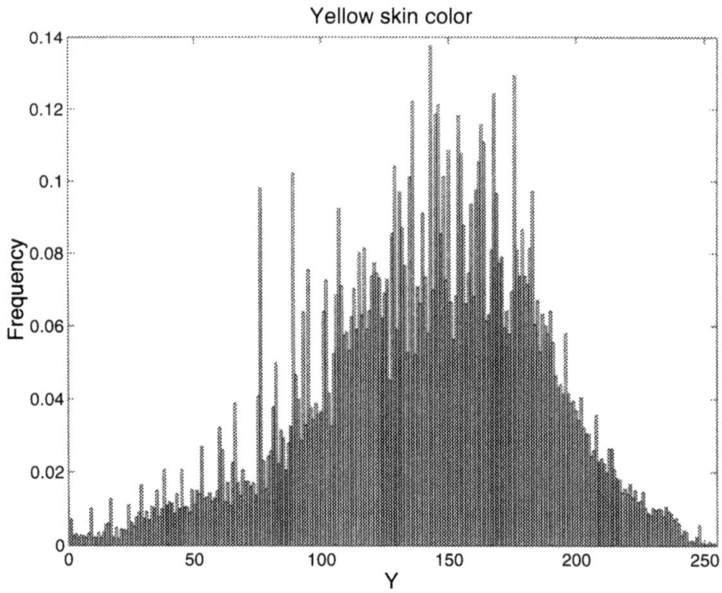

Figure 2.22: The histograms of Y, Cr and Cb values for yellow skin color.

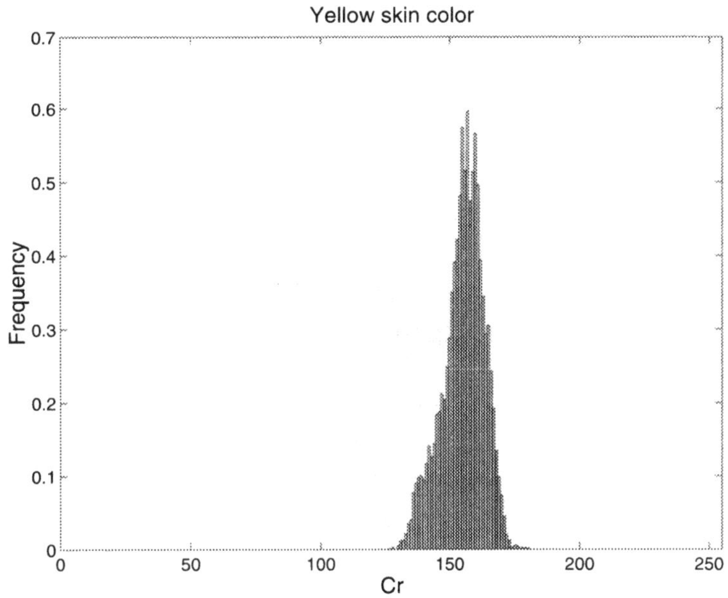

Figure 2.22: *Cont.*

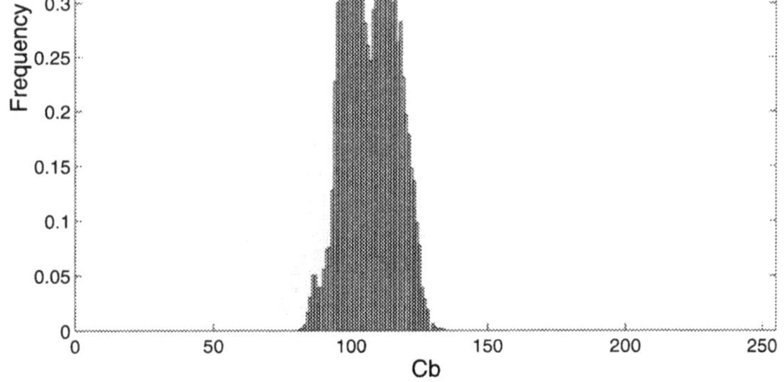

Figure 2.22: *Cont.*

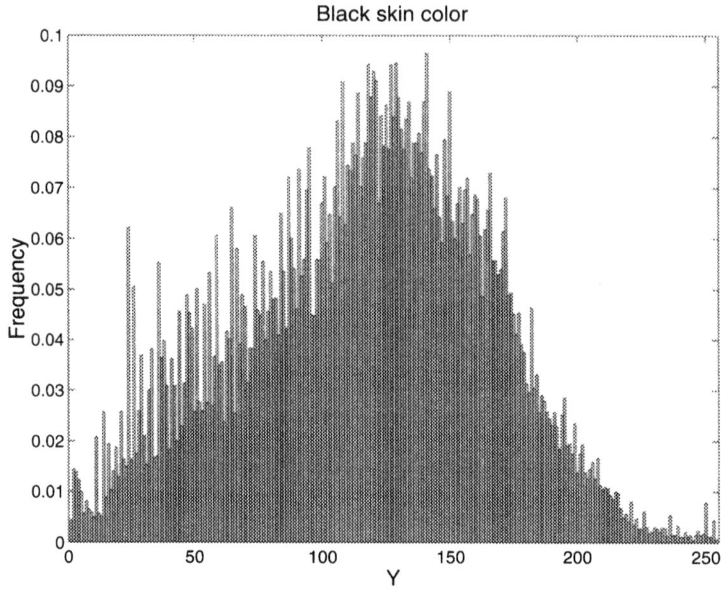

Figure 2.23: The histograms of Y, Cr and Cb values for black skin color.

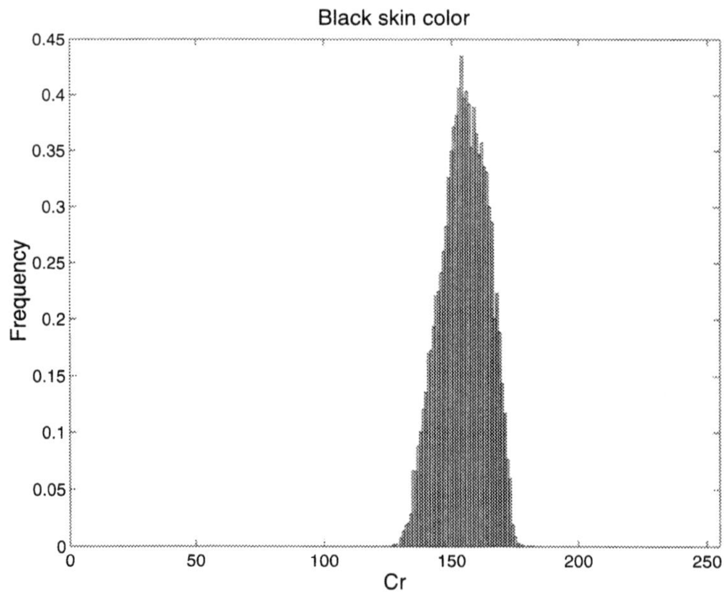

Figure 2.23: *Cont.*

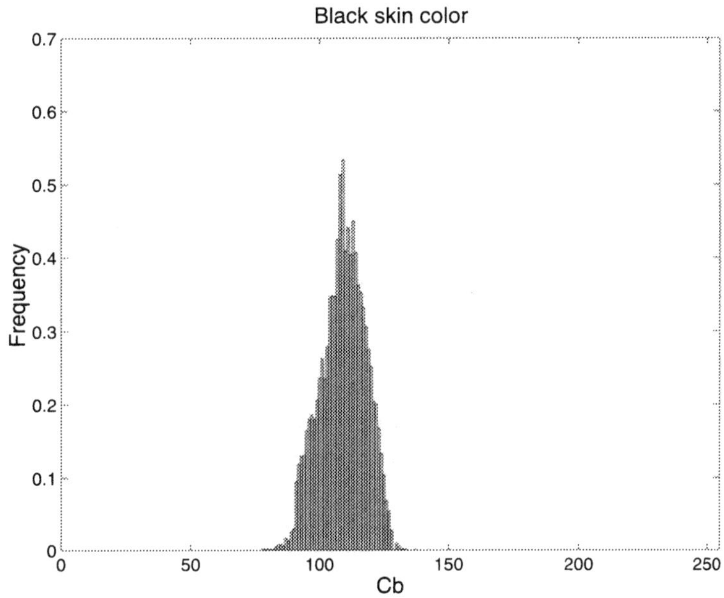

Figure 2.23: *Cont.*

Table 2.1: The results obtained from a test set of 60 images of different subjects, background complexities and lighting conditions. The correct localization is in terms of obtaining the correct position and contour of the person's face.

Test Set	**Success Rate**	**Failure Rate** - due to		
Number of Faces	Correct Localization	Incorrect Localization	Partial Localization	Incorrect and Partial Localization
60	49 (82%)	7 (12%)	2 (3%)	2 (3%)

2.5.7.2 Face Segmentation Results

The face segmentation algorithm with this universal skin-color reference map was tested on many head-and-shoulders images. Here, the emphasis is on the design of a completely automatic face segmentation process, and therefore the same design parameters and rules (including the reference skin-color map and the heuristic) were applied to all the test images. The test set now contained 20 images from each class of skin color. Therefore, a total of 60 images of different subjects, background complexities and lighting conditions from the three classes were used. Using this test set, a success rate of 82% was achieved. The results are shown in Table 2.1. The algorithm has performed successful segmentation of 49 out of 60 faces. Out of the 11 unsuccessful cases, 7 cases have incorrect localization, 2 partial localization and 2 cases with both incorrect and partial localization. The terms incorrect and partial localization will be explained later.

The representative results shown in Fig. 2.24 illustrated the successful face segmentation achieved by the algorithm on two images with different background complexities. The edges of the facial regions were accurately obtained with no noise appearing on either the facial region or the background. Moreover, the results were obtained in real-time as it took a SUN SPARC 20 computer less than 1 microsecond to perform all the computations required on a CIF-size input image.

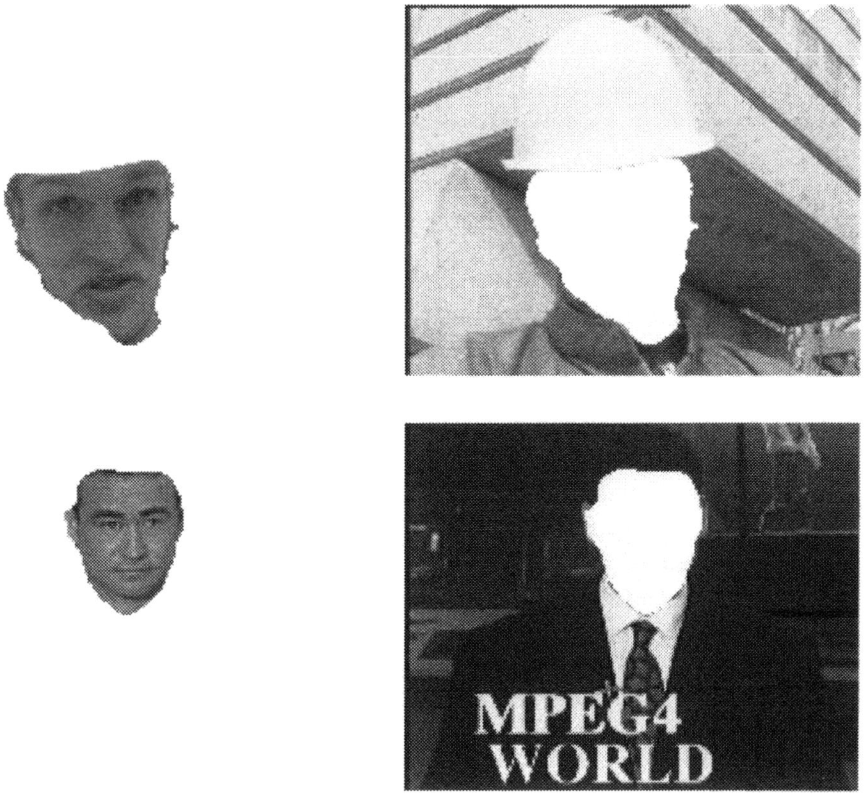

Figure 2.24: Successful segmented facial regions and the remaining background scenes.

Figure 2.25: The facial region is considered as incorrect localized if the result also includes the subject's hair.

In all 7 incorrect localization cases, the segmentation results did contain the complete facial regions but they also included some background regions. In 4 out of 7, the subject's hair, which is considered as background region, was falsely identified as facial region. One such case is shown in Fig. 2.25. Partial localization occurred in 2 cases and resulted in the localization of incomplete facial region. The 2 cases with both incorrect and partial localization have facial regions partially localized and the results also contained some background regions.

Note that of all cases in the experiment the facial regions were always located, whether they be completely or partially. The results and findings of the face segmentation process described in this chapter will be used in the foreground/background video coding scheme in Chapter 3.

References

[1] A. Eleftheriadis and A. Jacquin, "Model-assisted coding of video teleconferencing sequences at low bit rates," in *IEEE International Symposium on Circuits and Systems*, London, Jun. 1994, vol. 3, pp. 177–180.

[2] A. Eleftheriadis and A. Jacquin, "Automatic face location detection and tracking for model-assisted coding of video teleconferencing sequences at low-rates," *Signal Processing: Image Communication*, vol. 7, no. 4-6, pp. 231–248, Nov. 1995.

[3] A. Eleftheriadis and A. Jacquin, "Automatic face location detection for model-assisted rate control in H.261-compatible coding of video," *Signal Processing: Image Communication*, vol. 7, no. 4-6, pp. 435–455, Nov. 1995.

[4] S. Shimada, "Extraction of scenes containing a specific person from image sequences of a real-world scene," in *IEEE Region Ten Conference*, Melbourne, Australia, Nov. 1992, pp. 568–572.

[5] A. V. Nefian, M. Khosravi, and M. H. Hayes, "Real-time detection of human faces in uncontrolled environments," in *SPIE Visual Communications and Image Processing*, San Jose, California, USA, Feb. 1997, vol. 3024, pp. 211–219.

[6] K. Sobottka and I. Pitas, "Extraction of facial regions and features using color and shape information," in *Proceedings of the 13th International Conference on Patterm Recognition*, Vienna, Austria, Aug. 1996, vol. 3, pp. 421–425.

[7] K. Sobottka and I. Pitas, "Face localization and facial feature extraction based on shape and color information," in *Proceedings of the IEEE International Conference on Image Processing*, Sep. 1996, vol. III, pp. 483–486.

[8] K. Sobottka and I. Pitas, "Segmentation and tracking of faces in color images," in *Proceedings of the Second International Conference on Automatic Face and Gesture Recognition*, Vermont, USA, Oct. 1996, pp. 236–241.

[9] M. Hunke and A. Waibel, "Face locating and tracking for human-computer interaction," in *Proceedings of the 28th Asilomar Conference of Signals, Systems and Computers*, California, USA, Nov. 1994, vol. 2, pp. 1277–1281.

[10] M. Collobert, R. Feraud, G. Le Tourneur, and O. Bernier, "Listen: A system for locating and tracking individual speakers," in *Proceedings of the Second International Conference on Automatic Face and Gesture Recognition*, Vermont, USA, Oct. 1996, pp. 283–288.

[11] H. P. Graf, E. Cosatoo, D. Gibbon, M. Kocheisen, and E. Petajan, "Multi-modal system for locating heads and faces," in *Proceedings of the Second International Conference on Automatic Face and Gesture Recognition*, Vermont, USA, Oct. 1996, pp. 88–93.

[12] A. Neri, S. Colonnese, and G. Russo, "Automatic moving object and background segmentation by means of higher order statistics," in *SPIE Visual Communications and Image Processing*, San Jose, California, USA, Feb. 1997, vol. 3024, pp. 257–262.

[13] A. Neri, S. Colonnese, and G. Russo, "Video sequence segmentation for object-based coders using higher order statistics," in *IEEE International Symposium on Circuits and Systems (ISCAS'97)*, Hong Kong, Jun. 1997, vol. II, pp. 1245–1248.

[14] T. F. Cootes and C. J. Taylor, "Locating faces using statistical feature detectors," in *Proceedings of the 2nd International Conference on Au tomatic Face and Gesture Recognition*, Vermont, USA, Oct. 1996, pp. 204–209.

[15] A. J. Colmenarez and T. S. Huang, "Maximum likelihood face detection," in *Proceedings of the Second International Conference on Automatic Face and Gesture Recognition*, Vermont, USA, Oct. 1996, pp. 307–311.

[16] H. Li and R. Forchheimer, "Location of face using color cues," in *Proceedings of Picture Coding Symposium*, Lausanne, Switzerland, Mar 1993, paper 2.4.

[17] S. Matsuhashi, O. Nakamura, and T. Minami, "Human-face extraction using modified HSV color system and personal identification through facial image based on isodensity maps," in *Proceedings of the Canadian Conference on Electrica l and Computer Engineering*, Montreal, Canada, 1995, vol. 2, pp. 909–912.

[18] Q. Chen, H. Wu, and M. Yachida, "Face detection by fuzzy pattern matching," in *Proceedings of the Fifth International Conference on Computer Vision*, Cambridge, MA, USA, Jun. 1996, pp. 591–596.

[19] D. Saxe and R. Foulds, "Towards robust skin identification in video images," in *Proceedings of the Second International Conference on Automatic Face and Gesture Recognition*, Vermont, USA, Oct. 1996, pp. 379–384.

[20] R. Kjeldsen and J. Kender, "Finding skin in color images," in *Proceedings of the Second International Conference on Automatic Face and Gesture Recognition*, Vermont, USA, Oct. 1996, pp. 312–317.

[21] D. Chai and K. N. Ngan, "Automatic face location for videophone images," in *IEEE Region Ten Conference*, Perth, Australia, Nov. 1996, vol. 1, pp. 137–140.

[22] T. Cornall and K. Pang, "The use of facial color in image segmentation," in *Australia Telecommunication Networks and Applications Conference*, Melbourne, Australia, Dec. 1996, pp. 351–356.

[23] Y. J. Zhang, Y. R. Yao, and Y. He, "Automatic face segmentation using color cues for coding typical videophone scenes," in *SPIE Visual Communications and Image Processing*, San Jose, California, USA, Feb. 1997, vol. 3024, pp. 468–479.

[24] M. J. T. Reinders, P. J. L. van Beek, B. Sankur, and J. C. A. van der Lubbe, "Facial feature localization and adaptation of a generic face model for model-based coding," *Signal Processing: Image Communication*, vol. 7, no. 1, pp. 57–74, Mar. 1995.

[25] D. Chai and K. N. Ngan, "Locating facial region of a head-and-shoulders color image," in *Third IEEE International Conference on Automatic Face and Gesture Recognition (FG'98)*, Nara, Japan, Apr. 1998, pp. 124–129.

[26] D. Chai and K. N. Ngan, "Foreground/background video coding scheme," in *IEEE International Symposium on Circuits and Systems*, Hong Kong, Jun. 1997, vol. II, pp. 1448–1451.

[27] M. Menezes de Sequeira and F. Pereira, "Knowledge-based videotelephone sequence segmentation," in *SPIE Visual Communications and Image Processing (VCIP'93)*, Cambridge, MA, USA, Nov. 1993, vol. 2094, pp. 858–869.

[28] J. Luo, C. W. Chen, and K. J. Parker, "Face location in wavelet-based video compression for high perceptual quality videoconferenc-

ing," in *Proceedings of the International Conference on Image Processing (ICIP'95)*, Oct. 1995, vol. II, pp. 583–586.

[29] J. Luo, C. W. Chen, and K. J. Parker, "Face location in wavelet-based video compression for high perceptual quality videoconferencing," *IEEE Transactions on Circuits and Systems for Video Technology*, vol. 6, no. 4, pp. 411–414, Aug. 1996.

[30] D. Chai and K. N. Ngan, "Coding area of interest with better quality," in *IEEE International Workshop on Intelligent Signal Processing and Communication Systems (ISPACS'97)*, Kuala Lumpur, Malaysia, Nov. 1997, pp. S20.3.1–S20.3.10.

[31] D. Chai and K. N. Ngan, "Foreground/background video coding using H.261," in *SPIE Visual Communications and Image Proceeding (VCIP'98)*, San Jose, California, USA, Jan. 1998, vol. 3309, pp. 434–445.

[32] R. P. Schumeyer and K. E. Barner, "A color-based classifier for region identification in video," in *SPIE Visual Communications and Image Processing (VCIP'98)*, San Jose, California, USA, Jan. 1998, vol. 3309, pp. 189–200.

[33] MPEG AOE Sub Group, "MPEG-4 proposal package description (PPD) - revision 3," Document ISO/IEC JTC1/SC29/WG11 MPEG95/N0998, Jul. 1995.

[34] D. Chai and K. N. Ngan, "Extraction of VOP from videophone scene," in *International Workshop on Coding Techniques for Very Low Bit-rate Video*, Linkoping, Sweden, Jul. 1997, pp. 45–48.

[35] R. L. Rudianto, "Automatic 3-D wire-frame model fitting and adaptation to frontal facial image in model-based image coding," Honours thesis, Department of Electrical and Electronic Engineering, University of Western Australia, 1995.

[36] K. N. Ngan and R. L. Rudianto, "Automatic face location detection and tracking for model-based video coding," in *Proceedings of the Third Conference on Signal Processing (ICSP'96)*, Beijing, China, Oct. 1996, vol. 2, pp. 1098–1101.

[37] S. Satyanarayana and S. Dalal, "Video color enhancement using neural networks," *IEEE Transactions on Circuits and Systems for Video Technology*, vol. 6, no. 3, pp. 295–307, Jun. 1996.

[38] R. Chellappa, C. L. Wilson, and S. Sirohey, "Human and machine recognition of faces: a survey," *Proceedings of the IEEE*, vol. 83, no. 5, pp. 705–740, May 1995.

[39] J. Zhang, Y. Yan, and M. Lades, "Face recognition: eigenface, elastic matching and neural nets," *Proceedings of the IEEE*, vol. 85, no. 9, pp. 1423–1435, Sep. 1997.

[40] M. A. Turk and A. P. Pentland, "Face recognition using eigenfaces," in *Proceedings of the IEEE Conference on Computer Vision and Pattern Recognition (CVPR'91)*, Jun. 1991, pp. 586–591.

[41] Zhujie and Y. L. Yu, "Face recognition with eigenfaces," in *Proceedings of the IEEE International Conference on Industrial Technology*, Dec. 1994, pp. 434–438.

[42] S. McKenna and S. Gong, "Tracking faces," in *Proceedings of the Second International Conference on Automatic Face and Gesture Recognition*, Vermont, USA, Oct. 1996, pp. 271–276.

[43] H. Wu, T. Yokoyama, D. Pramadihanto, and M. Yachida, "Face and facial feature extraction from color image," in *Proceedings of the Second International Conference on Automatic Face and Gesture Recognition*, Vermont, USA, Oct. 1996, pp. 345–350.

[44] M. J. T. Reinders, F. A. Odijk, J. C. A. van der Lubbe, and J. J. Gerbrands, "Tracking of global motion and facial expressions of a human face in image sequences," in *SPIE Visual Communications and Image Processing (VCIP'93)*, Cambridge, MA, USA, Nov. 1993, vol. 2094, pp. 1516–1527.

[45] M. Okubo and T. Watanabe, "Lip motion capture and its application to 3-D molding," in *Proceedings of the Third IEEE International Conference on Automatic Face and Gesture Recognition (FG'98)*, Nara, Japan, Apr. 1998, pp. 187–192.

[46] E. Yamamoto, S. Nakamura, and K. Shikano, "Lip movement synthesis from speech based on hidden markov models," in *Proceedings of the Third IEEE International Conference on Automatic Face and Gesture Recognition (FG'98)*, Nara, Japan, Apr. 1998, pp. 154–159.

[47] Y. Ariki, Y. Sugiyama, and N. Ishikawa, "Face indexing on video data - extraction, recognition, tracking and modeling," in *Proceedings of the*

Third IEEE International Conference on Automatic Face and Gesture Recognition (FG'98), Nara, Japan, Apr. 1998, pp. 62–69.

[48] P. E. Mattison, *Practical digital video with programming examples in* C, John Wiley & Sons Inc., 1994.

[49] I. Pitas, *Digital image processing algorithms*, Prentice Hall, New York, USA, 1993.

[50] D. Chai and K. N. Ngan, "Face segmentation using skin color map in videophone applications," *to appear in IEEE Transactions on Circuits and Systems for Video Technology*, 1999.

[51] R. M. Haralick, S. R. Sternberg, and X Zhuang, "Image analysis using mathematical morphology," *IEEE Transactions on Pattern Analysis and Machine Intelligence*, vol. PAMI-9, no. 4, pp. 532–550, Jul. 1987.

[52] G. A. Baxes, *Digital image processing: principles and applications*, John Wiley & Sons, 1994.

Chapter 3

Foreground/Background Coding

3.1 Introduction

The current research activities in very low bit rate video coding have been commonly classified into two approaches. While one approach is heading towards the long-term goal of discovering new coding concepts, the other is concerned with the near-term goal. In the latter approach, the research activities have encompassed the modification and optimization of some conventional low bit rate video coding algorithms for use in the very low bit rate environment. Although this research has been pursued with impressive results, these hybrid algorithms still suffer from some inherent problems. Hence they have to compromise significantly on the image quality in order to cope with lower rates. As a result, they produce visual artifacts throughout the coded images. For example, it is well known that the hybrid predictive-transform coding scheme of the H.263 suffers from blocking effects at low bit rates. The effects are even more objectionable at very low bit rates. These artifacts are particularly annoying when they occur in areas of the picture that are of importance to viewers. Hence this shortcoming has motivated researchers to provide a practical solution to protect the important area of interest from visual artifacts.

A video coding scheme that treats the area of interest with higher priority and codes it at a higher quality than the less relevant background scene is presented here. The main objective is to achieve an improvement in the perceptual quality of the encoded picture; in other words, it is to provide a better subjective viewing quality. Furthermore, the intention is to achieve this at the encoder, rather than the decoder as a post-process image enhancement task.

Therefore the initial step for such an encoding approach is to identify and then segment out the viewer's area of interest from the less relevant background scene. Each frame of the input video sequence is to be separated into two non-overlapping regions, namely, the foreground region that contains the area of interest and the complementary background region. This step would involve some image scene analysis operations. These regions are then encoded using the same coder but with different encoding parameters. Bit allocation and rate control are assigned not only according to the buffer fullness but also on the importance of the coded region. In this way, we can redistribute the bit allocation for these regions that we have defined and encode each of them at different bit rate and quality. More important, the image quality of the more important foreground region can be improved by encoding it with more bits at the expense of background image quality. This approach is referred to as the Foreground/Background (FB) video coding scheme [1].

A block diagram of a basic FB coding scheme is depicted in Fig. 3.1. The figure shows that the input video data is first fed into the video content analyzer, also known as region classifier. Then the defined foreground and background regions, generated from the video content analyzer, become the inputs of the same source encoder. Although both regions are to be encoded with the same coding technique, their encoding parameters can be different. Depending on the source coding technique and the syntax of its video stream, the region classification information may or may not have to be transmitted. This is because the source decoder may or may not require the explicit knowledge of region location to decode a FB video stream.

The FB coding scheme has three major benefits:

1. It provides a short term solution to improve the subjective visual quality of an encoded image by selectively reducing the coding artifacts that typically arises from the current near-term approach to very low bit rate coding such as the H.263 coding technique.

2. The knowledge gained from the study of FB coding scheme can contribute to the long-term goal of searching for new coding concepts for very low bit rate video coding. As FB coding scheme and the other newly proposed coding concepts like object-based, content-based and model-based coding all share similar major coding problems. These problems include scene analysis, region/object segmentation and region/object/content-based (instead of frame-based), bit allocation and rate control strategies.

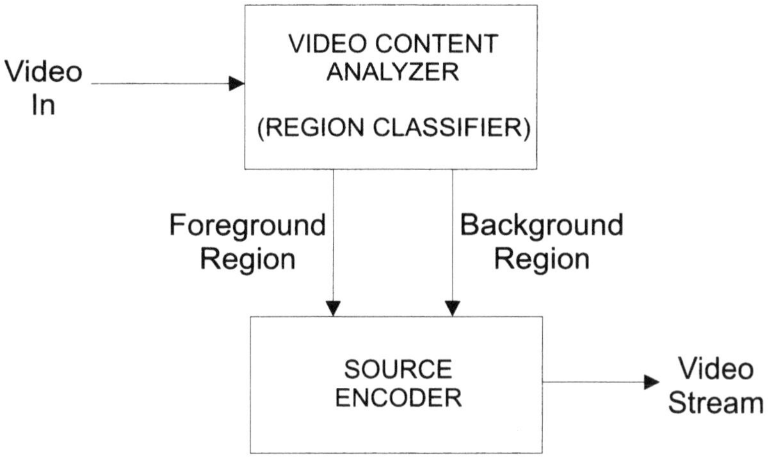

Figure 3.1: Block diagram of a basic FB coding scheme.

3. The FB coding scheme introduces new functionalities to old video coding technology. It can provide some of the much talked about MPEG-4 content-based functionalities to classical motion compensated DCT video coders, which by definition belonged to frame-based coding approach. The FB coder offers region/object/content-based bit allocation and rate control strategies to frame-based source encoder such as the most widely used videoconferencing standard of H.261.

It is fair to say that most of the current researches on new video coding techniques has been focusing on videotelephony applications, and the study of the FB coding scheme is of no exception. A videophone or videoconferencing image typically consists of a head-and-shoulders view of a speaker in front of a simple or complex background scene. Hence, in such case, the face of the speaker is typically the most important image region to the viewer, and it is to be considered as the foreground region of the input image.

The concept of FB video coding scheme was initially proposed by Chai and Ngan, and reported in [1], [2] and [3]. They presented, in [1], not only the introduction of the FB coding scheme but also the implementation of this scheme as an additional encoding option for the H.263 codec. While in [2] and [3], the implementation of FB coding scheme on the H.261 framework was discussed.

3.2 Related Works

Video coding techniques that make use of face location information are relatively new and popular, and are gaining increasing attention. This section reviews some of the works done by other researchers that are related to this FB coding scheme. The concise descriptions of their works are given below.

Eleftheriadis and Jacquin

They proposed in [4], [5] and [6] a coding approach known as the *model-assisted video coding*, as it is a mixture of classical waveform coding and model-based coding. Therefore, instead of modeling the face itself as in the case of the generic model-based coding, they modeled only the location of the face. Their approach is to first locate the facial area of a head-and-shoulders input image, and then exploit the face location information in an object-selective quantizer control. The aim of their work is to produce perceptually pleasing videoconferencing image sequences whereby faces are sharper. So, they adopted a rate control algorithm that transfers a fraction of the total available bit rate from the coding of the non-facial area to that of the facial area. The model-assisted rate control consisted of two important components, namely, buffer rate modulation and buffer size modulation. The buffer rate modulation forces the rate control algorithm to spend more bits in regions of interest, while the buffer size modulation ensures that the allocated bits are uniformly distributed within each region.

The integration of their proposed model-assisted bit allocation and rate control scheme on the H.261 video coding system was reported in [6]. Some experimental results were shown, as the authors compared the model-assisted RM8 coder with the standard RM8 coder. Note that although their rate control scheme was proposed to cater for a number of regions of interest, only two regions being facial and non-facial regions were used in their experiments. Moreover, vital model-assisted coding parameters such as γ and μ, which represent the relative average quality and the modulation factor respectively, were empirically obtained. Nonetheless, in their experiments, two test image sequences called *Jelena* and *Roberto* at QCIF size were used, with target rates set at 48 kbps and 5 fps. With parameter γ and μ determined experimentally, the model-assisted RM8 coder was able to achieve the target bit rate, which was also close to the value achieved by the standard RM8. The results showed a 60-75% increase in bits spent in the facial area and a 30-35% decrease in bits spent in the non-facial area. Subjective evaluation of the encoded images was carried out. From the images selectively provided, some quality improvement was noticeable in terms of reduced coding artifacts in the facial area.

Note that they have also studied the integration with different coders besides the H.261. Their model-assisted coding concept, without the model-assisted rate control scheme, was reported in the context of a 3D subband-based video coder in [4] and [5].

Ding and Takaya

Several methods were proposed in [7] to improve the encoding speed of the H.263 coder that is used for coding facial images from videotelephony applications, as encoding speed is the biggest obstacle for real-time image communications. These methods include the improvements of the computational efficiency in motion vector search, DCT and quantization, since these encoding components are the heart of the H.263 coder. The main assumption of their work is that the input video scene is constrained to only facial images, which are composed of a moving head and one still background. Their proposal is based heavily on this assumption, and referred to, by the authors, as *face tracking*. This name was given because the attention of their proposed approach is focused on the subspace of an image frame where a face is residing, while regarding the rest of the frame as background. Since facial expressions and head movements are of viewer's primary interest, the movement of a face will be tracked and transmission of any changes in the head area, instead of the whole frame, will suffice. Nevertheless, their coding approach can be explained as follows.

Firstly, based on the above assumption, the motion vector search for the head area can be restricted to within a small search range while the motion vectors for the background can be set to zero. This will save time in searching procedure and reduce the computation time necessary for getting the motion vectors.

Secondly, it is observed that the smaller the distortion between the current block and the corresponding prediction block, the more zero coefficients are produced in the DCT process. Therefore the computation of DCT coefficients can be limited to only some while imposing the others to be zero. Instead of consistently using an 8×8 point DCT on all 8×8 blocks of an image frame, they suggested the use of 2×2, 4×4 or 6×6 points in the lower frequency for DCT calculation. The selection of which size to use is according to the magnitude of the distortion (although not mentioned in [7], this should be the expected distortion as the authors assumed the general scenario and no distortion measure was actually calculated before the DCT operation). Generally, smaller point DCT is performed on the less detailed region such as the background region, while larger point DCT is performed on more detailed region like the face. It is expected that this DCT approach will maintain the same image quality as compared to the computation for all the DCT coefficients, because the coefficients that are being omitted in their DCT calculation should be zero or close to zero.

Lastly, it is suggested that the quantization adjustment be dependent on the region that it is covering, whereby smaller quantization step-size should be used for the important areas while larger for the unimportant areas. It is, however, unclear as to how this strategy can improve encoding time. In addition to this strategy, the use of constant quantization step-size was also mentioned. The so-called *bypass bitrate control* is nothing more than just fixing the quantizer to a certain value for all pictures in the sequence, and therefore the quantization parameter need not to be updated, and thus saving time.

A small set of experimental results, which lacks many details, were shown in [7]. It showed that the use of the above mentioned techniques has resulted in a significant increase of frame rate, indicating that the encoding speed had improved. An approximate increase from 1 f/s to 8 f/s was achieved with bit rate control, while 30 f/s was achieved without bit rate control. However, the improvement came at the expense of having a decrease in SNR value - an objective measurement of image quality. In contrary to what was described in [7] as a little decrease in image quality, a drop of around 10 dB from 42.5 dB should be considered as significant.

Lin and Wu

The work of Lin and Wu, as reported in [8] and [9] involved the use of block-based MC-DCT hybrid coder to code head-and-shoulders (videophone type) images with benign background scene at very low bit rate. They proposed a coding approach for the H.263 coder that involves fixing the temporal frequency and the introduction of a simple content-based rate control scheme.

Based on common observation, it is found that viewers are more sensitive to the unsteady movement of objects, and that heavy moving regions are more critical than the lightly moving regions in the very low bit rate video applications. Furthermore, the picture quality of the facial area is more important and noticeable to viewers. Therefore the intentions of their proposal are to fix temporal frequency so that the movement of objects in the video sequence are smooth, and more importantly, to spend more bits on regions of the image frame that receive higher level of viewers' concentration

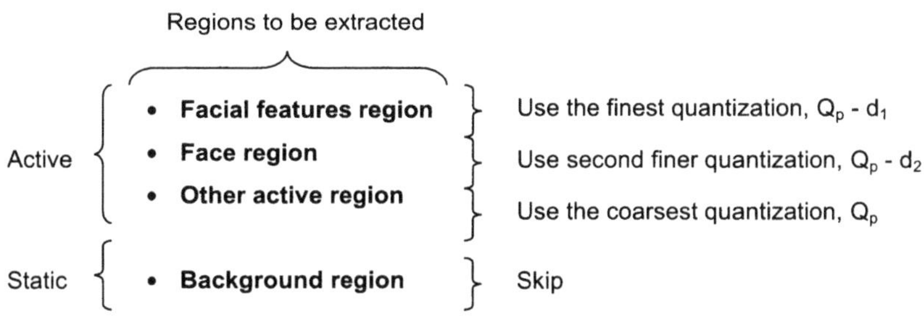

Figure 3.2: The regions to be extracted for the content-based bit rate control scheme proposed by Lin and Wu.

in order to improve the perceptual picture quality.

Hence, prior to the proposed encoding process, the contents of the input images are analyzed and then classified into different regions at macroblock level. As depicted in Fig. 3.2, there are four different regions to be extracted, namely, "facial features region" such as eyes and mouth, "face region", "other active region" such as shoulders, and "background region". The former three are considered as active regions while the latter is static.

The proposed rate control scheme adopts a quantization level adjustment based on not only the buffer fullness but also the content classification. Therefore the most active, and thus critical, facial features region is to be assigned with the finest quantization level of $Q_p - d_1$; face region with the second finer quantization level of $Q_p - d_2$; other active region with the coarsest quantization level of Q_p; and the static background region is to be directly skipped to save both bit rate and encoding time. Note that Q_p is the quantization parameter, and d_1 and d_2 are respectively selected as 4 and 2 in their implementation. Although content-based bit rate adjustment is introduced, the actual rate control scheme is rather restrictive and somewhat non-adaptive. The authors proposed the quantization parameter, Q_p to be identical for all macroblocks in the same picture, while the value of Q_p will only be updated at the start of each new picture that is to be encoded.

The content-based bit rate control scheme (CBCS) was implemented and embedded in an H.263 coder. It was then tested on the so-called *Miss America* and *Claire* video sequences at QCIF and against the reference coder that employs a frame-based control scheme (FBCS). The frame rate was fixed at 12.5 f/s, while the target bit rates were 8, 14.4 and 28.8 kb/s. A PSNR study was carried out, with results favoring the FBCS. A lower average PSNR values were resulted in the CBCS approach because, from observation, CBCS in overall reduced more bit rates from all the pixels in less critical image region than it injected bit rates into all other pixels in more critical image region. Therefore the authors have employed a weighted SNR (WSNR) evaluation function that takes the allocated bit counts of each region into account when calculating for mean-square-error (MSE). So each pixel that has been assigned with different number of bits will have different weight in this picture quality evaluation. With this evaluation, the CBCS was found to be slightly better than the FBCS in general. In addition, a MSE ratio graph, an average bit count ratio and a subjective evaluation of the results from CBCS and FBCS were carried out. The findings led to promising outcome that the CBCS could promote the perceptual picture quality of encoded pictures at very low bit rates.

Wollborn *et al.*

A content-based video coding scheme for the transmission of videophone sequences at very low bit rates was proposed by Wollborn *et al.* [10]. The suggested scheme was to use an MPEG-4 conforming codec to transmit the facial areas of the image in a better quality compared to the remaining image. Hence, a face detection algorithm was used to separate each input image into two video object planes (VOP). The facial area was to form the face VOP, while the remaining image was to form the residual VOP. Then, each image was coded and transmitted separately as two different VOPs. For this, the MPEG-4 video verification model (VM) version 6.0 [11] was used. The coder would code and transmit the shape, motion and texture parameters of the face VOP, whereas only the motion and texture parameters of the residual VOP. The shape parameters of the residual VOP was omitted because the residual VOP was to be coded and transmitted like the whole original image by using a lowpass extrapolation padding technique to fill/pad the hollow facial area of the residual VOP. The rationale behind this approach was that Woolborn *et al.* reported that coding of the padded area was less expensive in terms of bit rate than coding the shape information of the residual VOP. Nonetheless, the quality of the face VOP could be improved by spending a larger part of the bit rate on coding it, while only a small portion was used for the residual VOP. The bit rate allocation between the two VOPs was realized by setting the respective quantization parameter and/or frame rate differently, but it was done so manually. Moreover, the content-based rate control was not dealt with in [10]; therefore manual adjustment of quantization parameter was adopted in order to achieve the desired overall bit rate.

The proposed scheme of using the MPEG-4 VM6.0 for content-based coding was compared to the VM6.0 in frame-based mode. The so-called *Claire*, *Akiyo* and *Salesman* test sequences were used in their experiments. All sequences were coded at QCIF resolution with target bit rates ranging from 9 to 24 kb/s and two different frame-rates of 5 f/s and 10 f/s. The experimental testing showed two significant outcomes. Firstly, when coding sequences whereby motion was mainly occurring in the facial area, nearly no improvement for the facial area was achieved, while the quality of the remaining image is significantly decreased. Therefore frame rate for the residual VOP has to be reduced in order to achieve some improvement in the face VOP. Secondly, the experimental results showed that the improvement rises with increasing bit rate, since the overhead of coding two VOPs and the additional shape information has lesser impact.

Xie *et al.*

Xie *et al.* have presented in [12] and [13] a layered video coding scheme for very low bit rate videophone. Three layers are defined, and the different layers are basically pertaining to different coding modes. The first layer employs the standard H.263 coder, and this is considered as the basic coding mode of this proposed scheme. This basic layer will be used if there is no *a priori* knowledge of the image content. However, if this knowledge is available, the second layer is activated. The second layer assumes the input image as a head-and-shoulders type, and hence segments the image into two objects: the human face and everything else. This process produces a human face mask, which will be used to guide bit assignment in the encoder end. To maintain compatibility, this layer is restricted to the structure of the H.263 and the face mask is only required at macroblock resolution. If the face mask is also made available at the decoder end, by means of transmission along with the encoded bitstream as side information, then the scheme can be upgraded to its third layer. In this layer, pixel-level segmentation is required. The arbitrary-shaped face mask at pixel level will be used for motion estimation and the prediction error will be encoded by arbitrary-shaped DCT while the shape of the face mask will be encoded by B-spline (chain code was used in [12]). The aim of this layer is to further improve the subjective quality of the videophone by restructuring the boundary of the human face with higher fidelity.

The experimental results showed that the proposed approach of contour coding using B-spline with tolerable loss is much more efficient compared to the conventional chain-code and MPEG-4 M4R code. The system improvement was also shown when the motion estimation process makes use of the face mask to reduce searching scale.

There are two interesting points worth noting. One, the criterion to switch between different layers is reported to be based on subjective quality instead of a more objective and operable approach, and the switch is not done automatically. Two, their proposed methodology followed the Musmann's layered coding concept [14].

3.3 Foreground and Background Regions

Both the foreground and background regions are to be defined at macroblock level, since a macroblock is typically the basic processing unit of block-based coding systems such as the H.261 and H.263. Let α be a set of all macroblocks in an image frame, and let α_f and α_b be a set of all macroblocks that belong to the foreground and background regions, respectively. The relationship of these sets are illustrated in Fig. 3.3. Set α_f and α_b are non-overlapping, i.e.,

$$\alpha_f \cap \alpha_b = \oslash, \tag{3.1}$$

and the sum of these two sets forms the image frame, i.e.,

$$\alpha_f \cup \alpha_b = \alpha. \tag{3.2}$$

Note that the foreground region does not have to be in a rectangular shape as shown in Fig. 3.3. It can take on any arbitrary shape defined at macroblock level, while the background region will then take on the complementary shape of the foreground region. For instance, the identification and separation of α_f and α_b for videophone type images are done automatically and robustly according to the face segmentation technique as described in the previous Chapter. Fig. 3.4 shows a sample result produced from the *Carphone* image.

In some situations, the defined regions may consist of a physical object or a meaningful set of objects. Therefore the foreground region can also be appropriately referred to as the foreground object, and similarly, the background region as background object.

Furthermore, in terms of MPEG-4 Video Object (VO) definition, the foreground and background regions would then correspond to foreground and background VOs, respectively.

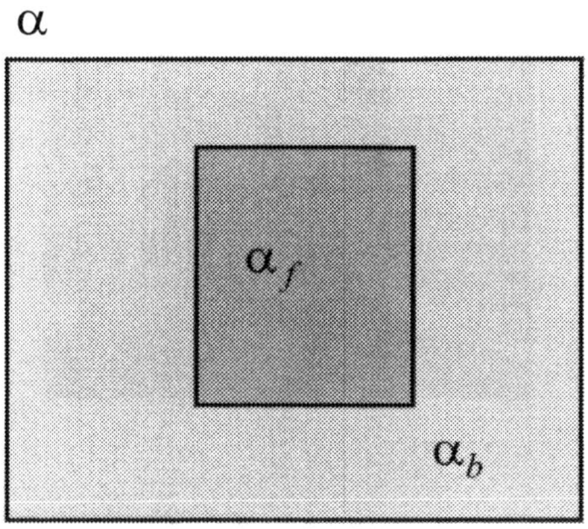

Figure 3.3: The relationship between α, α_f and α_b.

3.4 Content-based Bit Allocation

Our objective is to code α_f at a higher image quality but without increasing the overall bit rate. To do so, more bits are distributed to the coding of α_f while having less bits remained for α_b. Therefore this section explains two content-based bit allocation strategies for the FB coding scheme. The first strategy is known as Maximum Bit Transfer, while the second is known as Joint Bit Assignment.

3.4.1 Maximum Bit Transfer

The Maximum Bit Transfer (MBT) is a content-based bit allocation strategy that uses a pair of quantizers, one for the foreground region and one for the background region, to code a frame. It always assigns the highest possible quantization parameter to the background quantizer in order to facilitate maximum bit transfer from background to foreground region.

In this approach, the total number of bits spent on coding a frame, B_{MBT}, is computed as

$$B_{MBT} = B_{fg}(Q_f) + B_{bg}(Q_b) + h_{MBT} \tag{3.3}$$

where $B_{fg}(Q_f)$ and $B_{bg}(Q_b)$ represent, respectively, the number of bits spent on coding all foreground and background macroblocks, and h_{MBT} denotes the number of bits spent on coding all the necessary header information that are not directly associated to any specific macroblock. Both $B_{fg}(Q_f)$

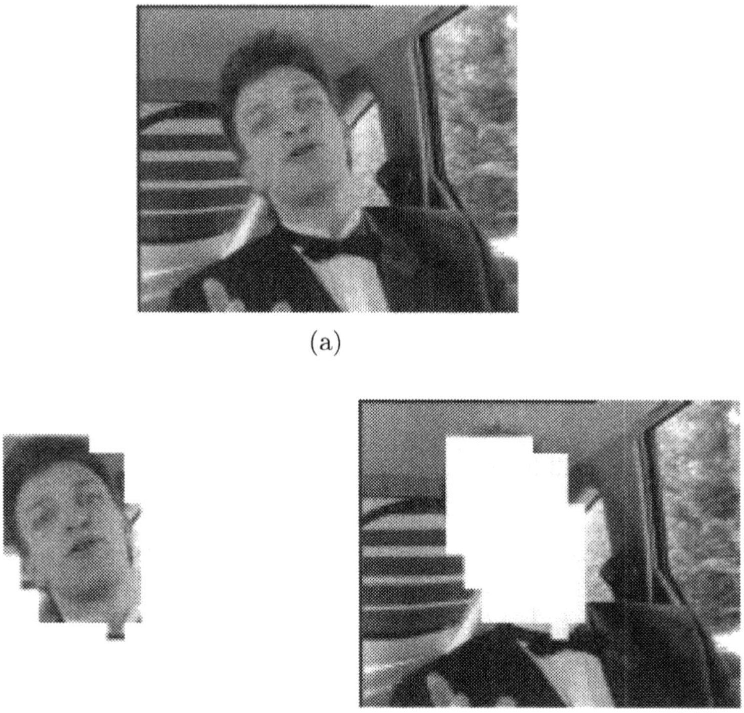

(a)

(b)

(c)

Figure 3.4: (a) α, (b) α_f and (c) α_b.

and $B_{bg}(Q_b)$ are a set of decreasing functions of quantization parameter. The foreground and background quantizers, which are represented by Q_f and Q_b respectively, can be assigned with quantization parameters (QP) that range from 1 to QP_{max}.

Typically, h_{MBT} is independent of $B_{fg}(Q_f)$ and $B_{bg}(Q_b)$, and it is fair to assume that h_{MBT} remains constant regardless of what values Q_f and Q_b have been assigned.

To maximize bit transfer, the texture information of the background region will be coded at the lowest possible quality. Hence, the largest possible quantization parameter of QP_{max} will be assigned to Q_b. As a consequence, this will reduce the size of B_{bg} and provide more bits for foreground usage. This extra resource will enable the use of finer quantizer for coding the texture information of the foreground region. The selection of the foreground quantizer, however, will be dictated by the given bit budget constraint. Let the target bits per frame be denoted by B_T, and define the difference between the target bits per frame and the actual output bit rate produced in this MBT approach as

$$\epsilon = B_T - B_{MBT}. \tag{3.4}$$

Ideally, ϵ should be zero. Practically, however, we can only obtain ϵ that is as close to zero as possible. Therefore we need to find Q_f such that $|\epsilon|$ is a minimum. If there exists two solutions, then the one that corresponds to a negative ϵ should be selected, as part of the aim to achieve minimum value of $|\epsilon|$ is to obtain the finest possible Q_f for foreground quantization.

Below we show how the MBT strategy can be used for coding the first picture of an input video sequence in intraframe mode.

Consider the following two coders: one is a reference coder while the other is a FB coder that uses the MBT strategy (FB-MBT). The purpose of the reference coder is to provide a reference for performance evaluation and comparison study. With the exception of the bit allocation strategy, both coders will have an identical encoding process.

In this case, the output bits per frame (b/f) of the reference coder, B_{REF}, will become the target bit rate (in terms of b/f) for the FB coder, i.e.,

$$B_T = B_{REF}. \tag{3.5}$$

Equation (3.4) now becomes

$$\epsilon = B_{REF} - B_{MBT}. \tag{3.6}$$

It is assumed that the reference coder adopts a "conventional" bit allocation technique, which uses only one fixed quantizer for coding the entire frame. Let Q be this quantizer, and similar to (3.3) we now have

$$B_{REF} = B_{fg}(Q) + B_{bg}(Q) + h_{REF}. \tag{3.7}$$

For FB-MBT coder to reallocate bits usage from background to foreground region, it will assign

$$Q_b = QP_{max} \geq Q, \tag{3.8}$$

so that

$$B_{bg}(Q_b) \leq B_{bg}(Q). \tag{3.9}$$

The reduction of bits spent on the background region will then be brought over for foreground usage so that

$$B_{fg}(Q_f) \geq B_{fg}(Q), \tag{3.10}$$

with

$$Q_f \leq Q. \tag{3.11}$$

We now have to find the value of Q_f such that $|\epsilon|$ is a minimum. Equation (3.6) can be rewritten as

$$\epsilon = B_{fg}(Q) + B_{bg}(Q) + h_{REF} - B_{fg}(Q_f) - B_{bg}(QP_{max}) - h_{MBT}. \tag{3.12}$$

At this stage, the values of $B_{fg}(Q)$, $B_{bg}(Q)$, h_{REF}, $B_{bg}(QP_{max})$ and h_{MBT} have all been obtained. Therefore let

$$A = B_{fg}(Q) + B_{bg}(Q) + h_{REF} - B_{bg}(QP_{max}) - h_{MBT} \tag{3.13}$$

so that (3.12) now becomes

$$\epsilon = A - B_{fg}(Q_f). \tag{3.14}$$

Using (3.14), Q_f can be decremented (starting from Q) in a recursive manner until the minimum value of $|\epsilon|$ is found. This numerical approach can be done using the C-code as shown below:

```
int Find_Qf(int Q, int QP_MAX)
{
 int Qf, Qb, finest_Qf;
 int A, diff, min_diff;

 Qf = finest_Qf = Q;
 Qb = QP_MAX;
 /* B_fg, B_bg, h_ref and h_mbt are        */
 /* functions that return integer values. */
 A = B_fg(Q) + B_bg(Q) + h_ref() - B_bg(Qb) - h_mbt();
 min_diff = A - B_fg(Qf);

 for (Qf=Q-1, Qf>=1, Qf--) {
   diff = A - B_fg(Qf);
   if ( abs(min_diff) > abs(diff) ) {
```

```
      min_diff = diff;
      finest_Qf = Qf;
    }
    else
      break;
  }
  return(finest_Qf)
}
```

Given the value of quantization used in the reference coder, the above C function determines the finest possible value of foreground quantizer that the FB-MBT coder can use and yet produces a bit rate similar (which is as close as possible) to the reference coder.

3.4.2 Joint Bit Assignment

In the Maximum Bit Transfer approach, the background region is always coded with the coarsest quantization level. However, it is not always desirable to have maximum bit transfer from background to foreground. Therefore, another bit allocation strategy termed as Joint Bit Assignment (JBA) is introduced. The JBA strategy performs bit allocation based on the characteristics of each region, such as size, motion and priority. The working of JBA is explained below.

Consider the two following approaches, namely, the proposed and reference approaches. The proposed approach employs the JBA strategy, while the reference (conventional) approach uses a generic strategy and its purpose is to provide a reference for the performance evaluation of the JBA strategy.

To maintain the same bit rate for both approaches, the number of bits spent on α_f, α_b and the overheads in the proposed approach should equal to the total number of bits spent on all macroblocks and the overhead information for a frame in the conventional approach. This equality condition can be mathematically expressed as

$$\beta_f N_f + \beta_b N_b + h_p = \beta N + h_c. \tag{3.15}$$

In this equation, β_f and β_b denote the average bits used per foreground and per background macroblock respectively, while β denotes the average bits used by the generic coder to code a macroblock. The parameters N_f, N_b and N represent the number of macroblocks in α_f, α_b and α, respectively. The amount of bits used in the overheads are represented by the parameter h_p in the proposed approach and h_c in the conventional approach.

Typically, $h_p = h_c$ or $h_p \approx h_c$, therefore (3.15) can be simplified as

$$\beta_f N_f + \beta_b N_b = \beta N. \tag{3.16}$$

The value of N is determined by the size of the input image frame, whereas the value of N_f and N_b are known once α_f and α_b have been defined. For instance, Fig. 3.4(a) shows a CIF size image with 352×288 dimension, which has $N = 396$ macroblocks. The defined α_f as shown in Fig. 3.4(b) contains $N_f = 77$ macroblocks, while α_b as shown in Fig. 3.4(c) contains $N_b = 319$ macroblocks. The value of β is obtained by dividing the total number of bits required for coding all the macroblocks in a frame using the generic coder by the number of macroblocks in a frame. Once the above values are obtained, the value for β_f and β_b can then be determined. To achieve higher quality coding for the foreground region, each foreground macroblock will use more bits and therefore β_f will be greater than β. Note that the parameter β_f has a maximum value of N/N_f times greater than β; this is the case when β_b is set to zero. Nonetheless, once a value for β_f is chosen, the value of β_b can be computed as

$$\beta_b = \frac{\beta N - \beta_f N_f}{N_b}. \tag{3.17}$$

where $N_b > 0$.

The amount of bits to be spent on α_f can be determined in a number of ways, and one of them is the *user-defined* approach. As the name suggested, in this approach β_f is set by the user using a scale s that ranges from 0 to N/N_f, and is defined as

$$\beta_f = s\beta. \tag{3.18}$$

If the user selects a value of s that is within $(0, 1)$, then less bits per macroblock will be spent on the foreground region as compared to the background region. Consequently, the quality of the foreground region will be worse than the background region. On the other hand, if a value within $(1, N/N_f)$ is chosen then more bits per macroblock will be spent on the foreground region as compared to the background region; thus the quality of the foreground region will be better than the background region. However, if $s = 0$ (lower bound) then the foreground region will not be coded; if $s = 1$ then the amount of bits spent on per foreground macroblock and on per background macroblock will be the same; and if $s = N/N_f$ (upper bound) then all the available bits will be spent on the foreground region while none will be allocated to the background region.

Hence the *user-defined* approach facilitates user interactivity in the video coding system. The user can control the quality of the foreground and background regions through the adjustment of the bit allocation for these image regions.

However, a bit allocation strategy that is content-based and can be carried out in an automatic and operative manner is also highly desired. Therefore, an alternative approach can be used, whereby bit allocation is determined based on the characteristics of the defined image regions. Each of these characteristics, including size, motion and priority is explained below.

- **Size**. In the size dependent approach, the amount of bits to be allocated to an image region is dependent on its size. The normalized size of the foreground region, S_{fg}, and the background region, S_{bg}, are respectively determined by

$$S_{fg} = \frac{N_f}{N} \tag{3.19}$$

and

$$S_{bg} = \frac{N_b}{N}, \tag{3.20}$$

where N_f, N_b and N denote the number of macroblocks in α_f, α_b and α respectively, and that

$$S_{fg} + S_{bg} = 1. \tag{3.21}$$

- **Motion**. Bit allocation can also be performed according to the activity of each region. The activity of a region can be measured by its motion. A region with high activity will yield more motion vectors. Let M_{fg} and M_{bg} be the normalized motion parameters for α_f and α_b respectively, and are derived as

$$M_{fg} = \frac{\sum_{\alpha_f} \left| MV \right|}{\sum_{\alpha_f + \alpha_b} \left| MV \right|} \tag{3.22}$$

and

$$M_{bg} = \frac{\sum_{\alpha_b} \left| MV \right|}{\sum_{\alpha_f + \alpha_b} \left| MV \right|}, \tag{3.23}$$

where $|MV|$ is the absolute value of the motion vector of a macroblock, and that

$$M_{fg} + M_{bg} = 1. \tag{3.24}$$

Note that large motion vectors are typically assigned to longer codeword representations, and therefore the transmission of these motion vectors will consume more bits; this is reflected in (3.22) and (3.23).

- **Priority**. The priority specifies the relative subjective importance of α_f and hence provides privilege to the foreground. After the available bits have been allocated to α_f and α_b based on their size and/or motion, we can selectively transfer a portion of the bits that has already been assigned to the background over to the foreground. Let P be the priority parameter that specifies the percentage of bit transfer. $P = 0\%$ signifies that no subjective preference is given to α_f, while $P = 100\%$ implies that 100% of the available bits are to be spent on α_f.

Now suppose B_T is the amount of bits available for a frame, and is defined as

$$B_T = \beta N. \tag{3.25}$$

Let B_{fg} and B_{bg} are the amount of bits to be spent on α_f and α_b, and are defined as

$$B_{fg} = \beta_f N_f \tag{3.26}$$

and

$$B_{bg} = \beta_b N_b, \tag{3.27}$$

respectively. Then, (3.16) can be rewritten as

$$B_T = B_{fg} + B_{bg}. \tag{3.28}$$

Subsequently, the amount of bits assigned to the α_f, based on size and motion, is given as

$$B_{fg} = (\omega_S S_{fg} + \omega_M M_{fg}) B_T, \tag{3.29}$$

where ω_S and ω_M are weighting functions of the respective size and motion parameters, and $\omega_S + \omega_m = 1$. Similarly, for α_b,

$$B_{bg} = (\omega_S S_{bg} + \omega_M M_{bg})B_T, \tag{3.30}$$

or simply

$$B_{bg} = B_T - B_{fg} \tag{3.31}$$

if B_{fg} has already been calculated from (3.29).

However, when the priority parameter is used, the amount of bit allocated to the foreground region becomes

$$B'_{fg} = B_{fg} + PB_{bg}, \tag{3.32}$$

while for the background region,

$$B'_{bg} = B_{bg} - PB_{bg}, \tag{3.33}$$

or

$$B'_{bg} = B_{bg}(1 - P), \tag{3.34}$$

3.5 Content-based Rate Control

For constant bit rate coding, a rate control algorithm is needed in an FB coding scheme to regulate the bitstream generated by the two image regions and to achieve an overall target bit rate. A content-based rate control strategy that not only takes the buffer fullness but also the content classification into account is typically required. The strategy can be classified into two general types, namely, independent and joint.

In an independent rate control strategy, the bit rate of each region is pre-assigned and two separate rate control algorithms are performed independent of each other. The output bit rate, R, is the sum of the individual bit rates for the foreground region, R_{fg}, and background region, R_{bg}, i.e.,

$$R = R_{fg} + R_{bg}. \tag{3.35}$$

On the other hand, in a joint rate control strategy, the controlling of the bit rates generated from both regions is carried out as a joint process. Since in FB coding scheme, the foreground and background regions are to be coded at different bit rates as defined by B_{fg} and B_{bg} bits per frame (or, β_f and β_b

bits per macroblock), a virtual content-based buffer is introduced. During the encoding of a frame, the virtual content-based buffer will be drained at two different rates depending on which region it is currently coding. The actual buffer will, however, still be physically emptied at a rate of B_T bits per frame in order to maintain a constant overall target bit rate. For instance, when the FB coder is coding a foreground macroblock, the virtual content-based buffer will be drained at a rate of β_f bits per macroblock, while physically the buffer is drained at a rate of β, which is lower than β_f. The effect of increasing the draining rate is that the virtual buffer occupancy level will be lower than the actual level. Therefore, it tricks the coder to encode the next foreground macroblock at a lower than actual quantization level. Similarly, when coding a background macroblock, the virtual content-based buffer will switch to a lower draining rate of β_b bits per macroblock. Since β_b is lower than the actual rate of β, the virtual buffer occupancy level will be higher than the actual level. As a result, this tricks the coder to use a higher quantization level for the next background macroblock. This quantization approach is known to us as the *discriminatory quantization process*.

The implementation of the joint content-based rate control algorithm depends much on the structure and bitstream syntax of the coder. In the next two sections, the implementations that suit the H.261 and H.263 coders will be discussed.

3.6 H.261FB Approach

The foreground/background coding scheme can be integrated into the H.261 framework. This is referred to as the H.261FB approach. As it is the case for the H.261, the work on the H.261FB coding approach is also focused on the application of personal-to-personal communications such as videotelephony. In this application, the face of the speaker is typically the most concerned image region for the viewer. Therefore the facial area is to be separated from its background to become the foreground region. This can be achieved using the automatic face segmentation algorithm. However, since the lowest possible quantization adjustment of the H.261 is at the macroblock level, the foreground and background regions are only to be identified at macroblock, instead of pixel, resolution. The significance of the lowest possible quantization adjustment lies in the fact that a discriminatory quantization process is used to transfer bits from background to foreground. In the encoding process, fewer bits will be allocated for encoding the background region and in doing so, it frees up more bits that can then be used for encoding the foreground region. This bit transfer will lead to a better quality encoded facial region at the expense of having lower quality background image. Furthermore, based on the premise that the background is usually of less significance to the viewer's perception, the overall subjective quality of the image will be perceptively improved and more pleasing to viewer.

An overview on the H.261 video coding system is first presented before the detailed explanation of the H.261FB implementation.

3.6.1 H.261 Video Coding System

The CCITT[1] Recommendation H.261 [15] is a video coding standard designed for video communications over ISDN[2]. It can handle $p \times 64$ kbps (where $p = 1, 2, \ldots, 30$) video streams and this matches the possible bandwidths in ISDN.

3.6.1.1 Video Data Format

The H.261 standard specifies the YCrCb color system as the format for the video data. The Y represents the luminance component while Cr and Cb represent the chrominance components of this color system. The Cr and Cb are subsampled by a factor of 4 compared to Y since the human visual system is more sensitive to the luminance component and less sensitive to the chrominance components.

The video size formats supported by the H.261 standard are CIF and QCIF. The Common Intermediate Format, CIF in short, has a resolution of 352×288 pixels for the luminance (Y) component and 176×144 pixels for the two chrominance components (Cr and Cb) of the video stream (see Fig. 3.5). The Quarter-CIF or QCIF contains a quarter size of a CIF, and therefore the luminance and chrominance components have a resolution of 176×144 pixels and 88×72 pixels, respectively.

3.6.1.2 Source Coder

The H.261 video source coding algorithm employs a block-based motion-compensated discrete-cosine transform (MC-DCT) design. Fig. 3.6 shows a block diagram of an H.261 video source coder.

The coder can operate in two modes. In the intraframe mode, an 8×8 block from the video-in is DCT-transformed, quantized and sent to the video multiplex coder. In the interframe mode, the motion compensator is used for

[1] CCITT is a French acronym for Consultative Committee on Telephone and Telegraph.
[2] ISDN is short of Integrated Services Digital Network.

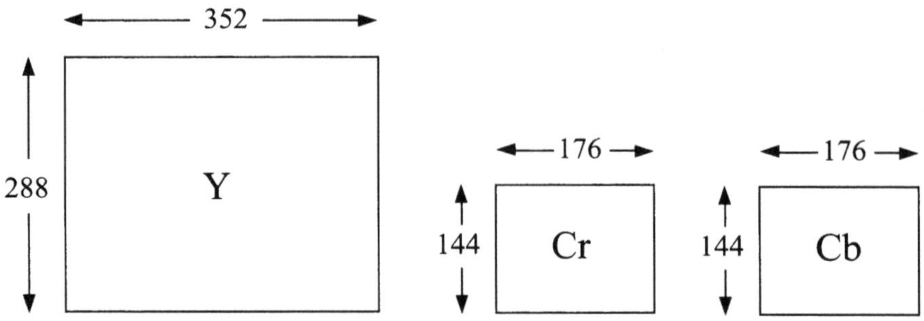

Figure 3.5: A CIF-size image in the YCrCb format with a spatial sampling frequency ratio of Y, Cr and Cb as 4:1:1.

comparing the macroblock of the current frame with blocks of data from the previous frame that was sent. If the difference, also known as the prediction error, is below a pre-determined threshold, no data is sent for this block, otherwise, the difference block is DCT-transformed, quantized and sent to the video multiplex coder. Note that if motion estimation is used then the difference between the motion vector for the current and the previous macroblocks is sent. A loop filter is used for improving video quality by removing high frequency noise, while the coding control is used for selecting intraframe or interframe mode and also for controlling the quantization step-size. At the video multiplex coder, the bitstream are further compressed as the quantized DCT coefficients are scanned in a zigzag order and then run-length and Huffman coded. The output of the video multiplex coder is placed in a transmission buffer. Then a rate control strategy that controls the quantizer will be used to regulate the outgoing bitstream.

3.6.1.3 Syntax Structure

The compressed data stream is arranged hierarchically into four layers, namely,

- Picture;
- Group of blocks;
- Macroblock; and
- Block.

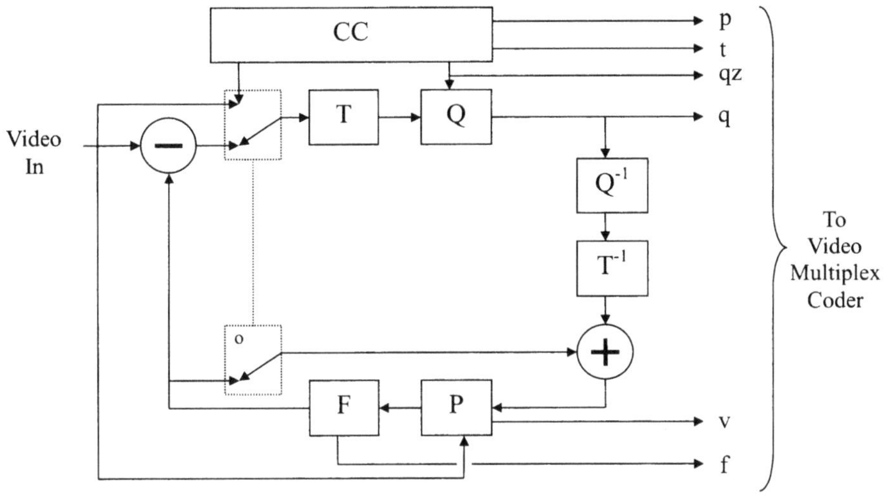

CC: Coding control
T: Transform
Q: Quantizer
F: Loop filter
P: Picture memory with motion compensated variable delay

p: Flag for INTRA/INTER
t: Flag for transmitted or not
qz: Quantizer indication
q: Quantizing index for transform coefficients
v: Motion vector
f: Switching on/off of the loop filter

Figure 3.6: Block diagram of an H.261 video source coder [15].

A picture is the top layer, it can be in QCIF or CIF. Each picture is divided into groups of blocks (GOBs). A CIF picture has 12 GOBs while a QCIF has 3. Each GOB is composed of 33 macroblocks (MBs) in an 3×11 array, and each MB is made up of 4 luminance (Y) blocks and 2 chrominance (Cr and Cb) blocks. A block is an 8×8 array of pixels. This hierarchical block structure are illustrated in Fig. 3.7.

The transmission of an H.261 video data starts at the picture layer. The picture layer contains a picture header followed by GOB layer data. A picture header contains a picture start code, temporal reference, picture type and other information. A GOB layer contains a GOB header followed by MB layer data. The GOB header includes a GOB start code, group number, GOB quantization value and other information. A MB layer has a MB header followed by block layer data. A typical MB header consists of a

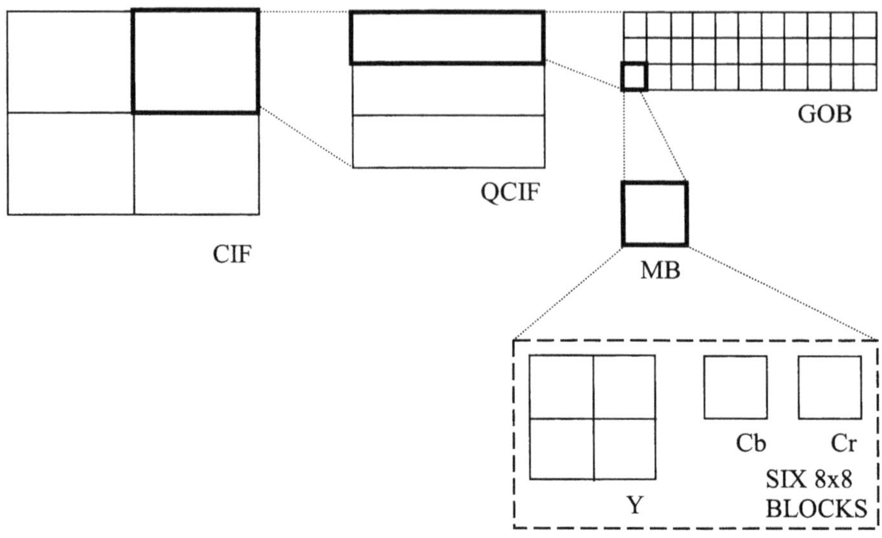

Figure 3.7: The hierarchical block structure of the H.261 video stream.

MB address, type, quantization value, motion vector data and coded block pattern. A block layer data contains quantized DCT coefficients and a fixed length EOB codeword to signal end of block. Fig. 3.8 depicts a simplified syntax diagram of the data transmission at the video multiplex coder. Note that, within a MB, not every block needs to be transmitted, and within a GOB, not every MB needs to be transmitted.

Readers can refer to the CCITT Recommendation H.261 document [15] for the detailed syntax diagram and the complete data structure information.

3.6.1.4 Unspecified Encoding Procedures

The H.261 standard is a decoding standard as it focuses on the requirements of the decoder. Therefore, there are a number of encoding decisions not included in the standard. The major areas left unspecified in the standard are:

- the criteria for choosing either to transmit or skip a macroblock;
- the control mechanism for intraframe or interframe coding;
- the use and derivation of motion vector;

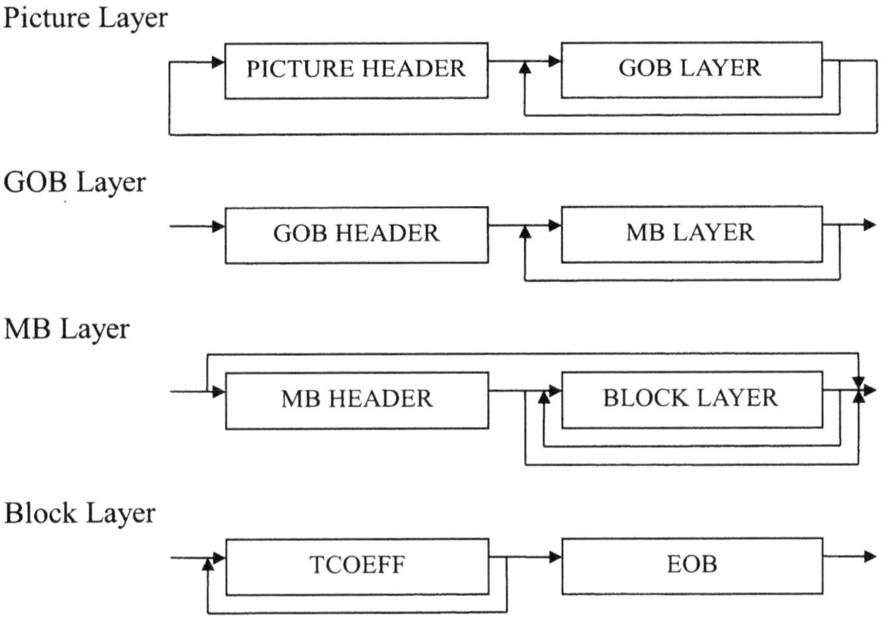

Figure 3.8: A simplified syntax diagram of the H.261 video multiplex coder.

- the option to apply a linear filter to the previous decoded frame before using it for prediction;
- the rate control strategy, and hence the quantization step-size adjustment.

By not including them in the standard, it provides the manufacturer of the encoder the freedom to devise its own strategy - as long as the output bitstream conforms to the H.261 syntax.

3.6.2 Reference Model 8

The Reference Model 8 [16], or RM8 in short, is a reference implementation of an H.261 coder. It was developed by the H.261 working group with the purpose of providing a common environment in which experiments could be carried out.

In the RM8 implementation, a motion vector $\vec{v}_m$ of macroblock m is determined by full-search block matching. The motion estimation compares only the luminance values in the 16×16 macroblock m with other nearby

16×16 arrays of luminance values of the previously transmitted image. The range of such comparison is between ± 15 pixels around macroblock m. The sum of the absolute values of the pixel-to-pixel difference throughout the 16×16 block (SAD in short) is used as the measure of prediction error. The displacement with the smallest SAD which indicates the best match is considered the motion compensation vector for macroblock m, i.e., $\vec{v}_m$. The difference (or error) between the best-match block and the current to-be-coded block is known as the motion compensated block.

Several heuristics are used to make the coding decisions. If the energy of the motion compensated block with zero displacement is roughly less than the energy of the motion compensated block with best-match displacement, $\vec{v}_m$, then the motion vector is suppressed and resulted in zero displacement motion compensation. Otherwise motion vector compensation is used.

The variance V_P of the motion compensated block is compared against the variance V_Y of the luminance blocks in macroblock m to determine whether to perform intraframe or interframe coding. If intraframe coding mode is selected then no motion compensation is used, otherwise motion compensation is used in interframe coding. The loop filter in interframe mode is enabled if V_P is below a certain threshold. The decision of whether to transmit a transform-coded block is made individually for each block in a macroblock by considering the sum of absolute values of the quantized transform coefficients. If the sum falls below a preset threshold, the block is not transmitted. All the above heuristics, threshold functions and default decision diagrams can be found in the RM8 document [16].

Quite often video coders have to operate with fixed bandwidth limitation. However, the H.261 standard specifies entropy coding that will ultimately result in video bitstream of variable bit rate. Therefore some form of rate control is required for operation on bandwidth-limited channels. For instance, if the output of the coder exceeds the channel capacity then the quality can be decreased, or vice versa. The RM8 coder employs a simple rate control technique based on a virtual buffer model in a feedback loop whereby the buffer occupancy controls the level of quantization. The quantization parameter QP is calculated as

$$QP = \min\left\{ \left\lceil \frac{buffer\_occupancy}{200p} + 1 \right\rceil, 31 \right\}. \tag{3.36}$$

Note that p was previously used in the definition of bit rate that the H.261 coder operates in, i.e., $p \times 64$ kbit/s. The quantization parameter QP has an integral range of $[1, 31]$. This equation can be redefined as a function of the normalized buffer occupancy level. Assuming that the buffer size is only related to the bit rate and defined as a quarter of a second's worth of information, i.e.,

$$\begin{aligned} buffer\_size &= \frac{bitrate}{4} \\ &= \frac{p \times 64000}{4} \quad \text{bits}, \end{aligned} \tag{3.37}$$

then the normalized buffer occupancy is

$$buffer\_occupancy' = \frac{buffer\_occupancy}{buffer\_size}. \tag{3.38}$$

Therefore (3.36) becomes

$$QP = \min\left\{ \left\lfloor 80 \times buffer\_occupancy' + 1 \right\rfloor, 31 \right\}. \tag{3.39}$$

This function is plotted in Fig. 3.9.

3.6.3 Implementation of the H.261FB Coder

The H.261FB coder utilizes the segmentation information to enable bit transfer between the foreground and background macroblocks. This redistribution of bit allocation is simply attained by controlling the quantization level in a discriminatory manner. In addition, a new rate control is devised in order to regulate the bitstream generated by this discriminatory quantization process. For proper evaluation of the foreground/background bit allocation, the discriminatory quantization process and the foreground/background rate control, all other coding decisions of the H.261FB coder are to be based on the RM8 implementation.

The implementation of the H.261FB coder will be carried out in such a way that the generated bitstream will still conform to the H.261 standard. The reasons that this can be done so are:

- The bit allocation strategy is not part of the standard;
- The new quantization process does not involve in any modification of the bitstream syntax, as it merely performs the allowable quantization step size adjustment;
- There are no standardized technique for rate control;

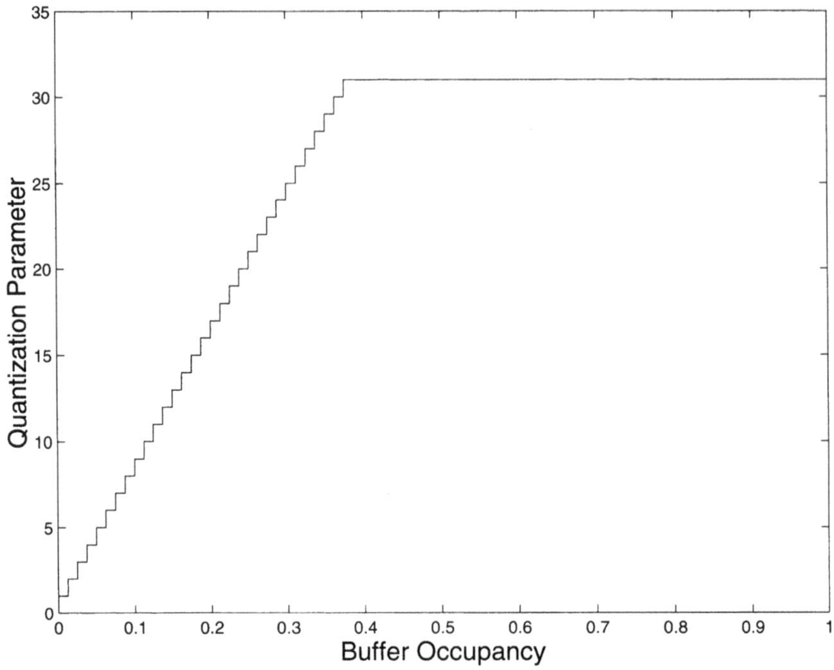

Figure 3.9: Quantization parameter adjustment based on the normalized buffer occupancy.

- The sequential processing structure defined in the standard is still maintained, i.e., macroblocks are still coded in their regular left to right and top to bottom order within each group of block;
- The segmentation information does not need to be transmitted to the decoder as it is only used in the encoder.

As a result, a full H.261 decoder compatibility is maintained.

3.6.3.1 Foreground/Background Bit Allocation

The foreground and background regions can be assigned to a certain amount of bits so that they can be coded at different quality and bit rate. Two types of foreground/background bit allocation strategies are introduced to the H.261FB coder, and they are the Maximum Bit Transfer and the Joint Bit Assignment as discussed in Section 3.4. A brief summary of each strategy is provided below.

The Maximum Bit Transfer (MBT) approach always assigns the highest possible quantization parameter, QP_{max}, to the background quantizer in order to facilitate maximum bit transfer from background to foreground region. The quantization parameter of the foreground region, on the other hand, is dictated by the given bit budget constraint. From (3.4) we know that ϵ is denoted as the difference between the target bits per frame, B_T, and the actual output bit rate produced in this MBT approach, i.e.,

$$\epsilon = B_T - B_{MBT}.$$

This can be expanded to become

$$\epsilon = B_{fg}(Q) + B_{bg}(Q) + h_{REF} - B_{fg}(Q_f) - B_{bg}(QP_{max}) - h_{MBT},$$

where $B_{fg}(Q)$ and $B_{bg}(Q)$ are the number of bits spent on coding all foreground and all background macroblocks respectively, at quantization level of Q, and h_{REF} and h_{MBT} are the number of bits spent on coding all the necessary header information that are not directly asociated to any specific macroblock in the reference and MBT approach, respectively. Now the objective is to find the value of the foreground quantizer, Q_f, such that $|\epsilon|$ is a minimum. See Section 3.4.1 for more details.

In the Joint Bit Assignment approach, the bit allocation is based on the characteristics of each image region, such as size, motion and priority. The amount of bits to be assigned to the foreground (B_{fg}) and background (B_{bg}) region are given as

$$B_{fg} = \left[\omega_S\left(S_{fg} + S_{bg}P\right) + \omega_M\left(M_{fg} + M_{bg}P\right)\right]B_T, \quad (3.40)$$

$$B_{bg} = \left(\omega_S S_{bg} + \omega_M M_{bg}\right)\left(1 - P\right)B_T, \quad (3.41)$$

where

B_T	:	the amount of bits available for the frame,
ω_S, ω_M	:	weighting functions of the size and motion parameters,
S_{fg}, S_{bg}	:	normalized size parameters of the foreground and background,
M_{fg}, M_{bg}	:	normalized motion parameters of the foreground and background,
P	:	priority parameter that specifies the % of subjective bit transfer.

See Section 3.4.2 for more details on this Joint Bit Assignment approach.

3.6.3.2 Discriminatory Quantization Process

The foreground/background bit allocation strategy distributes two different bit rates to the foreground and background regions, and therefore two quantizers, instead of one, are used in the H.261FB coder. We assign Q_f and Q_b to be the quantizers for the foreground and background macroblocks, respectively. The H.261FB coder uses the *MQUANT* header to switch between these two quantizers as shown in (3.42). The *MQUANT* header is a fixed length codeword of 5 bits that indicates the quantization level to be used for the current macroblock.

$$MQUANT = \begin{cases} Q_f, & \text{if current macroblock belongs to foreground,} \\ Q_b, & \text{if current macroblock belongs to background.} \end{cases} \tag{3.42}$$

It is, however, not necessary for the encoder to send this header for every macroblock. In fact, the transmission of *MQUANT* header is only required in one of the following cases:

- When the current macroblock is in a different region to the previously encoded macroblock; i.e., a change from foreground to background macroblock or vice versa;
- When the rate control algorithm updates the quantization level in order to maintain a constant bit rate.

Naturally, this approach has to sustain a slight increase in the transmission of *MQUANT* header. However the benefit easily outweighs this overhead cost. This will be demonstrated in the experimental results.

3.6.3.3 Foreground/Background Rate Control

A rate control algorithm is needed to regulate the bitstream and achieve an overall target bit rate. Here, a joint foreground/ background rate control strategy that is based on the RM8 rate control [16] is devised.

Suppose the source video sequence has L number of frames with frame index l starting from 1 to L, and has a frame rate of F_S frame per second (f/s). Each frame is partitioned into N number of macroblocks with macroblock index n starting from 1 to N. And suppose this source material is to be coded at a target bit rate of R_T bits per second (b/s) and a target frame rate of F_T f/s.

The target frame rate of F_T can be equal or less than the frame rate of the source material, and it can be achieved by skipping the appropriate number of frames, i.e.,

$$F_T = \frac{F_S}{F_{skip}} \quad \text{f/s} \tag{3.43}$$

where F_{skip} denotes the constant number of frames to be skipped. As a result, let K be the number of frames that will be coded (i.e., $K = L/F_{skip}$, where / is an integer division with truncation towards zero) and k be the frame index of the coded frames starting from 1 to K.

Let $buffer\_occupancy_k$ be the amount of information stored in the buffer prior to coding frame k, in unit of bits. The buffer occupancy at the start of the video sequence is initialized to zero:

$$buffer\_occupancy_1 = 0. \tag{3.44}$$

The very first frame of the sequence is intraframe coded with constant quantization parameter and no rate control is performed during this frame. After the first frame is coded, the buffer is assumed half full. Therefore the buffer occupancy prior to coding of the second frame is

$$buffer\_occupancy_2 = \frac{buffer\_size}{2}. \tag{3.45}$$

The rate control starts at the second coded frame and the buffer occupancy is updated according to the following equation:

$$buffer\_occupancy_{k,n} = buffer\_occupancy_k + B_{k,n} - buffer\_drain_{k,n}, \text{for } k \geq 2, \tag{3.46}$$

where $buffer\_occupancy_{k,n}$ denotes the amount of bits currently in the buffer after coding macroblock n of frame k, $buffer\_occupancy_k$ represents, as before, the buffer occupancy at the start of frame k, $B_{k,n}$ denotes the number of bits spent since the start of frame k and until after macroblock n of frame k, and $buffer\_drain_{k,n}$ represents the amount of bits to be emptied from the buffer after macroblock n of frame k is coded.

In the RM8 approach, the buffer is emptied at a constant rate of B_T/N bits per macroblock, whereby B_T is derived from

$$B_T = \frac{R_T}{F_T} \quad \text{b/f.} \tag{3.47}$$

Therefore the buffer drain for RM8 is

$$buffer\_drain_{k,n} = \frac{n}{N} B_T. \tag{3.48}$$

For the H.261FB joint foreground/background rate control, however, (3.48) becomes

$$buffer\_drain_{k,n} = \frac{n_f}{N_f} B_f + \frac{n_b}{N_b} B_b. \tag{3.49}$$

where n_f and n_b are the macroblock index for the respective foreground and background regions. During the encoding of a frame, the buffer will be drained at two rates depending on which region it is currently coding and therefore (3.49) is used as a virtual buffer drain. Note that the physical buffer will still be emptied at a rate of B_T b/f in order to maintain a constant overall bit rate of R_T b/s. This is based on the content-based joint rate control concept as discussed in Section 3.5.

Let QP be the quantization parameter with an integer range from 1 to 31. It is updated periodically according to the following equation:

$$QP = \frac{buffer\_occupancy_{k,n}}{Q_{division}} + Q_{offset} \tag{3.50}$$

The DCT coefficients of the foreground and background macroblocks will be quantized differently according to their assigned bit rates. When coding a foreground macroblock,

$$Q_{division} = \frac{N B_f F_T}{320 N_f}, \tag{3.51}$$

while when coding a background macroblock,

$$Q_{division} = \frac{N B_b F_T}{320 N_b}, \tag{3.52}$$

and, in both cases, $Q_{offset} = 1$.

Note that if the foreground/background regions are not defined, then (3.51) or (3.52) will become

$$\begin{aligned} Q_{division} &= \frac{N B_T F_T}{320 N} \\ &= \frac{R_T}{320}, \end{aligned} \tag{3.53}$$

which is the definition for the RM8 rate control.

The joint foreground/background rate control maintains the two individual bit rates of the foreground and background regions and also the sequential processing structure of the H.261 video coding system by switching between the buffer drain rates and the $Q_{division}$ parameters.

Figure 3.10: The original, first image frame of the *Foreman* sequence and its foreground and background macroblocks.

3.6.4 Experimental Results

The H.261FB coder was tested on several videophone image sequences. The H.261FB coder with the Maximum Bit Transfer (MBT) approach is examined first. For this, two standard CIF-size video sequences, namely, *Foreman* and *Miss America* were used. The face segmentation algorithm was employed to separate each frame of the input sequences into foreground and background regions at macroblock resolution. The segmentation results for the first frame of each sequence are shown in Figs. 3.10 and 3.11, and the number of foreground and background macroblocks identified in these frames are given in Table 3.1. Note that a CIF-size image has 396 macroblocks.

These images were encoded using the reference coder RM8, and the proposed coder H.261FB. The H.261FB coder made use of the segmentation results and adopted the MBT approach. Other than these inclusions, the rest of the encoding processes of the H.261FB were implemented in the same

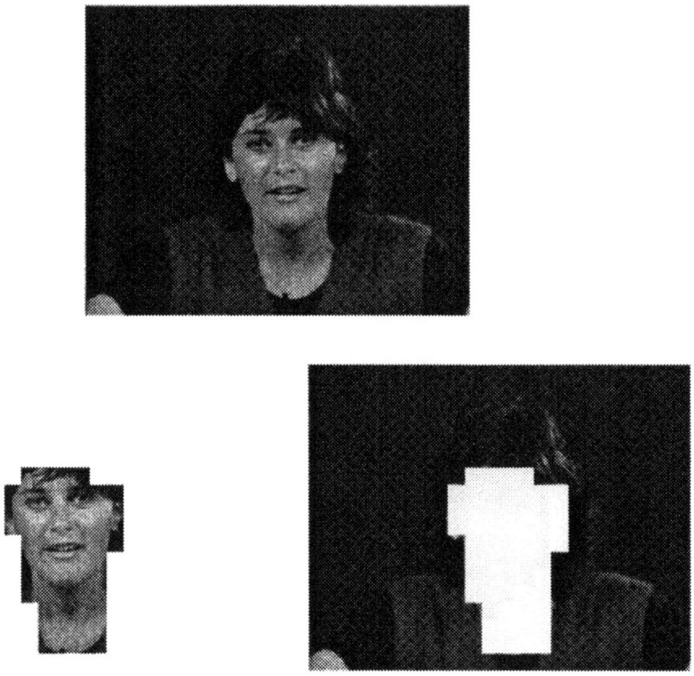

Figure 3.11: The original, first image frame of the *Miss America* sequence and its foreground and background macroblocks.

way as the RM8 so that a proper evaluation of the new coding scheme could be carried out.

Intraframe coding was first performed on these images. The quantizer, Q, of the RM8 coder was arbitrarily set to 25 for the *Foreman* image and 24 for the *Miss America* image. As for the H.261FB coder, the MBT bit allocation strategy forced the background quantizer, Q_b, to the maximum value of 31 for both images, while the value of the foreground quantizer, Q_f, was calculated to be 11 for the *Foreman* image and 21 for the *Miss America* image. These values are shown in Table 3.2 and note that they were fixed to their given values throughout the entire intraframe coding process.

With these settings, both coders spent approximately 39 kb/f on the *Foreman* image and 28 kb/f on the *Miss America* image. The encoded images are shown in Figs. 3.12 and 3.13, while their peak-signal-to-noise-ratio (PSNR) values can be found in Table 3.3.

Table 3.1: The number of foreground and background macroblocks in the *Foreman* image and the *Miss America* image.

Image	Number of Foreground Macroblocks, N_f	Number of Background Macroblocks, N_b
Foreman	72	324
Miss America	58	338

Table 3.2: The quantization parameters selected for the RM8 and H.261FB coders.

Image	RM8	H.261FB
Foreman	$Q = 25$	$Q_f = 11$, $Q_b = 31$
Miss America	$Q = 24$	$Q_f = 21$, $Q_b = 31$

Table 3.3: Objective quality measures of the encoded foreground (FG) and background (BG) regions and also of the whole frame (showing only the luminance component).

	Foreman		*Miss America*	
	RM8	H.261FB	RM8	H.261FB
PSNR_Y (dB)	29.68	29.11	35.37	35.25
PSNR_Y_FG (dB)	30.91	34.87	30.11	30.65
PSNR_Y_BG (dB)	29.45	28.45	37.61	36.94

(a)

(b)

Figure 3.12: *Foreman* image encoded by (a) RM8 and (b) H.261FB.

(a)

(b)

Figure 3.13: *Miss America* image encoded by (a) RM8 and (b) H.261FB.

(a)

Figure 3.14: Magnified images of Fig. 3.12, (a) is encoded by RM8 and (b) is encoded by H.261FB.

By comparing the two encoded *Foreman* images shown in Figs. 3.12(a) and 3.12(b), it can be clearly seen that the quality of facial region was much improved in the H.261FB-encoded image as a result of the bit transfer from background to foreground region, while the consequent degradation in the background region was less obvious. Moreover, based on the premise that the background is usually of less significance to the viewer's perception, the overall quality of Fig. 3.12(b) was subjectively better and more pleasing to the viewer. The improvement can be further illustrated by magnifying the face region of the images as shown in Fig. 3.14. Objectively, the overall PSNR of the luminance (Y) component of the H.261FB-encoded image was less than that of the RM8-encoded image by 0.57 dB. However, if two separate PSNR measurements were used for the encoded foreground and background regions, then the objective quality of the facial region would have improved by 3.96 dB, whereas the background image quality would have degraded by only 1.00 dB.

(b)

Figure 3.14: *continued.*

For the encoded *Miss America* images shown in Figs. 3.13(a) and 3.13(b), the improvement achieved by the H.261FB coder was harder to notice, even when the area of interest is magnified as displayed in Fig. 3.15. Note that, however, the subjective improvement is more visible when the image is displayed on monitor screen than when it is printed on paper. Nevertheless, the two similar results produced by the RM8 and the H.261FB coders were also evident from their comparably PSNR values. The H.261FB coder did not achieve significant quality improvement of the facial region in its encoding process because it was unable to free up substantial bits by coarse quantization of the background region. This explanation can be illustrated in Fig. 3.16, whereby the bit usage per foreground and per background macroblock are plotted against different quantization parameters. The diagram on the right shows that, unlike the *Foreman* image, we could not transfer significant amount of bits by encoding the background region of the *Miss America* image at higher quantization level. It was because the discrete cosine transform (DCT) could compress a smooth, uniform and low-

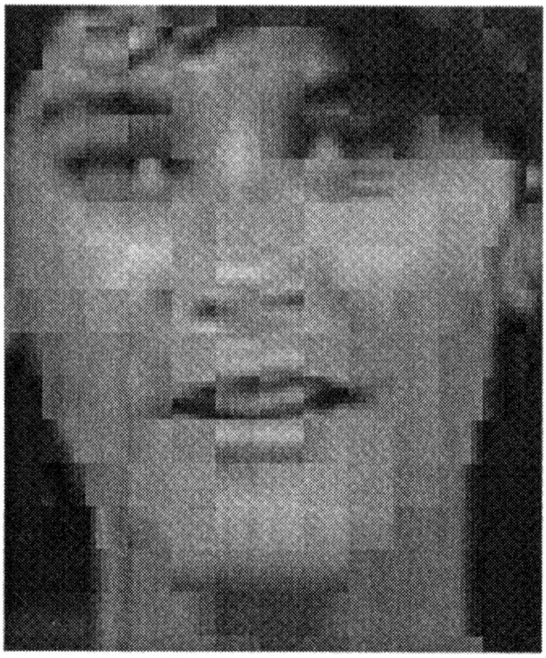

(a)

Figure 3.15: Magnified images of Fig. 3.13, (a) is encoded by RM8 and (b) is encoded by H.261FB.

texture background image of *Miss America* with great efficiency. Hence, the H.261FB coder could not reduce on what was already a minimal amount of bits used for the background and therefore the transfer of the bit saving to the foreground was small.

Furthermore, the bit usage for coding the facial region were quite similar, as can be seen in Fig. 3.16. Also from both these diagrams we can determine what value of Q_f will be selected for the H.261FB coder under the MBT strategy when the value of Q for the RM8 coder is other than the one we have previously chosen, for the *Foreman* and *Miss America* images.

The H.261FB coder was tested with the Joint Bit Assignment (JBA) approach and the joint rate control strategy. For comparison purpose, the CIF-size *Foreman* video sequence was encoded at 192 kb/s and 10 f/s using a conventional RM8 coder. Fig. 3.17 depicts the bits per frame (b/f) and PSNR values achieved by the RM8 coder. The coder spent on average 18,836 b/f and achieved an average PSNR value of 31.00 dB.

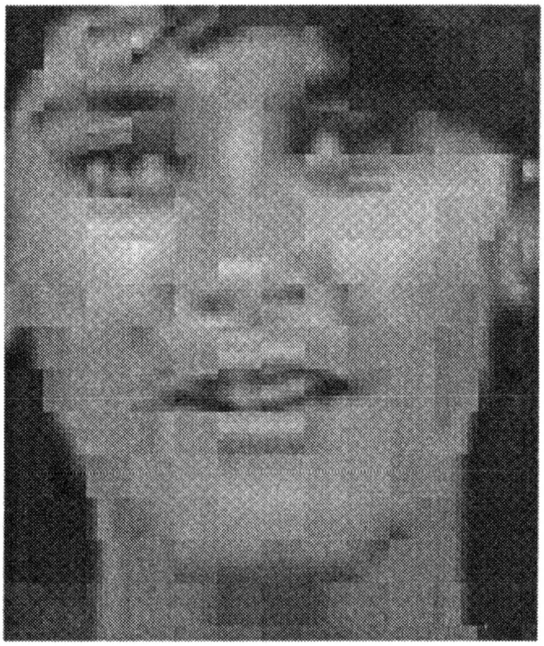

(b)

Figure 3.15: *continued.*

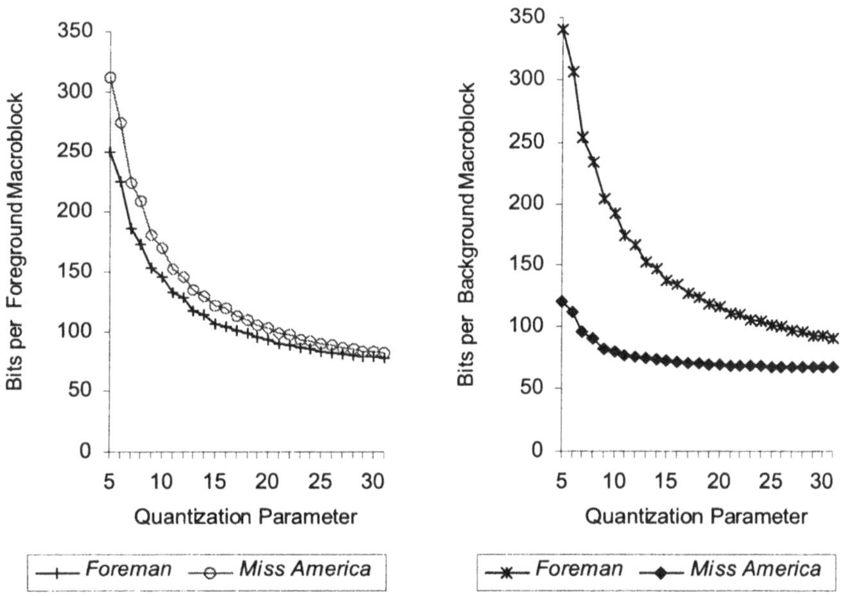

Figure 3.16: The average bits used per foreground and per background macroblock at different quantization parameters.

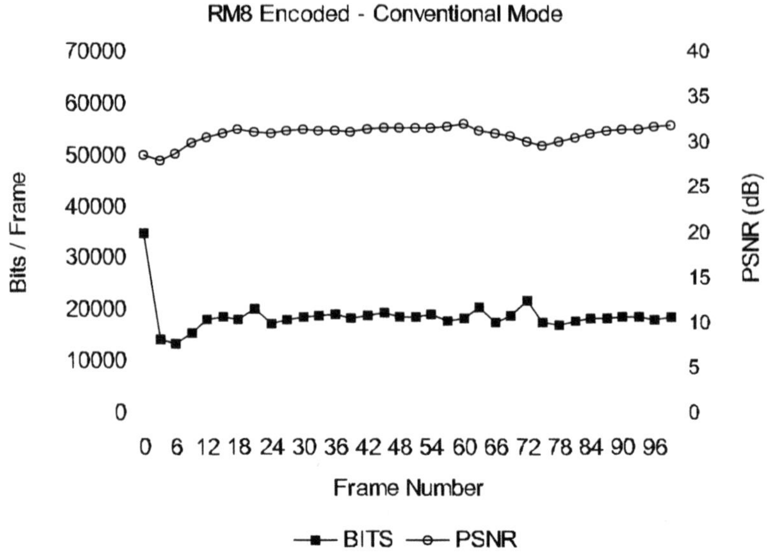

Figure 3.17: Bits/frame and PSNR values of the RM8-encoded *Foreman* sequence.

The normalized size and motion parameters of the foreground region of the *Foreman* video sequence are plotted as shown in Fig. 3.18. Since the values are normalized, the parameters for the background region are simply the complementary values. The figure shows a slow increase in the size of the foreground region, and that the background has higher activity than the foreground at most time.

Three sets of experiments were carried out on the H.261FB coder using the *Foreman* sequence with target bit rate of 192 kb/s and target frame rate of 10 f/s (i.e., same rates as those used in the RM8 coder). The first experiment was to test the bit allocation strategy based on size parameter only. This was done by setting P to 0%, ω_M to 0, and ω_S to 1 in (3.40) and (3.41). The input sequence was encoded with this bit assignment by the H.261FB coder. Fig. 3.19 depicts the coding results for the foreground and background regions. The H.261FB coder spent an overall[3] average of 18,843 b/f and achieved an overall average PSNR value of 30.99 dB - a result similar to what the RM8 has achieved (i.e., 18,836 b/f and 31.00 dB). It can be said that the proposed joint foreground/background rate control is

[3]The term overall here refers to the whole image instead of sub-region.

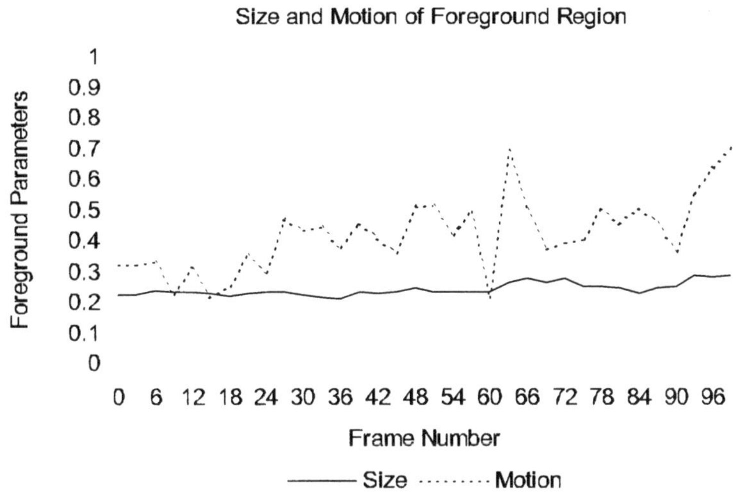

Figure 3.18: The characteristics of the foreground region of the *Foreman* sequence.

as accurate as the RM8 rate control. The bit difference between the above two cases (i.e., the RM8 and the H.261FB coder), as shown in Fig. 3.20, is indeed very small. Note that a positive bit difference in Fig. 3.20 indicates that the H.261FB is spending more bits per frame than the RM8 and vice versa. Nonetheless, the total difference after encoded 100 frames was only 7 bits.

In the second experiment, bit allocation based on size and priority parameters was performed. Therefore ω_M was set to 0 and ω_S to 1. With $P = 50\%$, the algorithm was transferring half the bits allocated to the background based on size parameter over to the foreground. The increase in the amount of bits eventually assigned to the foreground has led to an upward shift in the quality of the encoded foreground region, as depicted by the PSNR values in Fig. 3.21. By comparing the first and second experiments, the PSNR of the foreground region has increased from an average value of 31.91 dB to 35.58 dB, while the degradation of the background region from an average of 30.74 dB to 28.38 dB has resulted. As expected, the 50% drop in the amount of bits assigned to the background is evidenced by comparing the bits per background region values between Figs. 3.19 and 3.21.

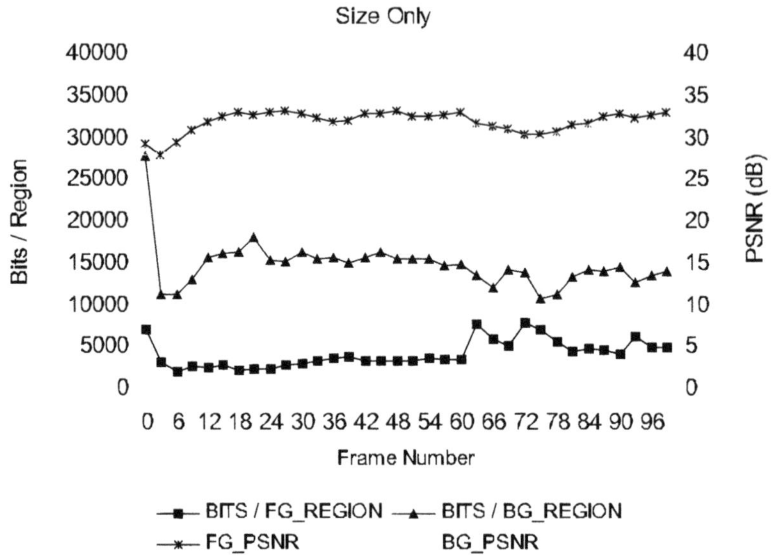

Figure 3.19: H.261FB encoded sequence with joint foreground/background bit allocation based only on the size of the region.

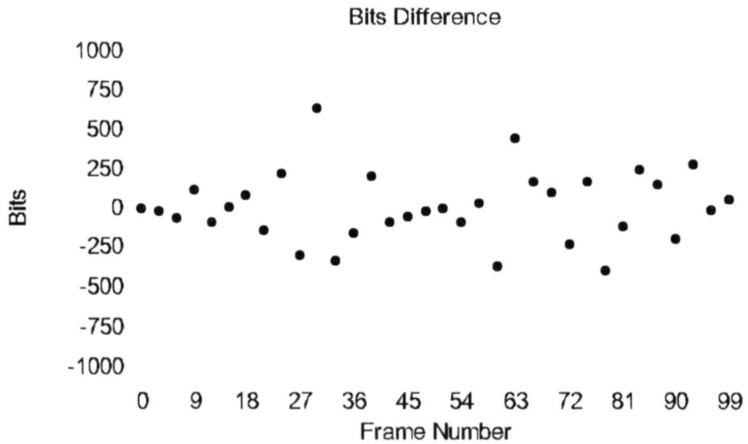

Figure 3.20: The difference in bit consumption per coded frame between the RM8 and the H.261FB at 192 kb/s and 10 f/s.

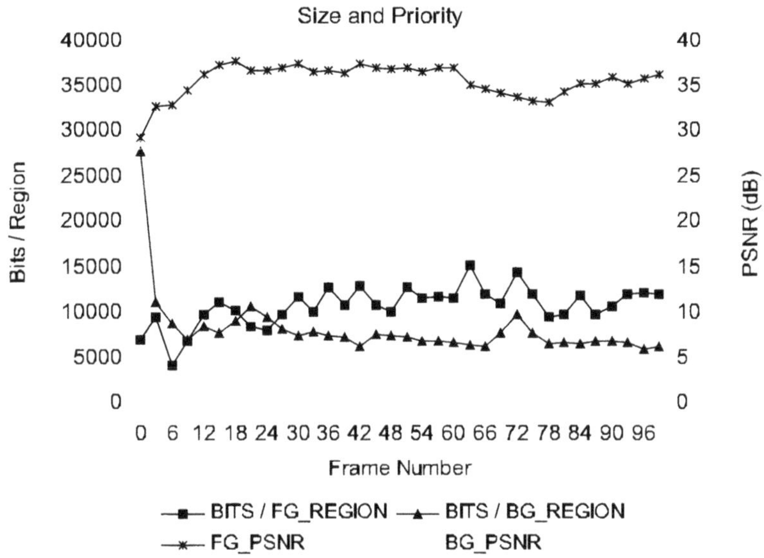

Figure 3.21: H.261FB encoded sequence with joint foreground/background bit allocation based on the size and priority of the region.

In the final experiment, the bit allocation was performed based on size and motion parameters. These two parameters were to have an equal influence to the bit allocation and therefore the weighting functions for both parameters were set at a constant value of 0.5. The coding results are shown in Fig. 3.22. It is evident from the figure that the inclusion of motion parameter in bit allocation has provided more bits to region with higher activity.

To show a sample of the subjective image quality achieved from the different approaches, frame 51 (middle frame) of each encoded sequence is selected for display. It can be observed that the image quality between the conventional RM8 approach (see Fig. 3.23(a)) and the size-only JBA approach (see Fig. 3.23(b)) is quite similar. However, improvement can be clearly seen in Fig. 3.23(c) for the size-and-priority JBA approach and in Fig. 3.23(d) for the size-and-motion JBA approach. The PSNR values of frame 51 can be found in Table 3.4. Note that the two separate PSNR values for the conventional RM8 approach were obtained using the segmentation information.

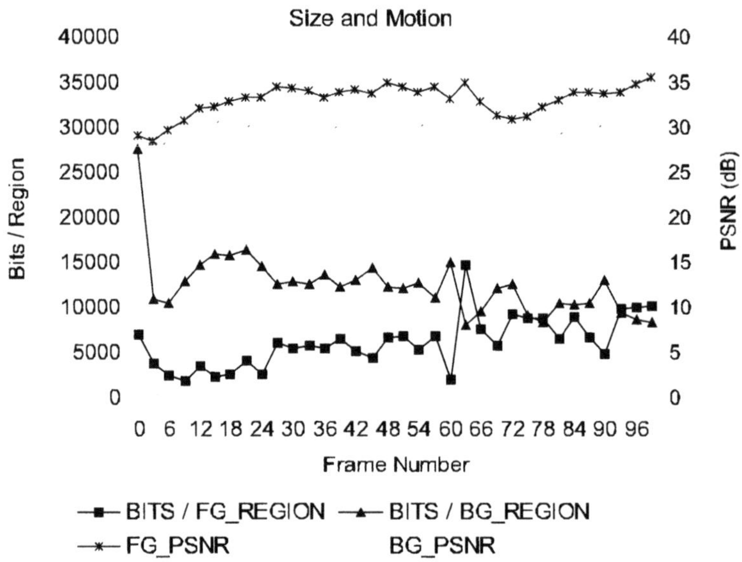

Figure 3.22: H.261FB encoded sequence with joint foreground/background bit allocation based on the size and motion of the region.

Table 3.4: PSNR values of Frame 51.

Approach	PSNR (dB) (Overall)	PSNR_FG (dB) (Foreground)	PSNR_BG (dB) (Background)
Conventional RM8	31.68	32.53	31.45
Size-only	31.58	32.51	31.33
Size-and-priority	29.59	37.07	28.62
Size-and-motion	31.03	34.68	30.33

(a)

(b)

Figure 3.23: Frame 51, encoded by (a) RM8 coder and H.261FB coder using (b) size-only JBA, (c) size-and-priority JBA and (d) size-and-motion JBA.

(c)

(d)

Figure 3.23: *continued.*

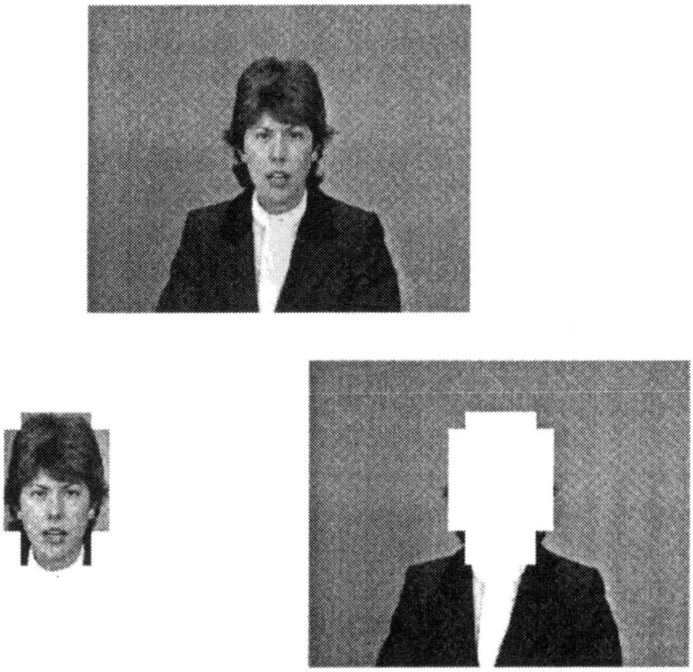

Figure 3.24: The original first frame of the *Claire* video sequence and its foreground and background regions at macroblock resolution.

The H.261FB was further tested on a different video sequence. Fig. 3.24 shows the original first frame and the foreground and background region of *Claire* sequence at CIF size. The normalized size and motion parameters of the foreground regions are shown in Fig. 3.25. The high values of the motion parameter signify that the main activity of the image is concentrated in the foreground region. The movement of the upper body of the speaker is the only activity in the background region.

This input sequence was coded using the RM8 coder at a target bit rate of 128 kb/s and a target frame rate of 10 f/s. Using the segmentation information, a separate set of PSNR values of the RM8-encoded foreground and background regions is plotted, as can be seen in Fig. 3.26. The figure exhibits a large difference in PSNR, with the quality of the background region being much higher than the foreground region as a large part of the background region is low in texture and motion.

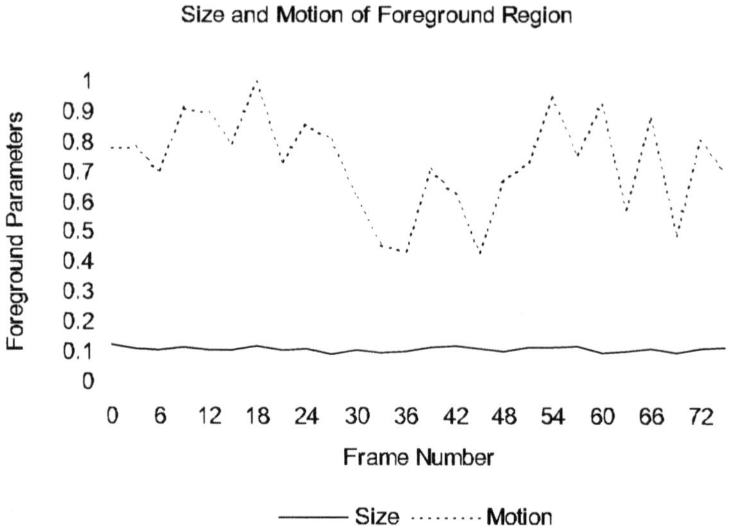

Figure 3.25: The characteristics of the foreground region of *Claire* sequence.

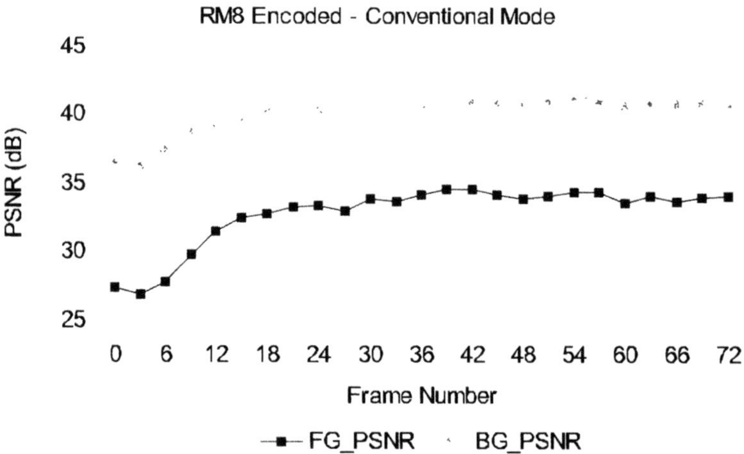

Figure 3.26: The PSNR values of the RM8-encoded foreground and background regions.

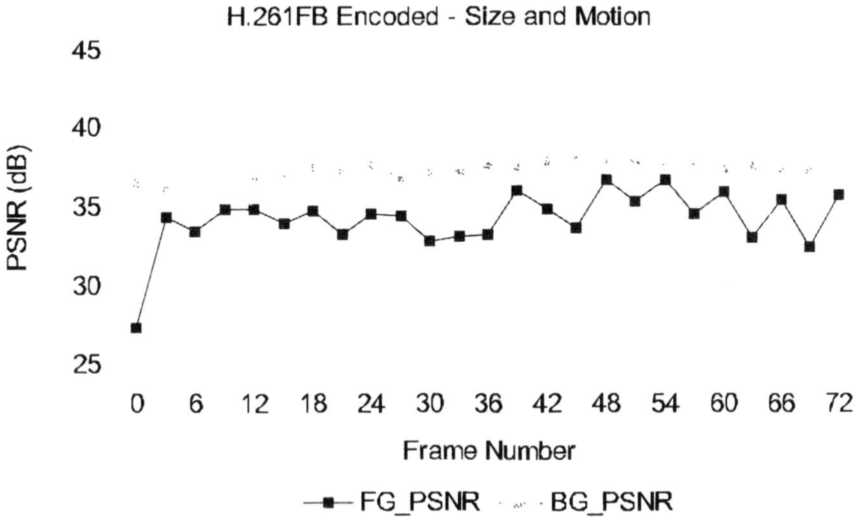

Figure 3.27: The PSNR values of the H.261FB-encoded foreground and background regions.

The same sequence was then encoded using the H.261FB coder with bit allocation based on the equal influence of the size and motion parameters. The coding results are shown in Fig. 3.27. The joint foreground/background bit allocation has resulted in higher PSNR values for the foreground region.

Both approaches used identical encoding parameters for intraframe coding of the first frame, and therefore the same results were produced as can be seen in Figs. 3.26 and 3.27. However, in the next encoded frame (interframe coding mode), the H.261FB coder allocated more bits to the foreground because it has detected a high foreground motion. Consequently, it improved the foreground image quality at a much quicker rate and also to a higher quality level. The first interframe coded images (i.e., Frame 3) are shown in Fig. 3.28.

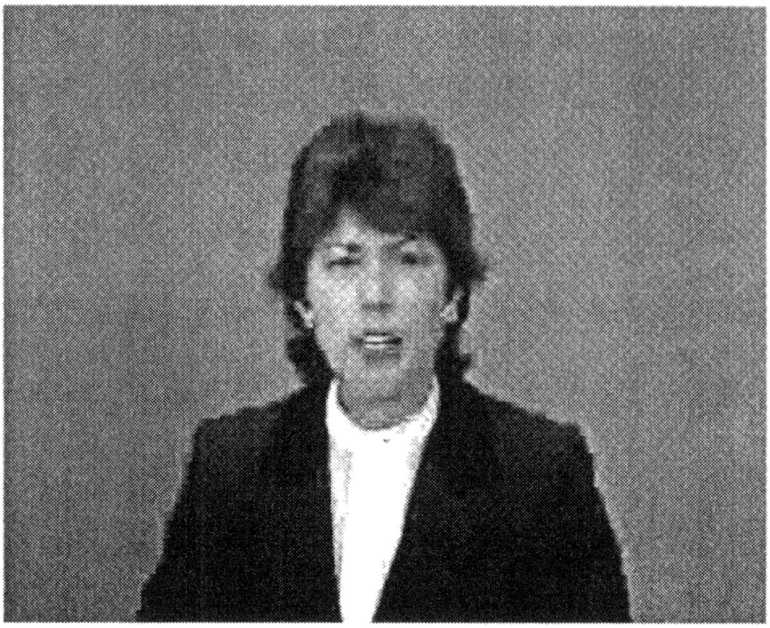

(a)

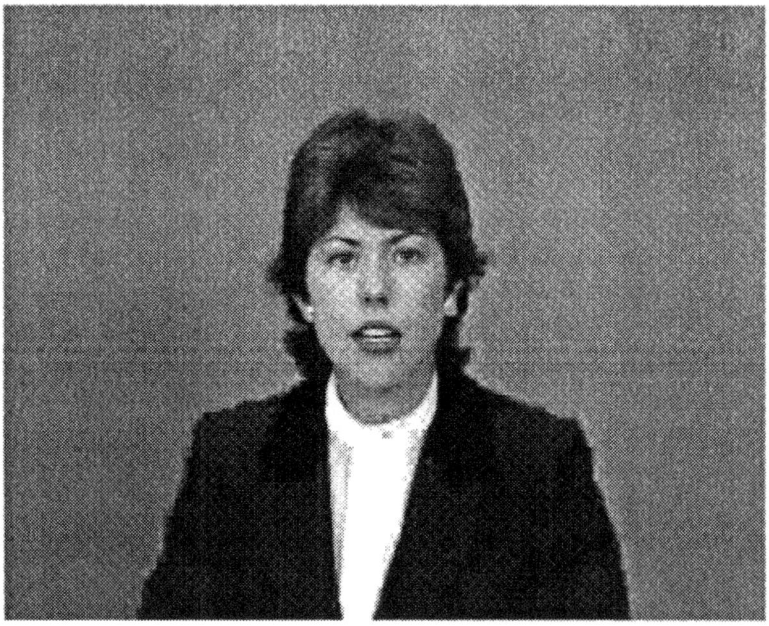

(b)

Figure 3.28: The first interframe coded images (i.e., Frame 3) by (a) RM8 coder and (b) H.261FB coder.

3.7 H.263FB Approach

The FB video coding scheme can also be integrated into the H.263 coder in a similar manner as with the H.261 coder. This is referred to as the H.263FB approach. Like the H.261 coder, the H.263 coder also focuses primarily on videotelephony applications, and the face of the speaker is typically the most concerned region by the viewers. For the H.263FB approach as discussed here, the facial area is to be separated from its background to become the foreground region. During the encoding process, more bits can be spent on the foreground at the expense of having fewer bits for the background. Hence it allows the facial region to be transmitted over a narrow-bandwidth data link with better subjective image quality, which in turn serves the main purpose of videotelephony better.

The implementation of such approach and the experimental results are presented in the following.

3.7.1 Implementation of the H.263FB Coder

Here, the implementation of FB video coding scheme on the H.263 framework is described. Similar to the H.261FB approach, the image segmentation of human face for the H.263 coder is achieved by the algorithm explained previously. Once again the final segmentation result is at macroblock resolution. This face segmentation algorithm is adopted here due to its appealing features. Firstly, it operates on the same source format as the H.263 coder does, i.e., a CIF or QCIF YUV411 format. Secondly, the segmentation process is mainly performed at block level, therefore it is fast in producing a result at resolution that is appropriate for the block-based H.263 coder. Finally, it is fully automatic and robust. It can cope with numerous types of videophone images without having to adjust any design parameter.

The face segmentation information enables bit transfer from background to foreground through the controlling of the quantization step-size. Since the lowest level that the H.263 coder can adjust its quantization parameter is at the macroblock level, the resolution of the segmentation results is set to the macroblock level. However, unlike the H.261 video coding system, the H.263 has a limited selection of quantization step-size for each macroblock. In any particular macroblock line, the quantization step-size for one macroblock can only be varied within the integral range of [-2, 2] from its previous value. This restricts the ability of bit transfer from one macroblock to another. Hence the H.263 bitstream syntax must be modified in order to perform bit transfer effectively. As a consequence, a full H.263 decoder compatibility can no longer be maintained. Below the modification of the H.263 coding

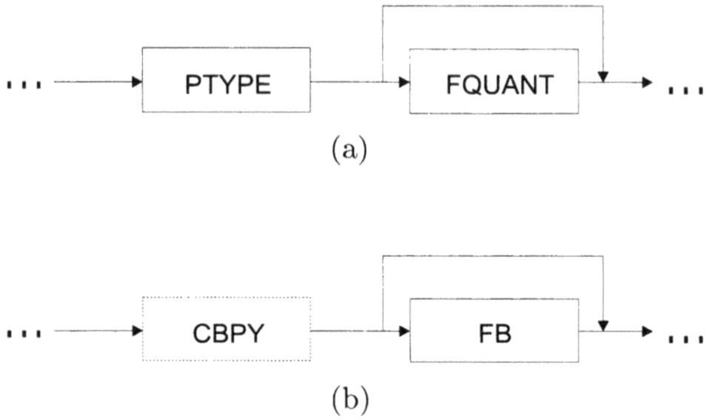

Figure 3.29: Syntax changes in H.263 video bitstream - (a) at the picture layer and (b) at the macroblock layer.

syntax is described. As a point to note, the changes in decoder are simply the reverse process, therefore they will not be discussed here. Readers are referred to [17] for the specifications of the H.263 codec.

The modification of the bitstream syntax involves only three headers, as illustrated in Fig. 3.29. The PTYPE header is modified and another header at the picture layer of the video bitstream is added; while at the macroblock layer, only one new header is introduced.

The use of FB coding scheme forms another negotiable option for the H.263 codec. This is referred as the FB coding mode. An extra bit is added to the PTYPE (Picture Type) header at the picture layer of the bitstream in order to indicate the use of this optional mode. This extra bit will become the bit 14 of the PTYPE header and be set to '0' if this mode is off, or '1' if it is on. If FB coding mode is off then the rest of the coding processes do not require any new syntax, or else further changes in syntax are required.

If the FB coding mode is in use, an additional header called FQUANT is sent before the PQUANT header at the picture layer of the bitstream. This new FQUANT header is a fixed length codeword of 5 bits that indicates the quantization level to be used for the foreground region. This leaves the PQUANT header for the background region. Instead of having only one quantizer for the entire picture, the FB coding mode requires two quantizers - one assigned to each region. Let Q_f and Q_b be the quantizers for the foreground and the background, respectively. The quantizer, Q_f, takes on the FQUANT value while Q_b is defined by PQUANT. Q_b, as the coarser quantizer, is used on macroblock that belongs to the background, while the finer quantizer Q_f is used on the foreground macroblock.

The final syntax change occurs at the macroblock layer of the bitstream. Here, a 1-bit header called FB is introduced to signify the region the coded macroblock is in; using '0' to indicate that it belongs to the background and '1' for otherwise. This header is required to be sent only if MCBPC and CBPY headers indicate that there is at least one non-INTRADC transform coefficient in any of the six blocks that needs to be transmitted. If so, the transmission of FB header occurs immediately after CBPY. For a QCIF size image, there are 99 macroblocks, hence the *maximum* number of transmissions of FB header in one frame is 99 times.

Therefore the overhead bits required by the FB coding mode is *at most* 105 bits per QCIF frame. This includes one compulsory extra bit in PTYPE header, five bits in FQUANT header and 99 bits from the transmission of 99 1-bit FB headers.

3.7.2 Experimental Results

The FB coding scheme was tested on a QCIF-size *Foreman* video sequence. The intraframe coding on the first frame with and without the use of the FB coding mode was tested, and the results are given in Figs. 3.30(a) and 3.30(b), respectively. Fig. 3.30(a) was coded using 15,502 bits with quantization step-size for the foreground and background set at 9 and 21 respectively, whereas Fig. 3.30(b) was coded using 15,796 bits with quantization step-size for the entire picture set at 16. The bit transfer of 2379 bits or 15% was achieved. The overall PSNR value for Fig. 3.30(a) is 30.701 dB; which is lower than the value for Fig. 3.30(b) by 0.766 dB. This is expected since the larger region of the background was coded at higher quantization step-size and therefore producing more noise. Subjectively, however, it can be observed that Fig. 3.30(a) is more pleasing to view as it has less noise in the facial region, while the increase in noise at the background is less noticeable and annoying.

(a)

(b)

Figure 3.30: Intraframe coded images - (a) with the FB coding mode and (b) without the FB coding mode.

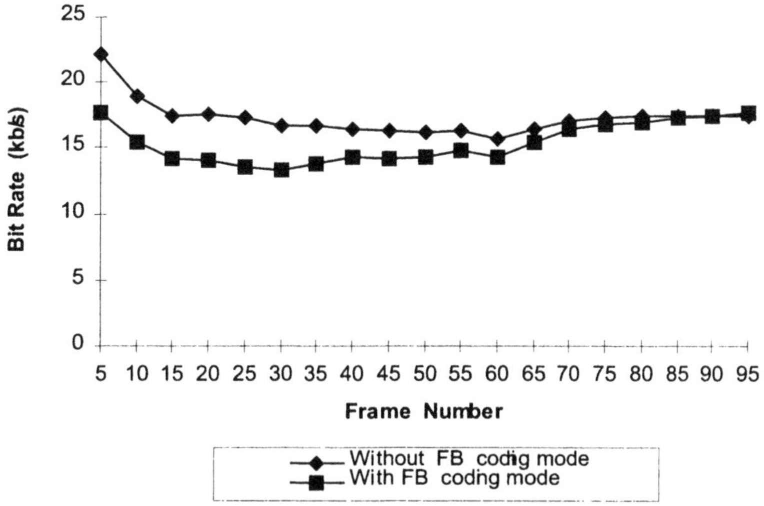

Figure 3.31: A plot of bit rate against frame number at 5.0 f/s.

The performance of the H.263FB coding scheme was then tested on interframe coding. One hundred frames of the *Foreman* video sequence were coded at variable bit rate with fixed quantization step-size and fixed frame rate of 5.0 f/s. In FB coding mode, the quantizers for the foreground and background were set at 9 and 28 respectively, while the quantizer for the case of without FB coding mode was set at 16. For proper comparison of interframe coding, the first frame was intraframe coded entirely with quantization step-size at 16 for both cases. A plot displaying the bit rates achieved is provided in Fig. 3.31. Notice that up to Frame 30, the bit rate obtained in FB coding mode is a few kb/s lower than that of without the FB coding mode. After that, the bit rate climbs steadily to match its counterpart due to rapid motion in the facial region and hence more finely quantized transformed coefficents are coded from the foreground regions.

To illustrate the subjective image improvement, Frame 90 from the coded sequence is shown in Fig. 3.32. It is observed that the image in Fig. 3.32(a) has a better perceived quality than Fig. 3.32(b) due to the improvement in the rendition of facial features when the FB coding mode is used. Note that the subjective improvement has been achieved even though its overall average PSNR value is 1 dB lower, at 28.10 dB, and about 10% below its average bit rate.

(a)

(b)

Figure 3.32: Interframe coded images - (a) with the FB coding mode and (b) without the FB coding mode.

3.8 Towards MPEG-4 Video Coding

Both H.261FB and H.263FB coders can be considered as frame-based video coders that imitate, to some extent, the object-based video coding approach that is much talked about in the MPEG-4 standard [18]. A traditional frame-based video coding system is blind to image content and therefore treats all parts of an image with equal importance. However, by integrating the FB coding scheme into the H.261 and H.263 coders, we are able to tune the encoder parameters for each video object, like an MPEG-4 coder.

Unlike the MPEG-4 approach, the H.261FB and H.263FB coders are, however, limited to two image regions (or video objects) decomposition. Furthermore, these coders are restricted by the sequential processing structure of the traditional frame-based video coding system, i.e., a top-bottom, left-right processing order of image blocks, and the basic processing unit is in an 8×8 block or 16×16 macroblock. This is followed in order to conform with the existing H.261 and H.263 video coding standards.

In contrast to the multitude of functionalities that the MPEG-4 standard is set to provide, the objective of the FB coder is to only provide spatially variable reconstuction quality and bit rate in relation to the foreground and background regions of an image. In particular, it is to protect the area of interest, i.e., the foreground, from visual artifacts and to code this area at a better quality (and thus at a higher bit rate) than the background. Therefore the above mentioned restrictions do not hamper the FB coder from achieving its objective.

Nevertheless, the FB coder serves a good platform to further research on the implementation of MPEG-4 codec. Firstly, the face segmentation technique used in the FB coder can be brought over to an MPEG-4 codec. Secondly, the block-based DCT operation employed in the FB coder can be replaced with shape adaptive DCT [19] for arbitrarily shaped video objects. Thirdly, the FB content-based bit allocation strategies can be extended to multiple-object content-based bit allocation. The only aspect of the FB coder that cannot be used in a MPEG-4 codec is the FB content-based rate control strategy. This is because this strategy adapts specifically to the fundamental sequential processing structure of a frame-based video coding system whereby the foreground and background regions are coded jointly, whereas the video objects in a MPEG-4 approach are coded separately.

3.8.1 MPEG-4 Coder

The performance study on the MPEG-4 coder is presented here with the following questions in mind.

- How does it perform in frame-based and object-based coding?
- How much overheads are required to use object-based mode as compared to frame-based?
- What is the capability of bit/quality transfer among video objects?
- What difference does it make if the video objects were segmented at different resolutions?

Four sets of experiments were carried out in search of these answers. The aim, procedure, results and discussion for each experiment are presented below.

3.8.1.1 Experiment 1

The aim of the first experiment was to run the MPEG-4 coder in a rectangular frame-based and variable bit rate (VBR) video coding mode, and then to measure its performance in terms of bit consumption and output image quality.

For this, *Foreman* was selected as the source sequence, with 100 CIF-size frames at 30 f/s. The alpha channel was set to rectangular mode, and rate control disabled. The entire sequence (100 frames) was encoded at constant quantization parameter (QP) of 16 and at constant target frame rate of 10 f/s.

A total of 34 frames (i.e., Frame 0, 3, 6, 9, ... , 99) were encoded. A plot of bit consumption against frame number is shown in Fig. 3.33, while a plot of output image quality against frame number is shown in Fig. 3.34.

It was found that the coder spent approximately 10,300 b/f to encode the *Foreman* sequence at a frame rate of 10 f/s, using a constant QP of 16 throughout. The average output image quality was measured at a PSNR value of 31.39 dB.

3.8.1.2 Experiment 2

The objective of the second experiment was to test the MPEG-4 coder in object-based mode and observe how it compares against frame-based and how much overheads are required.

The same source sequence was used as before, but the alpha channel was switched to binary mode. The *Foreman* sequence was decomposed into two video objects (VOs), i.e., a foreground (VO0) and a background (VO1), using the face segmentation algorithm as described in Chapter 2.

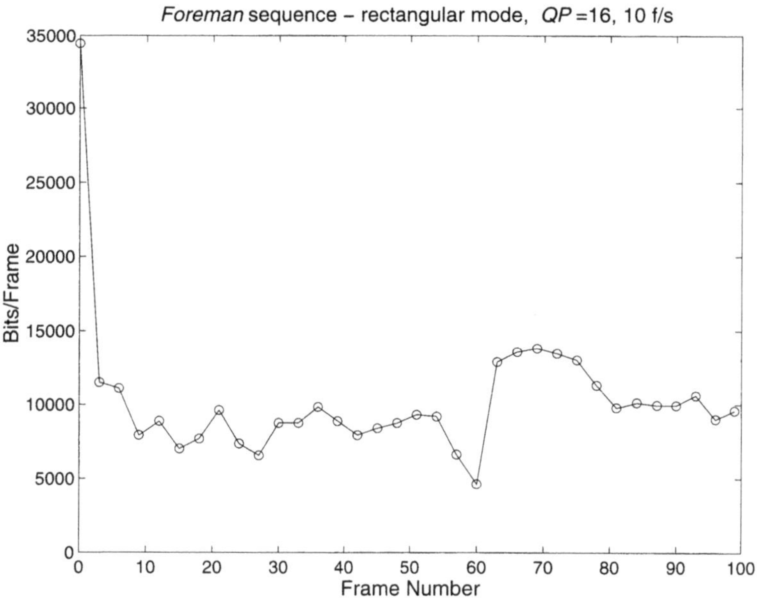

Figure 3.33: Experiment 1 - VBR coding of *Foreman* sequence, a plot of bits/frame against frame number.

The foreground contained only the facial region. For each VO, a set of alpha maps were generated at MB resolution. Then, both VOs (2×100 video object planes (VOPs)) were encoded at constant QP of 16 and at constant target frame rate of 10 f/s. Note that the rate control was not needed.

The experimental results are presented in Table 3.5. The average PSNR values for the foreground (FG) and background (BG) video objects were found to be 31.11 dB and 32.14 dB, respectively. However, note that since both experiment 1 and 2 have used the same QP value, the output image quality of the whole scene in this experiment would be the same as in experiment 1. In terms of bit consumption, the total bits spent on coding both VO0 and VO1 were $271,144 + 133,904 = 405,048$ in the object-based coding mode.

As compared to the frame-based mode, the coder in binary alpha channel mode spent an extra 54,728 bits, or approximately 15.6% more bits, to encode 100 frames of the *Foreman* sequence at the same image quality. This is quite an expensive overhead cost. Note that this overhead cost is incurred from the transmission of additional header information, alpha

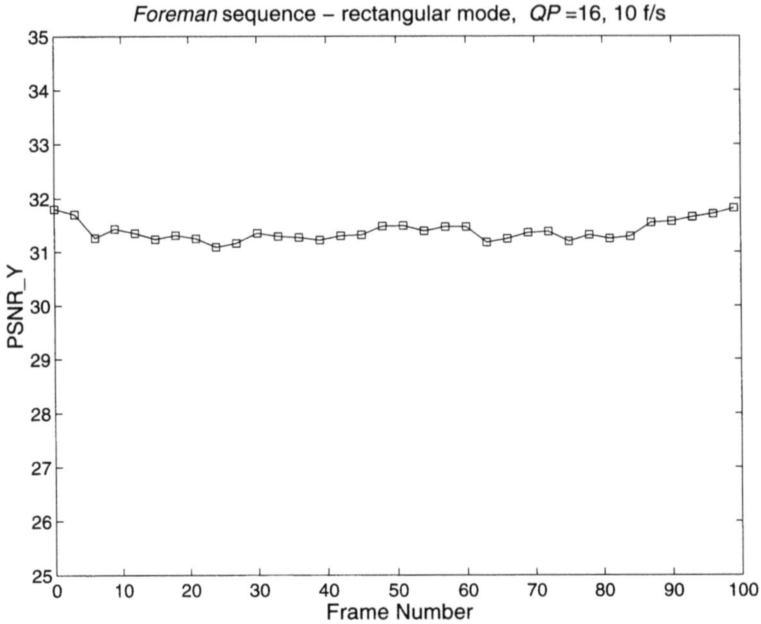

Figure 3.34: Experiment 1 - VBR coding of *Foreman* sequence, a plot of PSNR against frame number. Note that these are the PSNR values of luminance (Y) component only.

Table 3.5: Results from coding 100 frames of *Foreman* sequence in rectangular and also binary alpha channel modes, all using constant QP of 16.

	Expt #1	Expt #2	
	Rect. Frame	VO0 (BG)	VO1 (FG)
Total bits	350320	271144	133904
Av. bits/VOP	10299.76	7972.00	3935.53
Av. PSNR_Y_BG (dB)	31.39	31.11	32.14

channel, shape information, etc. Therefore, the use of binary alpha channel must be justified by the additional content-based functionalities that it provides.

Table 3.6: Coding VO0 (background region) at various QPs.

QP	Total bits	PSNR (dB)
16	271,144	31.11
22	227,392	30.15
23	226,904	30.04
24	224,168	29.91
25	219,288	29.82
26	217,096	29.73
28	215,184	29.55
29	211,784	29.45
31	209,336	29.25

Table 3.7: Coding VO1 (foreground region) at various QPs.

QP	Total bits	PSNR (dB)
16	133,904	32.14
12	158,008	33.22
10	180,232	33.97
9	201,352	34.55
8	220,128	35.00

3.8.1.3 Experiment 3

The aim here was to encode the foreground and background regions of the input video at various quality in a VBR environment by adjusting the QPs, so that the capability of bit/quality transfer among VOs can be investigated.

Once again the same source sequence was selected, the alpha channel remained in binary mode and the rate control remained disable for VBR environment. Using the same sets of alpha maps as before, both VOs were encoded at various QPs but at constant target frame rate of 10 f/s.

The total amounts of bits spent on encoding 100 background VOPs and their average PSNR values under various QPs can be found in Table 3.6. Similarly, the results for the foreground VOPs are shown in Table 3.7. Note that lower QP values were chosen for the foreground VOPs since they are visually more important than the background VOPs.

This experiment considers the given bit constraint and the condition of

Table 3.8: A combination of VOs at different bit rate and quality.

VO1 (Face)			VO0 (Non-face)			Total bit
QP	Total bits	PSNR	QP	Total bits	PSNR	consumption
8	220,128	35.00	31	209,336	29.25	429,464
9	201,352	34.55	31	209,336	29.95	410,688
10	180,232	33.97	24	224,168	29.91	404,400

not spending more than the amount of bits used in Experiment 1. In other words, it is required to encode the same source sequence without consuming more than 350,320 bits. One way for achieving this is as follows. From Tables 3.6 and 3.7 it can be noticed that if VO0 was encoded at the maximum QP of 31 and VO1 at QP of 16, then the total bit consumption would be 343,240 (i.e., $209,336 + 220,128$) bits, which is 7080 bits under the bit budget. Therefore similar bit consumptions were achieved but at the expense of having to quantize the background video object at the coarsest level. Note that in Experiment 1, each frame was encoded using QP value of 16 throughout in the frame-based approach. This demonstrates and reinforces the finding in Experiment 2 that the overhead cost of encoding two separate VOs to be quite significant. Therefore the concept of transferring bits from one VO to another in order to encode one particular VO at a better quality is clearly not feasible in MPEG-4 object-based approach, due to the expensive overhead cost. This is unless, of course, the use of object-based approach is also to provide additional functionality such as content-based user interactivity.

Nevertheless, MPEG-4 coder is certainly capable of transferring bit/quality among video objects, but it comes at a cost. Table 3.8 shows some of the possibilities of encoding different VO at different bit rate and quality, and the cost is indicated by the total amount of bit consumption.

3.8.1.4 Experiment 4

An input video to the MPEG-4 coder can be decomposed into VOPs at pixel or macroblock (MB) resolution. In Experiment 2 and 3, VOPs at MB resolution were used. So, the aim of this experiment was to determine what difference does it make if the VOPs were defined at pixel resolution instead.

The source image as displayed in Fig. 3.35 was used. The source image was decomposed into two VOPs using the face segmentation algorithm at both pixel and MB resolution. VOP0 represents the non-facial region while

Figure 3.35: Source image.

Table 3.9: Overall bit rates and PSNR values achieved from using different binary alpha maps.

	Binary alpha maps					
	MB resolution			Pixel resolution		
	VOP0	VOP1	Overall	VOP0	VOP1	Overall
QP value	31	6	n/a	31	6	n/a
PSNR (dB)	28.42	37.84	30.61	28.41	37.92	30.40
Bits/VOP	16,896	12,912	29,808	18,808	9,600	28,408

VOP1 contains the facial region. The binary alpha maps at MB and pixel resolution are depicted in Figs. 3.36 and 3.37, respectively. Both VOPs were then encoded using the MPEG-4 coder. The statistics of the results are presented in Table 3.9, and the encoded images are shown in Fig. 3.38.

Note that the face segmentation algorithm will attempt to include all pixels in facial region to the foreground alpha map. So, to have it at MB resolution, it is inevitable that some non-facial-pixels will be included in this map. Therefore the size of the alpha map for the facial region in MB resolution will never be smaller than the map in pixel resolution. This is demonstrated in Figs. 3.36(b) and 3.37(b).

Hence, the reasons why more bits are required to encode VOP1 at MB resolution are twofold when compared against VOP1 at pixel resolution. Firstly, the area is larger, and this leads to greater bit consumption. Secondly, pixels in this VOP are encoded at finer QP value, and so the increase in bit consumption is even greater.

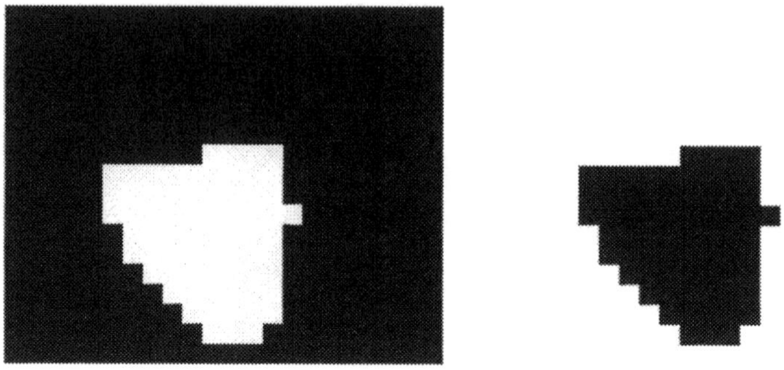

(a) (b)

Figure 3.36: Binary alpha maps at MB resolution for (a) VOP0 (non-face) and (b) VOP1 (face).

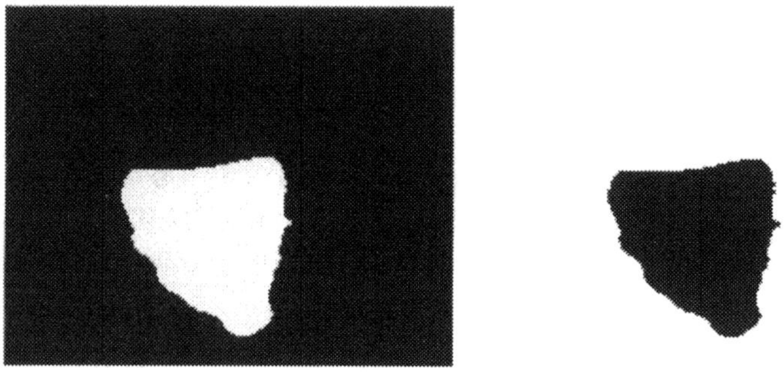

(a) (b)

Figure 3.37: Binary alpha maps at pixel resolution for (a) VOP0 (non-face) and (b) VOP1 (face).

However, as far as the quality of the encoded images are concerned, there is little difference in terms of objective and subjective quality.

(a)

(b)

Figure 3.38: Encoded images using binary alpha maps at (a) MB and (b) pixel resolution.

3.8.2 Summary

The performance of MPEG-4 coder was studied. It was found that the use of binary alpha channel mode incurs an expensive overhead cost. Therefore, the use of binary instead of rectangular alpha channel must be justified by the content-based functionalities that it provides. Note that, however, due to this overhead cost, the use of binary alpha channel mode solely for the purpose of transferring bits from one image region to another, as described in the FB coding scheme, is clearly not feasible in MPEG-4 coding system. Additionally, it was found that it does not make much difference whether the foreground and background VOs are defined in MB or pixel resolution.

References

[1] D. Chai and K. N. Ngan, "Foreground/background video coding scheme," in *IEEE International Symposium on Circuits and Systems*, Hong Kong, Jun. 1997, vol. II, pp. 1448–1451.

[2] D. Chai and K. N. Ngan, "Coding area of interest with better quality," in *IEEE International Workshop on Intelligent Signal Processing and Communication Systems (ISPACS'97)*, Kuala Lumpur, Malaysia, Nov. 1997, pp. S20.3.1–S20.3.10.

[3] D. Chai and K. N. Ngan, "Foreground/background video coding using H.261," in *SPIE Visual Communications and Image Proceeding (VCIP'98)*, San Jose, California, USA, Jan. 1998, vol. 3309, pp. 434–445.

[4] A. Eleftheriadis and A. Jacquin, "Model-assisted coding of video teleconferencing sequences at low bit rates," in *IEEE International Symposium on Circuits and Systems*, London, Jun. 1994, vol. 3, pp. 177–180.

[5] A. Eleftheriadis and A. Jacquin, "Automatic face location detection and tracking for model-assisted coding of video teleconferencing sequences at low-rates," *Signal Processing: Image Communication*, vol. 7, no. 4-6, pp. 231–248, Nov. 1995.

[6] A. Eleftheriadis and A. Jacquin, "Automatic face location detection for model-assisted rate control in H.261-compatible coding of video," *Signal Processing: Image Communication*, vol. 7, no. 4-6, pp. 435–455, Nov. 1995.

[7] L. Ding and K. Takaya, "H.263 based facial image compression for low bitrate communications," in *Proceedings of the 1997 Conference on Communications, Power and Computing (WESCANEX'97)*, Winnipeg, Manitoba, Canada, May 1997, pp. 30–34.

[8] C.-H. Lin and J.-L. Wu, "Content-based rate control scheme for very low bit-rate video coding," *IEEE Transactions on Consumer Electronics*, vol. 43, no. 2, pp. 123–133, May 1997.

[9] C.-H. Lin, J.-L. Wu, and Y.-M. Huang, "An H.263-compatible video coder with content-based bit rate control," in *IEEE International Conference on Consumer Electronics*, Jun. 1997, pp. 20–21.

[10] M. Wollborn, M. Kampmann, and R. Mech, "Content-based coding of videophone sequences using automatic face detection," in *Picture Coding Symposium (PCS'97)*, Berlin, Germany, Sep. 1997, pp. 547–551.

[11] MPEG-4 Video Group, "MPEG-4 video verification model version 6.0," Document ISO/IEC JTC1/SC29/WG11 N1582, Sevilla, Spain, Feb. 1997.

[12] T. Xie, Y. He, C.-J. Weng, and C.-X. Feng, "A layered video coding scheme for very low bit rate videophone," in *Picture Coding Symposium (PCS'97)*, Berlin, Germany, Sep. 1997, pp. 343–347.

[13] T. Xie, Y. He, C.-J. Weng, Y.-J. Zhang, and C.-X. Feng, "The study on the layered coding system for very low bit rate videophone," in *SPIE Visual Communications and Image Processing (VCIP'98)*, San Jose, California, USA, Jan. 1998, vol. 3309, pp. 576–582.

[14] H. G. Musmann, "A layered coding system for very low bit rate video coding," *coding*, vol. 7, no. 4-6, pp. 267–279, 1995.

[15] ITU-T Recommendation H.261, "Video coder for audiovisual services at $p \times 64$ kbit/s," Mar. 1993.

[16] CCITT Study Group XV, "Document 525, description of reference model (RM8)," Jun. 9, 1989.

[17] ITU-T Recommendation H.263, "Video coding for low bitrate communication," May 1996.

[18] ISO/IEC JTC1/SC29/WG11 N2323, "Overview of the MPEG-4 standard," Jul. 1998.

[19] T. Sikora and B. Makai, "Shape-adaptive DCT for generic coding of video," *IEEE Transactions on Circuits and Systems for Video Technology*, vol. 5, no. 1, pp. 59–62, Feb. 1995.

Chapter 4

Model-Based Coding

4.1 Introduction

Research into model-based image coding has intensified as studies into very low bit rate video coding have recently expanded. To represent and encode image signals efficiently, a suitable image model is required. Model-based image coding methods make use of variations of image source models taking into account the structural features of the image. There are two aspects of image source model used for image coding: segmentation model and motion model.

From the various proposals, 2 different approaches have emerged, the 2-D and 3-D model-based [1]. The 2-D model is a more general approach using deformable triangular segmentation of the image and affine transform based motion model. The 3-D model-based coding is more specific utilizing the 3-D properties of the objects in the scene. Table 4.1 shows the kind of image models used in various coding schemes.

4.1.1 2-D Model-Based Approaches

These coding methods exploit the important 2-D properties of the image such as edges, contours and regions. Two particular examples are contour-based coding and region-based coding. The first method extracts contours and encodes shapes and intensities of contours and reconstructs an image from them [2, 3]. The second method segments images into homogenous regions and encodes their shapes and intensities [2, 4]. These methods encode the images with natural intensity levels unlike the earlier works that only encode binary images.

For image sequences, the two successive frames are modeled and coded as

Table 4.1: Image coding techniques and their image source models. [1]

Image Source Models		Coding Schemes
Segmentation Model	**Motion Model**	
Pixel		PCM
Statistically dependent pixels block	2-D translation	MC-DCT etc.
2-D model-based approaches 2-D features such as edges, contours, 2-D rigid regions, deformable triangle blocks, deformable square blocks, etc.	translation, bilinear transform affine transform etc.	contour-based coding region-based coding object-based coding 2-D deformable triangle-based coding
3-D model-based approaches 3-D global surface model such as planes or geometric surfaces parameterized 3-D model	3-D global motion 3-D local motion	object-based coding 3-D model-based coding

arbitrarily shaped 2-D objects translating two-dimensionally [5]. Both rigid and flexible regions are used for modeling 2-D moving areas. The motion models can be described with an affine transform or a bilinear transform to better approximate the motion fields of a 3-D moving rigid object and linear deformations such as rotation and zooming. The affine transform motion model is used with triangular segmentation in the deformable triangular based motion compensation scheme [6].

4.1.2 3-D Model-Based Approaches

This is the more specific approach to model-based coding which utilizes 3-D structural models of the scenes. There are two kinds of approaches to 3-D model-based schemes. The first approach makes use of surfaces of the object modeled by general geometric models such as planes or smooth surfaces. The second approach utilizes parameterized model of the object. In order to distinguish between the two approaches, the first one is referred as 3-D feature-based approach and the second as 3-D model-based approach.

In 3-D feature-based approaches, information such as surface structure and motion information is estimated from image sequences and utilized in image coding. There are several different methods that have been proposed. Hotter *et al.* [5, 7] and Diehl [8] have proposed a method utilizing a segmented surface model, in which changing regions caused by object motion are detected and modeled by planar patches or parabolic patches. Ostermann *et al.* [7, 9], Morikawa et al. [10], and Koch [11] have proposed another method utilizing global surface models, in which a smooth surface model of the scene is estimated from an image sequence. These methods have also been applied with motion compensation and interpolation to improve the performance of conventional waveform coding methods.

In 3-D model-based coding, the parameterized models are usually given in advance. To obtain a 3-D model from a general scene is extremely difficult, but when the object to be coded is restricted to specific classes, such as human faces in videophone images, then a 3-D generic face model is sufficient for describing scene objects, since most of the images are head-and-shoulder images. The need for construction of a 3-D model from 2-D images is no longer necessary. Earlier work for this approach is the semantic coding as proposed by Fochheimer [12, 13]. This approach lacked in the reconstruction of the image, with the resulting images being too synthetic. More recent work include Aizawa, Harashima [14, 15] and Welsh [16, 17], utilizing a detailed parameterized 3-D model of a person's face. The emphasis is on human facial images, with the 3-D model given in advance.

Sometimes a combination of 3-D model-based/waveform hybrid coding is used to improve the fidelity of the reconstructed images. In these schemes, waveform coding is used to compensate errors which occur in the model-based coding process. Waveform coding methods used include MC/DCT [18], vector quantization [19], and contour coding [20].

Automatic modeling poses the biggest problem in 3-D model-based coding, as described later in Sections 4.3 and 4.4, and the other major problem is in analysis. Some automatic motion tracking has been reported, with the model made in advance and the initial position of the face is assumed. The face motion is tracked by using facial feature points which were detected by simple threshold logic. The direct estimation of face motion without using feature points has been reported by Choi *et al.* [21, 22] and Li *et al.* [23].

The method of 3-D model-based coding described in this chapter follows the work of K. Aizawa et al. [14, 15] This method utilizes a 3-D model of a human head for representation of facial images such as the ones used in videoconferencing. The encoder analyses the head motion and facial expression of the input images based on the common knowledge of the 3-D facial model, it then transmits these parameters. The decoder uses these parameters and synthesizes the images using the 3-D facial model. The image source model used is the 3-D facial model adjusted to the user's face. The original image texture is projected onto the 3-D model so that the intensity information is stored at each point on the 3-D model, enabling natural image reproduction.

4.1.3 Applications of 3-D Model-Based Coding

With the rapid advancement in the telecommunication, TV/Film entertainment and computer industry, a whole multitude of applications is emerging from these industries. Sound and video are being added to telecommunications and computer industries, interactive capability is being added to communications and entertainment, and networking is being added to computer and entertainment.

Due to its synthesis capability, that is, given the image model any desired scene can be described in a structural way into codes which can be easily operated on and edited, a new class of applications is emerging. New image sequences can be created by modeling and analyzing stored old image sequences. Such manipulations of image content may be the most important application of model-based coding. Thus, model-based coding has much wider range of applications than the conventional waveform coding techniques. One-way communication type applications may be important application areas, in which database applications, broadcasting type communication applications and machine-interface applications are included.

The following list describes several specific application examples of 3-D model-based coding:

1. Virtual Space Teleconference [24]: The idea is to incorporate 3-D computer graphics database with 3-D model-based coding to set up a virtual conference room. The other parties are coded by 3-D model-based coding and displayed using various computer graphics data. It will provide an advanced communication interface with realistic sensations.

2. Structured Video and Virtual Studio [25, 26, 27]: Because model-based coding is able to describe scenes in a structural way, new scenes can be created from pre-existing material using 3-D properties of the scene. Video modeling will provide a way to handle and edit video materials and compose new scenes employing common computer graphics technology. It can also provide the means for video indexing of video database applications. Virtual studio is a computer generated studio setting. It is generated with computer graphics and image analysis techniques to program production for broadcasting. The clipped images of persons and scenes are generated taking into account the camera motion which is detected by either mechanical sensors or analysis of an image sequence.

3. Speech/Text-Driven Facial Animation System for an Advanced Man-Machine Interface [28, 29, 30]: By having a friendly machine interface with speech or text-driven 3-D facial model, this represents an improvement to the interface between the human user and the computer. Applications that can benefit from this are current prerecorded message systems and voice activated databases.

4. Real-time Implementation of Model-Based Coding System: A prototype real-time system has been developed. The motion analysis method is rather simplistic, the model used in the images was pre-existent and the initial position of the face is known.

5. Synthesis of Facial Expressions for Psychological Studies [31, 32]: Facial synthesis techniques can be used to generate a variety of different facial expressions. This can be applied to psychological studies so that judgmental experiments can be performed using the facial images controlled by parameters as stimuli.

6. 2-D to 3-D Conversion of Images [33]: Another potential application is the conversion of 2-D images into stereo images by using a 3-D facial model. Stereo images can be viewed by just receiving 2-D image information.

The importance of 3-D model-based coding is underscored by the inclusion in the latest video coding standard for multimedia content, the MPEG-4, the syntax for the coding of human face and body using the 2-D meshes approach. The syntax contains the parametric descriptions of a synthetic description of human face and body and the animation streams of the face and body. It also includes the static and dynamic mesh coding with texture mapping, and texture coding for view dependent applications. The syntax allows the animation of face at the decoder upon receiving the Facial Definition Parameters and/or Facial Animation Parameters; and body animation when the corresponding body parameters are received. More details are contained in Section 6.7 (Coding of Synthetic Objects).

4.2 3-D Human Facial Modeling

The 3-D model-based coding system can be subdivided into 3 main components, a 3-D facial model, an encoder and a decoder as depicted in Fig. 4.1. The encoder separates the object from the background, estimates the motion of the person's face, analyzes the facial expressions, and then transmits

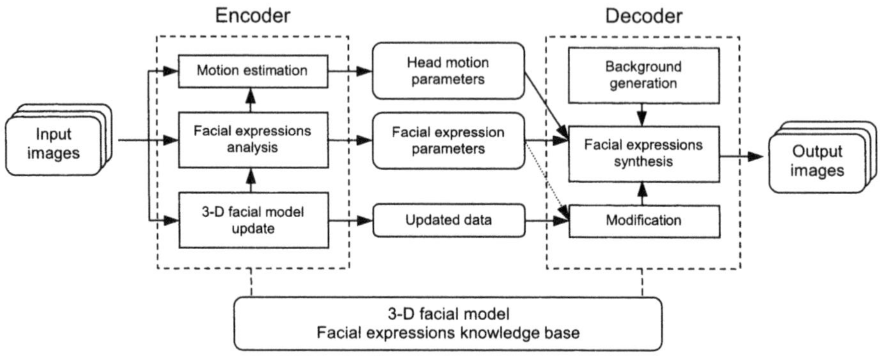

Figure 4.1: Model-based analysis and synthesis image-coding system. ©IEEE 1994

the necessary analysis parameters. Most information included in the facial image sequences are the 3-D head motions and the facial expressions. The head motion parameters (HMP) and the facial expression parameters (FEP) describe information in the model-based coding system. When necessary, the encoder will also add new depth information and initially unseen portion of the object into the model by updating and correcting it if required. During analysis and synthesis, the encoder and the decoder use the 3D facial model and prior knowledge on the facial muscular actions as the common knowledge. Since only analysis parameters are the information that needs to be transmitted, this results in a very low bit rate transmission.

This section gives an overview of the synthesis and analysis of facial image sequences from the point of view of model-based coding. The modeling of a person's face and the expected transmission rates of the model-based coding system is also discussed.

4.2.1 Modeling A Person's Face

Face modeling represents the most important part in 3-D model-based coding because the analysis and synthesis of facial images are strongly dependent upon it. The initial work on face modeling was started by Frederick Parke who developed the parameterized facial model [34, 30]. The model consisted of a human face with geometrical details of facial features such as eyes, mouth, and so forth. The work had drawbacks in image reconstruction, with lack of surface details and reality because the reconstruction is using only wire-frame models and shading techniques. Thus, the reconstructed images did not appear natural.

For image communication purposes, not only the reconstructed images must resemble as closely as possible to the original images, but they must also appear natural. For these particular reasons, texture mapping technique [14] is used to enhance the naturalness of the synthesized images, as with this technique the original intensities of the image is used. The human face is represented by a highly detailed generic 3-D wire-frame (WFM) model consisting of triangulated mesh of wire-frames. To fit an individual's face, the wire-frame model is scaled and adjusted to correctly fit the frontal facial image of that person. The original facial image is then texture mapped to the adjusted wire-frame model.

In most of model-based coding systems, face modelling has taken a similar approach: using a 3-D wire-frame model and texture mapping an original image to the model. Additional information such as side views of a face [35], continuous aspect view of a face [36] and range data [19] can be used for increasing the accuracy of the initial 3-D facial model. Recently, use of range data to generate a 3-D facial model has been attempted [37].

1. The 3-D wire-frame generic face model used for the 3-D model-based coding system consists of approximately 500 triangles. There are two different wire-frame models used depending on the method of image synthesis, namely, clip-and-paste synthesis and structure deformation synthesis method. In this chapter, image synthesis employing structure deformation method will be described. Figure 4.2 shows a wire-frame generic face model used for structure deformation synthesis method.

2. Four feature points are defined on the face as depicted in Fig. 4.3. The wire-frame is 3-D affine transformed to fit through the four feature point positions. Since the facial image is a 2-D image with no depth information, the depth of the four feature points are estimated using the general face model as follows:

$$Z_{face} = Z_{model} \frac{AD_{face}}{AD_{model}} \tag{4.1}$$

where Z_{face} is the depth information of the feature points on the 2-D image, Z_{model} is the depth information of the feature points on the 3-D generic WFM, AD_{face} is the length from A to D on the 2-D image, and AD_{model} is the length from A to D on the 3-D generic WFM.

3. The movement of points on the wire-frame model can be described as follows: the points on the lower face outline of the adjusted WFM

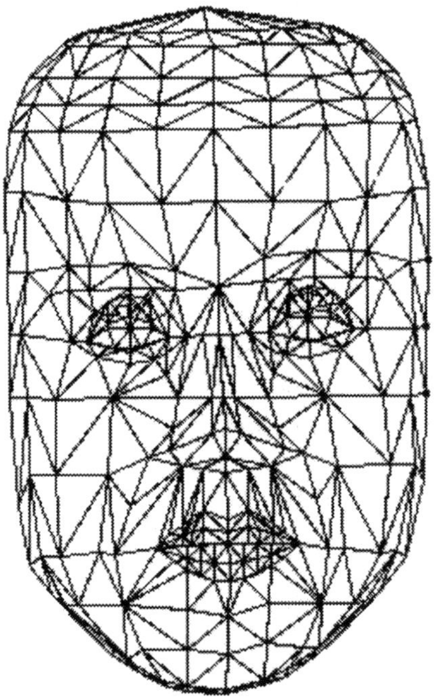

Figure 4.2: Wire-frame generic face model for structure deformation synthesis method.

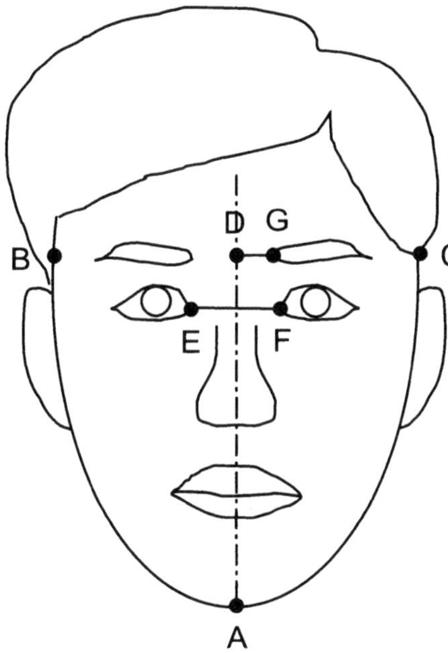

Figure 4.3: The four feature points (A, B, C, D) that are used to roughly fit the wire-frame generic face model to the full-face image. D is a point which equally divides line EF.

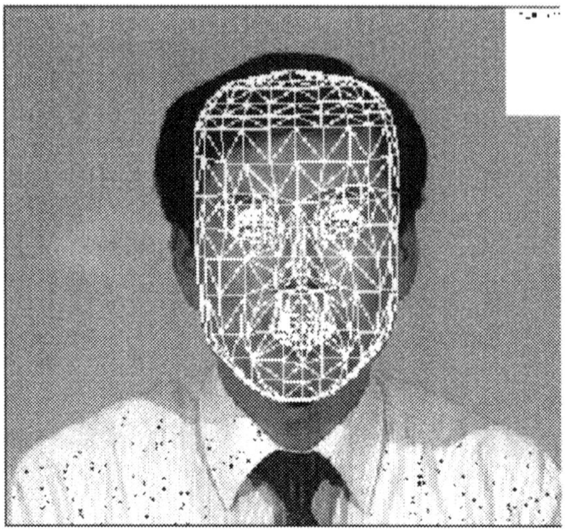

Figure 4.4: Adjusted 3-D general face model on the full-face image.

are moved so that they are located on that of the full-face image (see Fig. 4.4). The other points not on the lower face outline are moved towards the direction of the wire-frame center axis in proportion to the translation of the points on the lower face outline (see Fig. 4.5). Point P_0 is adjusted to the lower face outline and the other points (P_i) are moved such that

$$\hat{P}_i = P_i \left(1 - \frac{|\hat{P}_0 - P_0|}{|P_0|}\right) \tag{4.2}$$

After this step, the 3-D generic WFM is roughly scaled and fitted to the frontal facial image as shown in Fig. 4.4. The 3-D generic WFM fitting and adaptation is already completed at this stage for the clip-and-paste synthesis method.

4. For the detailed model used for structure deformation synthesis method, the facial features positions need to be located and the corresponding wire-frame features representing them need to be fitted to the frontal facial image. Four control points are defined for each component as defined in Fig. 4.6. These points are located on the facial image and the facial component models of eyebrows, eyes, lips and nose are then

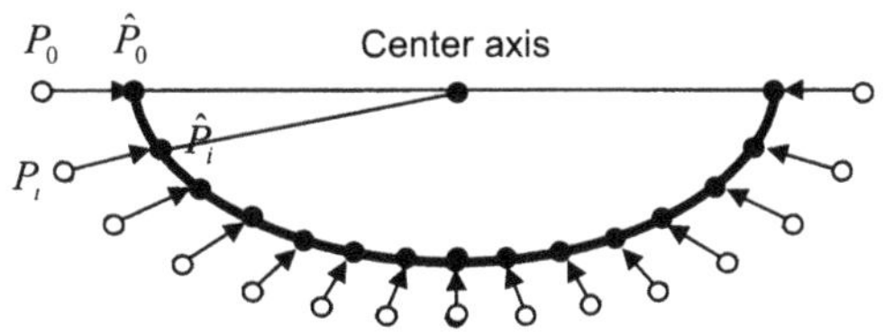

Figure 4.5: This figure shows a horizontal slice of the head. Adjustment of points excluding the lower face outline points [14].

3D affine transformed to harmonize each corresponding feature point positions.

5. After the accurate scaling and adjusting of the 3-D generic WFM of the face, the frontal facial image is then projected and mapped onto the adjusted WFM. A 3-D facial model is created which consists of points which have 3-D coordinate values and intensities. The block diagram representing the whole process for construction of the 3-D model of a person's face is given in Fig. 4.7.

   After the 3-D facial model representing a person's face has been constructed, the model can be moved or rotated in any direction. The synthesized image which has been rotated is given in Fig. 4.8 alongside the frontal image. It can be seen that the rotated image still appears natural since the texture from original image is used to synthesize the new image.

   Currently, the process of scaling and adjusting the wire-frame model to fit the frontal facial image is not yet fully automated, and this represents one of the biggest problems in face modeling. In the next section, a solution to fully automate this process will be described in detail.

4.3 Facial Feature Contours Extraction

Automatic fitting and adaptation of the generic 3-D WFM to the facial image requires the outline of facial features including eyebrows, eyes, nose, mouth and face profile to be located precisely. The nodes of the WFM can then be moved to their correct location to fit the facial image. The

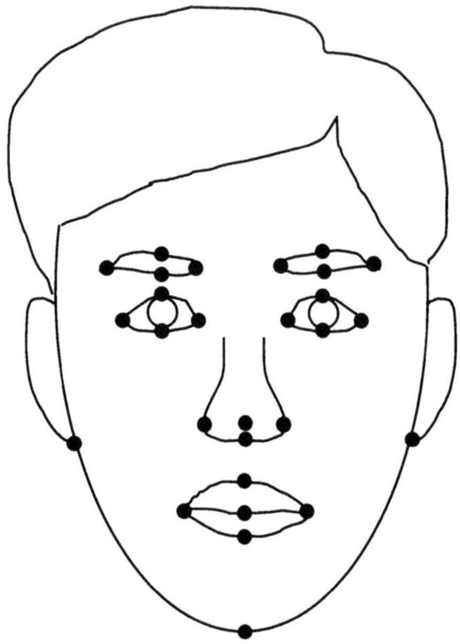

Figure 4.6: The facial component control points used to define the 3-D facial component models.

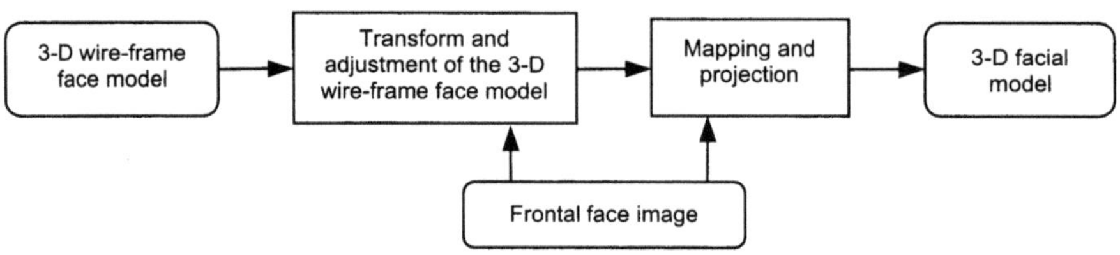

Figure 4.7: Block diagram illustrating 3-D facial model construction [14].

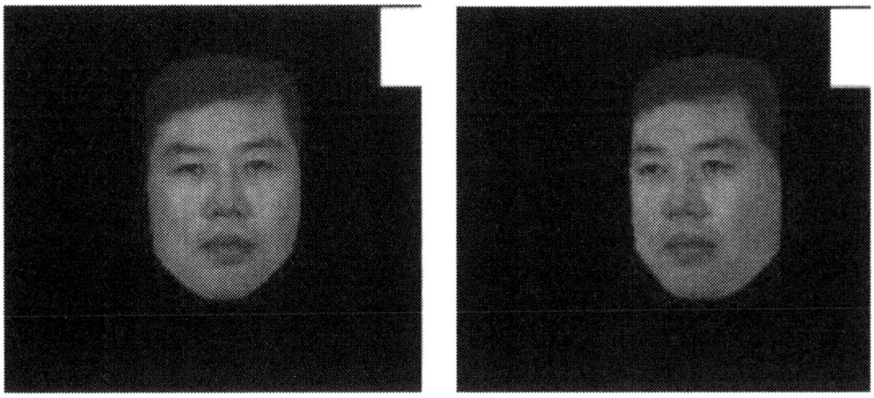

(a) (b)

Figure 4.8: Synthesize image by rotating 3-D model (a) Frontal view frame and (b) Side view frame.

outline of the face is used to adjust and scale the 3-D head model and the other facial features outline is used to adjust and scale the 3-D facial component models. The head model alone is used for the clip-and-paste synthesis method, and the head model with facial component models are used for the facial structural deformation synthesis method.

Facial feature contours extraction has applications in areas other than model-based coding. One important application is for face recognition and interpretation of human faces. Different methods for features extraction include extraction from both profile and side-view cases. In the profile case [38, 39], components of the feature vector are extracted which include the distances between feature points, areas, angles and curvatures. In the front-view case, Nakamura *et al.* [40] developed human face identification based on isodensity maps. Yuille *et al.* [41] developed a method to extract the eyes and mouth using deformable templates.

In this section, facial features contours extraction using active contour models (or snakes) [42] and deformable templates [41] is described. Active contour model is an energy minimizing spline guided by the external constraint forces and influenced by image forces that pull it toward features such as lines and edges. The deformable templates are specified by a set of parameters which use a priori knowledge about the expected shape of the features to guide the contour deformation process. The templates are flexible enough in shape and orientation to extract the desired contour. Both the active contour models and deformable templates require the contours/templates to be initially located roughly near the features to be extracted, and the procedure for initial estimates of the rough contour location is presented in the next section.

4.3.1 Rough Contour Location Finding

For correct contour extraction, it is necessary that the initial 'rough' contour is located near the features to be extracted. Otherwise wrong contour can be extracted. The initial 'rough' contour is located by localizing of the facial features components. Kanade [43] has pioneered the work in localization of facial feature points. Reinders *et al.* [44] proposed a method for facial feature localization through candidate region generation and feature selection. De Silva *et al.* [45] proposed an automated facial features detection using a method called edge pixel counting. In this section, a procedure for rough contour estimation routine by Huang *et al.* [46] is described.

The Rough Contour Estimation Routine (RCER) firstly locates the left eyebrow. With a priori knowledge of the position and the image gray-level of the left eyebrow, the rough contour can be extracted by RCER. From the rough contour of the left eyebrow, other rough contours including the left eye, right eyebrow, right eye, mouth, nose and face can be subsequently estimated.

There are no universal threshold values for the intensity values of the facial features since different portraits have varying brightness. The image gray-level of a facial feature such as the eyebrow, is derived using the scale space filter [47] to determine the zero-crossings of the intensity histogram at different scales. The histogram is partitioned into peaks, valleys and ambiguous regions. The positions of the major peaks are selected as the thresholds.

The following steps describes the procedure for rough contour location finding:

1. The background is presumed to have constant intensity values, RCER can estimate the left and right side of the face.

2. The left eyebrow is observed to be on average 1/4 of the facial width. RCER can then calculate the x-coordinate of a contour point of the left eyebrow.

3. Using the x-coordinate found in step 2, RCER goes downward from the top of the forehead to find the y-coordinate of the left eyebrow.

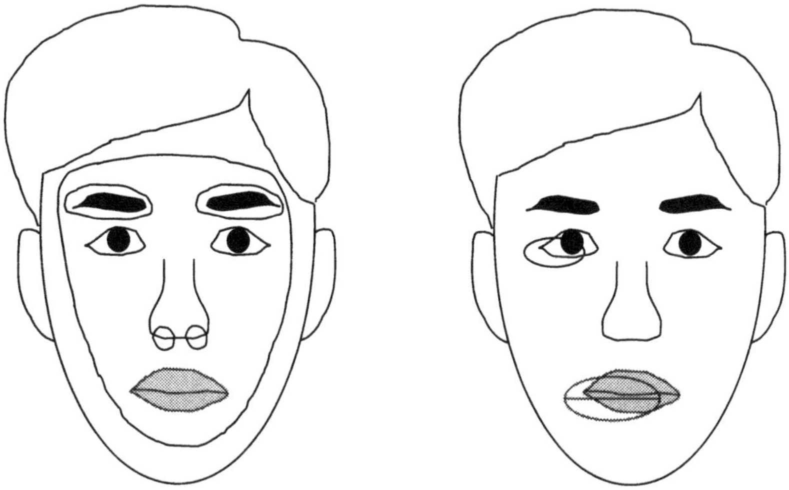

(a) (b)

Figure 4.9: Illustration of the initial facial contours and templates.

4. By using the Contiguous Object Region Finding (CORF) method [48] the rough contour of the left eyebrow is located.

5. Similar to steps 2-4, the y-coordinate of the right eyebrow is estimated and its rough contour is subsequently located.

6. From the left and right eyebrow respectively, and using CORF method the left and right eyes can be located.

7. Going downward from the center of the left and right eyes, and by using CORF method, the rough contours of the nose and mouth can be located.

8. A rough contour for the face is then located by enclosing all the contours derived in steps 1-7.

The RCER estimates all the rough contours to be larger than the precise contour of the features, except for the facial profile in which it is smaller as shown in Fig. 4.9. This presents no problem as the iteration process will shrink or expand the estimated contour to the precise one.

4.3.2 Image Processing

The deformable templates act on three representations of the image, as well as on the image itself. An energy function is defined which contains terms attracting the templates to salient features such as peaks, valleys in the image intensity, the edges, and the image intensity itself. The three image representations are therefore the peak, valley and edge images.

In active contours, the image forces defined in the energy function draw the contour to the edge in the image. Therefore the image representation used for this method is the edge image. In both cases, the image representations (peak, valley and edge) are smoothed to attract contour over longer distances.

4.3.2.1 Image Morphological Processing

Image morphology [49] pertains to the study of the structure of objects within an image. There are two forms of image morphological processing, binary and gray-scale. As the images used in model-based image coding systems have many intensity values, we will restrict ourselves to gray-scale morphological processing. Some more information on mathematical morphology can also be found in Section 1.3.1.

Morphological operations are similar to image convolution, where the morphological process moves across the input image, pixel by pixel, placing the resulting pixels in the output image. At each input pixel location, the input pixel and its neighbors are combined using a structuring element (or morphological mask) to determine the output pixel's brightness value. The structuring element is usually square in dimension, that is, 33 or 55 and so forth.

Erosion and Dilation

Erosion and dilation operations are the two most fundamental morphological operations. The erosion operation reduces the size of the objects relative to their background and conversely, the dilation expands the size of the objects. The erosion operation on a pixel of the input image is the minimum value of the pixel intensity and those of its neighboring pixels. That is,

$$\begin{aligned} O(x,y) = \min\{&I(x,y), I(x,y-1), I(x,y+1), I(x+1,y-1), \\ &I(x+1,y), I(x+1,y+1), I(x-1,y-1), \\ &I(x-1,y-1), I(x-1,y), I(x-1,y+1)\} \end{aligned} \tag{4.3}$$

This has the effect of darkening bright objects, and thus making them appear smaller. The overall image brightness is reduced as well.

The dilation operation is very similar to that of erosion, but instead the maximum value of a pixel and its neighbors is the value of the output pixel. That is,

$$\begin{aligned} O(x,y) = \max\{&I(x,y), I(x,y-1), I(x,y+1), I(x+1,y-1), \\ &I(x+1,y), I(x+1,y+1), I(x-1,y-1), \\ &I(x-1,y-1), I(x-1,y), I(x-1,y+1)\} \end{aligned} \tag{4.4}$$

Dilation has the effect of brightening bright objects, and thus making them appear larger. As a result, the overall image brightness is also increased.

Opening and Closing

Opening is an image morphological operation that darkens small objects and entirely removes single-pixel objects like noise spikes and small spurs. Objects tend to retain their original shapes and sizes. The opening operation is erosion followed by dilation. This operation can be applied numerous times to achieve the necessary effect. The multiple operations are performed by applying the erosion operations a number of times, followed by the same number of dilation operations.

Closing is the opposite of opening operation, whereby dilation is followed by erosion. The multiple operations are similar to the opening, with dilation performed a number of times, followed by the same number of erosion operations. The closing operation has the effects of brightening small objects and entirely filling in single-pixel objects like small holes and gaps while maintaining the original shapes and sizes of the objects.

4.3.2.2 Peak and Valley Images

Peak (or top-hat) image is one of the image representations used with the deformable templates for contour extraction. The image highlights the peaks in the image intensity such as the white of the eye. Derivation of the peak image is a variant of the opening operation described in the previous section. The opening is first performed on the image, then this image is subtracted from the original image using a dual image point process. The result is an image in which only bright peaks appear. The derivation of the peak image is illustrated in Fig. 4.10.

Valley image highlights dark areas within an image, such as the iris of the eye in a facial image. The derivation of this image is opposite to that of the peak image. That is, closing operation is first performed on the image, and the resulting image is subtracted from the original image. The result

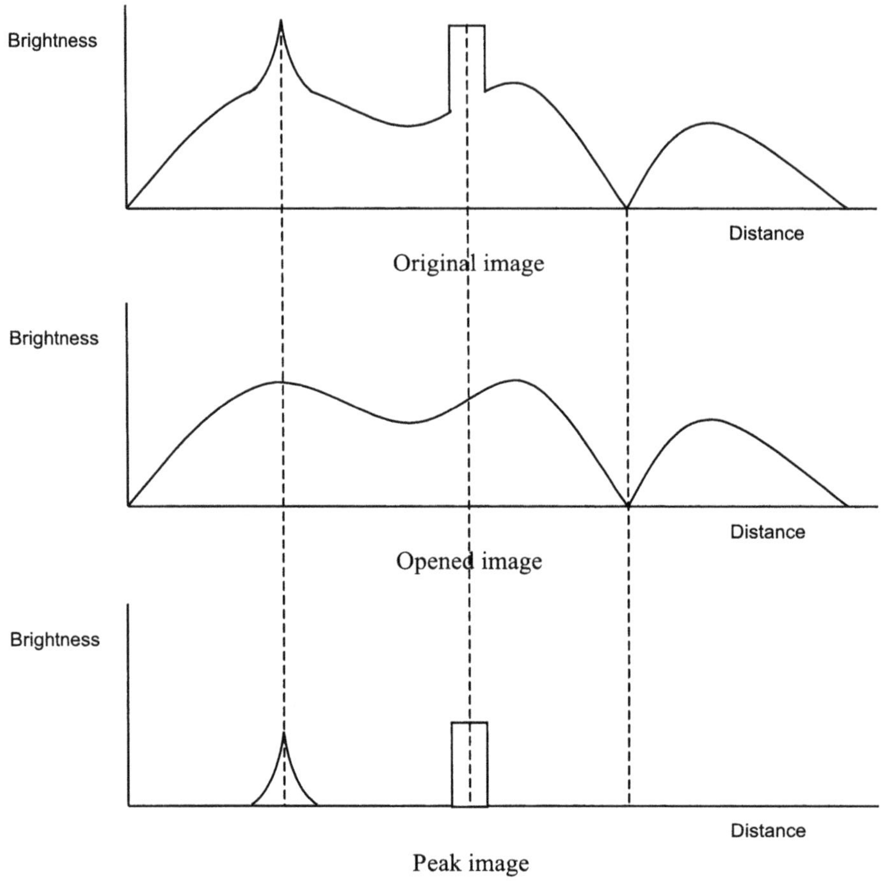

Figure 4.10: Derivation of peak (top-hat) image from original image.

is an image in which only the dark valleys appear. Figure 4.11 shows an original image with its associated peak and valley images.

4.3.2.3 Edge Image

Edges of an image correspond to areas where image intensities change rapidly. There exist many standard methods for edge extraction. Here Sobel operator (1.25) is used to extract the edge image. The image derived from application of the Sobel operation to the image in Fig. 4.11 (a) is shown in Fig. 4.12.

(a)

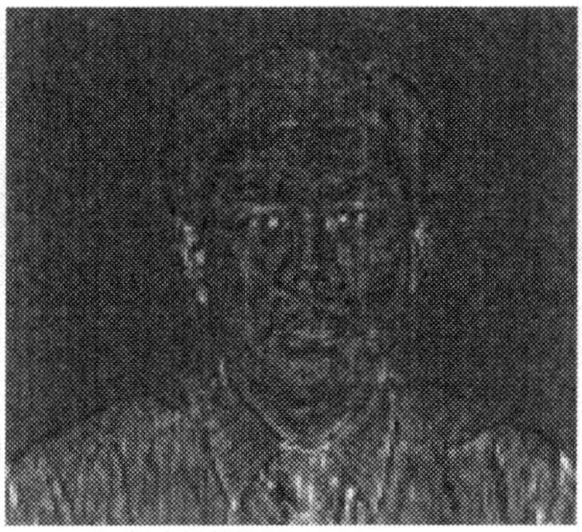

(b)

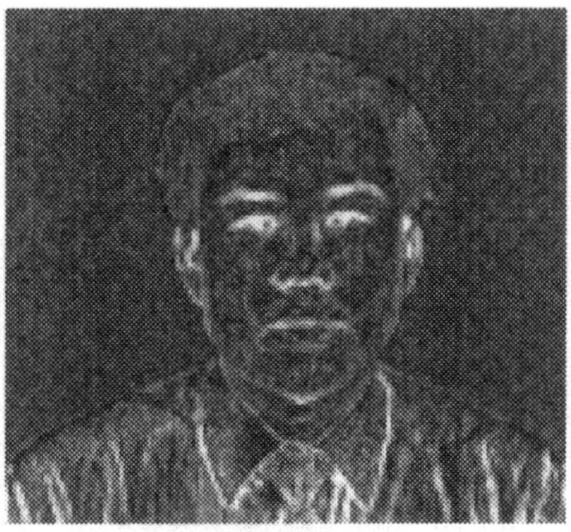

(c)

Figure 4.11: Morphological image processing (a) original image, (b) peak image, (c) valley image.

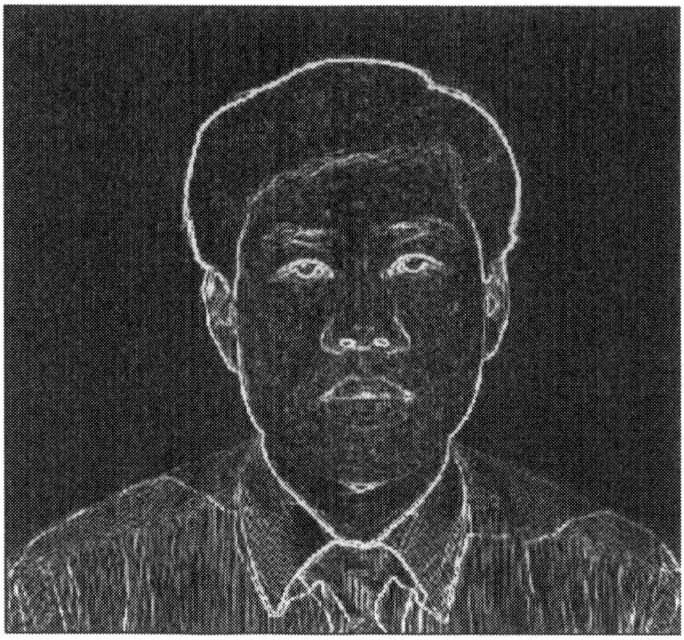

Figure 4.12: Edge image derived using Sobel operator.

4.3.2.4 Smoothing Operator

With the rough contour location finding procedure described in Section 4.3.1, the precise contour can be relatively distant from the initial contour. Smoothing the image representations enables the contours to be extracted at longer distances. The images are smoothed by using an averaging low-pass filter. This filter corresponds to a simple local average of the image elements inside the operator window of size 5x5, with constant weighing of 1/25. That is, the convolution mask used for the smoothing operation is given as

$$\frac{1}{25}\begin{bmatrix} 1 & 1 & 1 & 1 & 1 \\ 1 & 1 & 1 & 1 & 1 \\ 1 & 1 & 1 & 1 & 1 \\ 1 & 1 & 1 & 1 & 1 \\ 1 & 1 & 1 & 1 & 1 \end{bmatrix} \tag{4.5}$$

Images of Fig. 4.11 and Fig. 4.12 are smoothed using the above mask and the resulting images are given in Fig. 4.13.

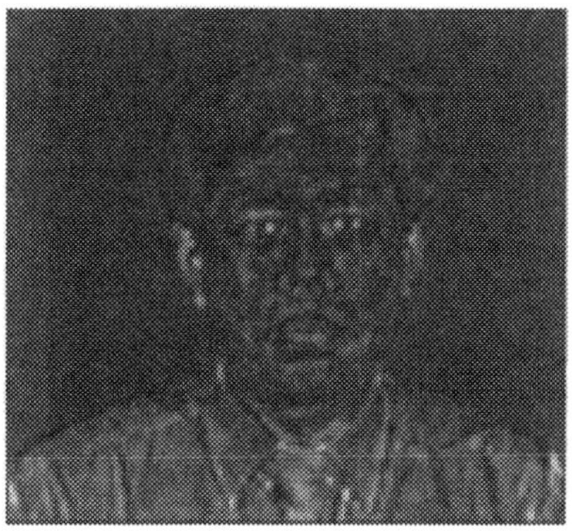

(a)

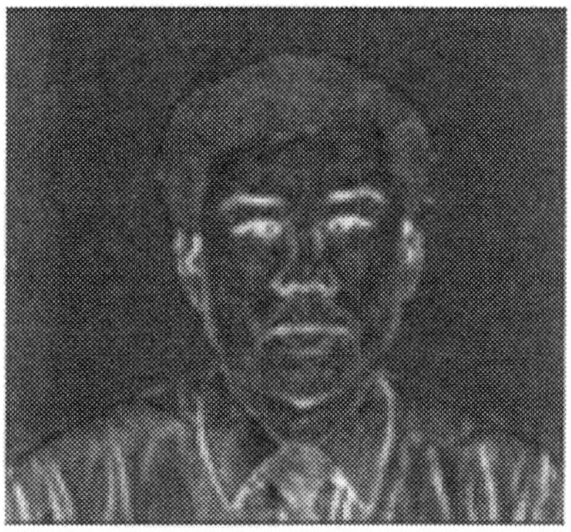

(b)

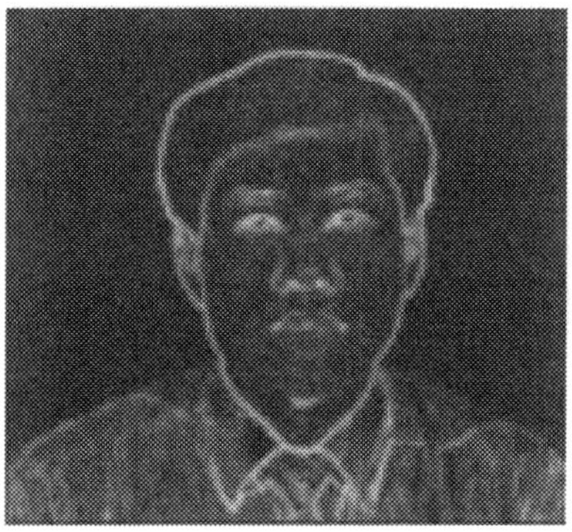

(c)

Figure 4.13: Image smoothing operator on (a) peak image, (b) valley image and (c) edge image.

4.3.3 Features Extraction Using Active Contour Models

Since the shape of human faces and eyebrows may vary quite significantly from one individual to another, a contour extraction technique that can capture contours that are flexible in shape and size is required. An active contour model or commonly known as snake is a method for contour extraction with the desired properties. Features extraction using active contour models has been developed by Huang *et al.* [46]. The active contours described in this section differs with the introduction of an external energy term for the face contour, namely the 'expansion' energy. This energy exerts forces to expand the initial contour enabling a more robust extraction of the fact at longer distances.

The initial contour is placed relatively near the feature, the image forces draw the contour to the edge of the image. For fast computation, the contours are extracted using the greedy algorithm [50].

4.3.3.1 Active Contour Model

Definition and Properties

An active contour model or more commonly known as *snake* is an energy minimizing spline guided by external constraint forces and influenced by image forces that pull it toward features such as lines and edges. The name arises from the behaviour that is similar to that of a snake, that is, it locks onto nearby edges and localizes them accurately.

A contour can be represented by a vector $v(s) = [x(s), y(s)]$, having the arc length s. With this definition, the energy functional of the active contour model is defined as

$$\begin{aligned} E_{total} &= \int_0^1 E_{snake}(v(s))ds \\ &= \int_0^1 [E_{internal}(v(s)) + E_{image}(v(s)) + E_{constraint}(v(s))]\, ds \end{aligned} \tag{4.6}$$

where $E_{internal}$ represents the internal energy of the contour due to bending or discontinuities, E_{image} is the image forces, and $E_{constraint}$ is the external energy due to other factors. The extracted contour corresponds to local minima of the energy functional.

Numerical Solution

The definition of active contour energy functional can be written as

$$E_{total} = \int_0^1 [\alpha(s)E_{continuity}(v(s)) + \beta(s)E_{curvature}(v(s)) + \gamma(s)E_{image}(v(s)) + \kappa(s)E_{constraint}(v(s))]\, ds \tag{4.7}$$

The form of this equation is similar to the previous (4.6), with $E_{continuity}$ and $E_{curvature}$ corresponding to internal energy. The first and second terms are the first- and second-order continuity constraints. The third term measures some image quantity such as edge strength, and the last term is a measure of other external constraints. The relative sizes of the coefficients α, β, γ, κ are more important, rather than their absolute sizes to balance the relative influences of the four terms.

The continuity term refers to the distances between the points and can be calculated as $|v_i - v_{i-1}|$, but this has the effect of shrinking the contour. It also contributes to the problem of points bunching up on strong portions of the contour. A more appropriate definition that still preserves the continuity constraints and encourages even spacing of points should be used. This definition gives the term as being the difference between the average distance between points, d, and the distance between the two points under consideration, which can be written as

$$E_{continuity} = |d - |v_i - v_{i-1}|| \tag{4.8}$$

With this definition, points having distances near the average will have the minimum value. This term is normalized by dividing the largest value in the neighborhood to which the point may move, giving a value in $[0, 1]$. After each iteration, a new value of d is computed.

Since the formulation of the continuity term causes the points to be evenly spaced, the curvature term is then defined as

$$E_{curvature} = |v_{i-1} - 2v_i - v_{i+1}| \tag{4.9}$$

This term is also normalized by dividing the largest value in the neighborhood.

The image force represented by the third term in (4.7) corresponds to the measure of the edge strength of the image. The image representation used for this is the smoothed edge image as derived previously in Section 4.3.2.4. Smoothed edge image can attract the contour over longer distances. For eight-neighbors, we have nine energy measurements. The image energy is normalized using the following equation

$$E_{image} = \frac{Min - Mag}{Max - Min} \tag{4.10}$$

where Mag is the edge intensity value of the point being considered, Min is the minimum of the 9 energy measurements, and Max is the maximum of the 9 energy measurements.

The above equation gives a negative value, so points with strong edges will have small values. Now, if $(Max - Min) < 5$ then Min is defined as

$$Min = Max - 5 \tag{4.11}$$

This is to prevent large differences in image areas where the gradient magnitude is nearly uniform. For example, in a neighborhood of points where the image energy values are 50, 51 or 52, using (4.10) will give normalized values of 0, -0.5, or -1. If (4.11) is incorporated, then this will give -0.6, -0.8, or -1 which is a more accurate representation. Near an edge, this situation does not normally arise.

The last term in (4.7) corresponds to the external constraint. The constraints can be due to any external factors, and may not exist for some contours. In fact, this term is only used for extraction of the face profile and not for the eyebrow. This will be described in more details later.

At the end of each iteration, the curvature at each point is determined. Points which meet specific criteria are considered as corner points and their β values are set to 0. The criteria for a corner point are, if the curvature is larger than some threshold, the curvature is larger than the two neighboring points, and the edge strength is above some threshold. The curvature can be calculated as follows

$$curvature = \left| \frac{\vec{u}_i}{|\vec{u}_i|} - \frac{\vec{u}_{i+1}}{|\vec{u}_{i+1}|} \right|^2 \tag{4.12}$$

where $\vec{u}_i = (x_i - x_{i-1}, y_i - y_{i-1})$ and $\vec{u}_{i+1} = (x_{i+1} - x_i, y_{i+1} - y_i)$.

Implementation

The greedy algorithm [50] is used for fast computation of the active contour, being of $O(nm)$ where n is the number of points and m is the neighborhood size. The O-notation refers to the proportionality of the computation of the algorithm, that is, $O(x)$ means the speed of computation is proportional to the variable x. The energy function is computed for each point and each of its neighbors. The neighbor having the smallest value is chosen as its new position. The pseudo-code for greedy algorithm is given below.

PSEUDO CODE FOR GREEDY ALGORITHM

Index arithmetic is modulo n.
Initialize α_i, β_i, γ_i, κ_i to some values for all i.

```
do    /* loop to move points to new locations */
   for i=0 to n  /* point 0 is first and last one processed */
      E_min = BIG
      for j=0 to m-1    /* m is size of the neighborhood */
         E_j = α_i E_continuity,j + β_i E_curvature,j + γ_i Eimage,j
              +κ_i E_constraint,j
         if E_j < E_min then
            E_min = E_j
            j_min = j
         move point v_i to location j_min
         if j_min not current location
            ptsmoved += 1    /* count points moved */
   /* process determines where to allow corners in the next
   iteration */
   for i=0 to n-1
      c_i = | u_i/|u_i| - u_{i+1}/|u_{i+1}| |^2
   for i=0 to n-1
      if c_i > c_{i-1} and c_i > c_{i+1}
      \* if curvature is larger than neighbors */
      and c_i > threshold1
      \* and curvature is larger than threshold 1 */
      and mag(v_i) > threshold2
      \* and edge strength is above threshold 2 */
         β_i = 0
until ptsmoved < threshold3
```

4.3.3.2 Eyebrow Extraction

To locate the contour of the eyebrow using active contour, the energy of the function is defined as:

$$\begin{aligned} E_{brow} &= \sum_{i-1}^{n} [\alpha_i(s)E_{continuity}(v(s)) + \beta_i(s)E_{curvature}(v(s)) \\ &\quad +\gamma_i(s)E_{image}(v(s))] \\ &= \sum_{i-1}^{n} \left[\alpha_i \frac{|d-|v_i'-v_{i-1}||}{Max_{icon}} + \beta_i \frac{|v_{i-1}+2v_i'+v_{i+1}|}{Max_{icur}} \right. \\ &\quad \left. +\gamma_i \left(\frac{Min_{iEdge}-Mag_{iEdge}}{Max_{iEdge}-Min_{iEdge}}\right)\right] \end{aligned} \tag{4.13}$$

where v_i' is the next location of v_i for the next iteration, d is the average distance between points, Max_{icon} is the maximum value of 9 measurements of $|d-|v_i-v_{i-1}||$, Max_{icur} is the maximum value of 9 measurements of $|v_{i-1}-2v_i+v_{i+1}|$, Mag_{iEdge} is the edge response of v_i', Max_{iEdge} is the maximum value of 9 measurements of Mag_{iEdge}, and Min_{iEdge} is the minimum value of 9 measurements of Mag_{iEdge}.

From the rough contour location finding procedure described in Section 3.1, the rough contour of the eyebrow is quite close to the precise contour. The contour extraction procedure should converge more quickly and precisely.

4.3.3.3 Face Profile Extraction

Unlike eyebrow extraction, the rough contour of the face profile is estimated more coarsely. The initial contour is also smaller than the actual contour as depicted in Fig. 4.9. The energy functional of the active contour for the face is defined as

$$\begin{aligned} E_{face} = \int_0^1 &[\alpha_i E_{continuity} + \beta_i E_{curvature} + \gamma_i E_{image} + \kappa_i E_{contraint1} \\ &+ \delta_i E_{constraint2}]ds \end{aligned} \tag{4.14}$$

The first three terms are defined as in previous section, the last two terms are external constraints imposed to expand the original contour. The 'expansion' energy corresponding to these terms are defined as follows, with the contour described as a set of points such as in Fig. 4.14. The final contour is assumed to maintain roughly the same shape as the original with

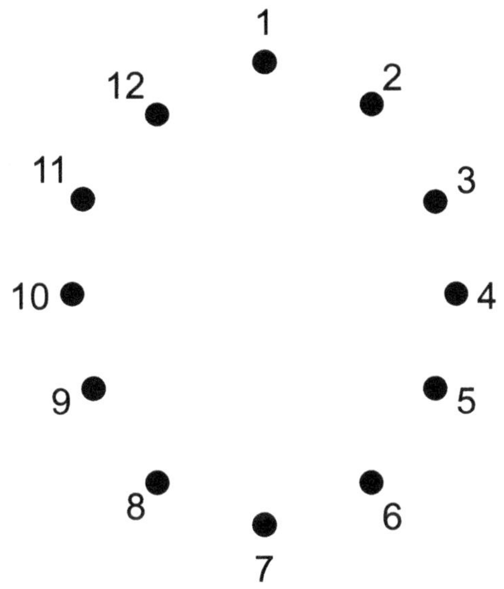

Figure 4.14: Points representing facial profile contour.

all points moved outward in a similar proportion. This is even more evident when a large number of points is used. Relatively, the opposite point with respect to the horizontal and vertical axis is the same point on the initial and final contour. That is, point 1 will have point 7 as opposite point with respect to horizontal axis on the initial and final contour. Similarly with respect to the vertical axis, point 4 will be opposite to point 10 before and after the snake's iterations. The external constraints are based on the separation of these 'opposite' points.

$E_{constraint1}$ and $E_{constraint2}$ are defined as the distances between the point being evaluated and its 'opposite' point with respect to the vertical axis and horizontal axis, respectively. Both terms are normalized in a similar way as the image energy term. Therefore large distances will have smaller values, in effect expanding the contour. The image force will ensure that the contour gets localized near the edges rather than expanding out of bound.

The facial profile can be distinguished into two parts with different characteristics, the lower face and upper face. The lower face include points from the base of one ear, around the chin to the other ear, and the upper face include the other points around the hair line. The lower face is roughly elliptical in shape, so the coefficient β of the curvature energy is set larger to get a smoother contour. For the upper face, due to the hairline having no predictable shape, the curvature coefficient β is set to smaller value and also the edge coefficient γ is set larger to ensure that the contour is localized on the edges of the hairline. The area between the chin and the neck usually give strong edge intensities because of the shadow cast on the neck. For this reason, the few points around the chin are given higher edge coefficient γ to ensure the localization of these points on the chin. The point on the tip of the chin is one of the control points used in WFM fitting, so it is important the chin contour is extracted accurately.

4.3.4 Features Extraction Using Deformable Templates

Even though human eyes and mouth vary from one person to another, the general shape of these features are quite fixed. Therefore a deformable template that is flexible enough in shape and size is a suitable representation for these facial features. Features extraction using deformable templates has been developed by Yuille *et al.* [41] and Huang et al. [46]. The technique described in this section differs as the number of template matching stages is less resulting in a faster extraction. With the initial template near the feature to be extracted, the template scales and orients itself to the final contour.

4.3.4.1 Deformable Templates

Deformable templates are specified by a set of parameters which utilizes *a priori* knowledge of the expected shape of the features to guide the contour deformation process. The templates are flexible enough to be able to change their size, and other parameter values, so as to match themselves to the data. The final values of these parameters can be used to describe the features. The method should work despite variations in scale, tilt, rotation of the head, and lighting conditions. Variations of these parameters should allow the template to fit any instance of the feature.

The templates interact with the image in a dynamic manner. An energy functional is defined which contains terms attracting the template to salient features such as peaks and valleys in the image intensity, edges, and the intensity itself. The final template corresponds to the local minimum of the energy function. The parameters are updated by method of steepest descent. Technique of using deformable templates for features extraction is described in the next two sections for the eye and mouth contours.

4.3.4.2 Eye Extraction

Definitions and Properties

The eye template is developed through observation of the different features of the eye. The eye template is decided to have all the important features of the eye, but not too complicated for the ease of computation. The template is developed to have the following features:

1. A circle of radius r representing the iris, centered on a point $\vec{x}_c$. The boundary of the circle is attracted to edges in the image intensity, while the interior of the circle is attracted to valleys in the image intensity.

2. Two parabolic sections representing the boundary of the eye. The parabolas have the point $\vec{x}_t$ as their center, width $2b$, maximum height a of the boundary above the center, and maximum height c of the boundary below the center. The eye contour has an angle of orientation θ. This bounding contour is attracted to edges in the image intensity.

3. Two points representing centers of the whites of the eye. These points are approximated by the points at half the distance between the center of the eye $\vec{x}_e$ and the corners of the eyes. These points are labeled $\vec{x}_e + p_1(\cos\theta, \sin\theta)$ and $\vec{x}_e + p_2(\cos\theta, \sin\theta)$, where $p_1 = 0.5b$ and $p_2 = -0.5b$. These points are attracted to peaks in image intensities.

4. The whites of the eye are the areas between the bounding contour of the eye and the iris. These regions are attracted to peaks in image intensity.

The above mentioned components are linked together by two types of forces, forces which encourage $\vec{x}_c$ and $\vec{x}_t$ to be close together, and forces which make the width $2b$ of the eye roughly four times the radius r of the iris.

The eye template is illustrated in Fig. 4.15. It has eleven parameters $\vec{x}_c$, $\vec{x}_t$, p_1, p_2, r, a, b, c and θ. All the parameters values can change during the iterations, with different variables allowed to changed at different stages of the matching as described later.

To make representation of parabolas as bounding contours of the eyes more explicit, two unit vectors are defined as follows

$$\vec{e}_1 = (\cos\theta, \sin\theta) \tag{4.15}$$

$$\vec{e}_2 = (-\sin\theta, \cos\theta). \tag{4.16}$$

Figure 4.15: Deformable template for the eye [41].

Using the above unit vectors, a point $\vec{x}$ in space can be represented by (x_1, x_2) where

$$\vec{x} = x_1\vec{e}_1 + x_2\vec{e}_2. \tag{4.17}$$

The top parabola representing the upper boundary of the eye can then be written as

$$x_2 = a - \frac{a}{b^2}x_1^2, \quad x_1 \in [-b, b] \tag{4.18}$$

Similarly for the bottom parabola representing the lower boundary of the eye can be written as

$$x_2 = c + \frac{c}{b^2}x_1^2, \quad x_1 \in [-b, b] \tag{4.19}$$

Energy Function for the Eye Template

Matching the initial template to the data requires the process to be divided into different stages. The energy functional of the eye template is defined accordingly at different stages of the template matching process to utilize the salient features of the image at each stage. The complete energy function is given as a combination of terms due to valley, peak, edge, image and internal potentials. The original image, and its smoothed peak, valley and edge representations are denoted by $\Phi_i(\vec{x})$, $\Phi_p(\vec{x})$, $\Phi_e(\vec{x})$, and $\Phi_v(\vec{x})$, respectively. The complete energy function E_c can be written as

$$E_c = E_v + E_e + E_i + E_p + E_{prior} \tag{4.20}$$

The valley potentials are given by the integral over the interior of the circle divided by the area of the circle. When the iris is partially hidden by the boundary of the eye, thus the part of the circle outside the boundary cannot be allowed to interact with the image. This is dealt by only considering the area of the circle inside the bounding parabolas. The valley potentials is given as

$$E_v = -\frac{c_1}{|R_b|}\int_{R_b}\Phi_v(\vec{x})dA \tag{4.21}$$

The edge potentials are given by the integrals over the boundaries of the circle divided by its length and over the parabolae divided by their lengths,

$$E_e = -\frac{c_2}{|\partial R_w|}\int_{\partial R_w}\Phi_e(\vec{x})ds - \frac{c_3}{|\partial R_w|}\int_{\partial R_w}\Phi_e(\vec{x})ds \tag{4.22}$$

The image potentials give contributions that attempt to minimize the total brightness inside the circle divided by its area, and maximize it between the circle and the parabolae (note the signs of c_4 and c_5).

$$E_i = -\frac{c_4}{|\partial R_w|}\int_{R_w}\Phi_i(\vec{x})dA - \frac{c_5}{|\partial R_w|}\int_{R_w}\Phi_i(\vec{x})dA \tag{4.23}$$

The peak potentials, evaluated at the two peak points, are given by

$$E_p = c_6[\Phi_p(\vec{x}_e + p_1\vec{e}_1) + \Phi_p(\vec{x}_e + p_2\vec{e}_1)] \tag{4.24}$$

The prior potentials are given by

$$\begin{aligned} E_{prior} =& \frac{k_1}{2}\|\vec{x}_e - \vec{x}_c\|^2 + \frac{k_2}{2}[p_1 - p_2 - (r+b)]^2 \\ &+ \frac{k_3}{2}(b-2r)^2 + \frac{k_4}{2}[(b-2a)^2 + (a-2c)^2] \end{aligned} \tag{4.25}$$

In the above equations, R_w and R_b are intensity regions containing the whites and dark center of the eye respectively. R_w is bounded by parabolic curves ∂R_w specified by parameters a, b, and c. R_b is bounded by a circle ∂R_b of radius r. The areas, or lengths, are given by $|R_b|$, $|R_w|$, $|\partial R_b|$ and $|\partial R_w|$. A and s correspond to area and arc-length, respectively.

Implementation

The eye template scales and orients itself to match the contour of the eye in the facial image, the circle in the template is also positioned accurately on the iris in the image. The implementation is done by firstly using the valley potential to find the iris, then the peaks to orient the template, and so on.

The final template corresponds to the minimum value of the energy function defined on the template. The implementation uses a search strategy that is divided into a number of distinct stages or epochs with different values of the parameters $\{c_i\}$ and $\{k_i\}$. The energy terms are written as explicit functions of the parameter values. For example, the sum over the boundary can be expressed as an integral function of $\vec{x}_e$, a, b, c and θ by

$$\frac{1}{|\partial R_w|}\int_{\partial R_w}\Phi_e(\vec{x})ds = \frac{c_3}{L(a,b)}\int_{x_1=-b}^{x_2=b}\Phi_e\left[\vec{x}_e + x_1\vec{e}_1 + \left(a - \frac{a}{b^2}x_1^2\right)\vec{e}_2\right]ds + \frac{c_3}{L(a,b)}\int_{x_1=-b}^{x_2=b}\Phi_e\left[\vec{x}_e + x_1\vec{e}_1 + \left(c - \frac{c}{b^2}x_1^2\right)\vec{e}_2\right]ds \tag{4.26}$$

where s corresponds to the arc length of the curves and $L(a,b)$ and $L(c,b)$ to their total length.

The parameters of the templates are updated using method of steepest descent, that is,

$$\frac{dr}{dt} = -\frac{\partial E}{\partial r} \tag{4.27}$$

where r is a parameter of the template.

To get the desired final eye template, some initial experimentation with the coefficients was needed. The relative sizes of the coefficients are more important, rather than their absolute sizes. Coefficients need to be carefully selected, otherwise problems can be encountered. For example, when trying to get iris of the eye, the intensity and valley terms over the circle attempt to find the maximum value of the potential terms over the circle attempt to find the maximum value *averaged* inside the circle. This led to the circle shrinking to one point at the darkest part (brightest valley intensity) of the circle. This problem is solved by strengthening the edge terms, therefore attracting the circle to the iris edge. Problem may also occur when the initial template is placed above the eye from the interaction with valleys from the eyebrows.

Four epochs of the implementation stage is defined. They are given as follows

1. Position of the iris is roughly located by using the valley terms to attract the circle. The variables used are the center of the iris $\vec{x}_t$ and the radius of the iris r. The center of the eye $\vec{x}_t$ is set equal to center of the iris $\vec{x}_c$ to drag the template toward the eye.

2. In this epoch the valley and the edge terms are used for a more precise extraction of the iris of the eye as it helps scale the circle to the correct the size of the iris. The parameters allowed to vary are $\vec{x}_c$, a, b, c, and r. The iris center is set equal to the eye center. After this, the position and size of the iris are considered essentially fixed.

3. Peak forces are used to get the correct orientation of the eye. Variables in this epoch are orientation of the eye θ, center of the eye $\vec{x}_t$ it is allowed to separate from the center of the iris $\vec{x}_e$. The template at this stage is roughly at its correct location.

4. This is a fine-tuning stage, where the eye contour is precisely located by incorporating the edge and other intensity fields. The parameters varied are orientation θ, length of the eye b, and the center of the eye $\vec{x}_t$. Edge and peak field are used to orient the template, and the original image is also used to minimize the brightness inside the circle. The prior potentials are also used to fine-tune the template in the final stage.

4.3.4.3 Mouth Extraction

Definition and Properties

Shape of the mouth is generally quite fixed with wide variations when the mouth is open or closed. Yuille *et al.* [42] has developed templates for both open and closed mouths. It is assumed that the face in the image is upright and neutral, so the mouth is horizontal and closed. Another assumption is that the mouth is vertically symmetric, these assumptions can be relaxed by using a more complicated template. Through observation of the different features of the mouth, it is decided to have the following features for the mouth template:

1. The mouth is centered on a point $\vec{x}_m = (M_{xc}, M_{yc})$. For a closed mouth the most salient feature is a deep valley in the image intensity where the lips meet as shown in Fig. 4.12. The edges at the top and bottom of the lips can also be used, but usually they are not as strong. The gap between the lips is represented by a parabola with the following equation

$$y = height_s - \frac{4height_s}{length^2} \times x^2, \tag{4.28}$$

$$x \in \left[-\frac{length}{2}, \frac{length}{2}\right] \tag{4.29}$$

where x is the x-coordinate of the point in the parabola, and y is the y-coordinate of the point in the parabola. The co-ordinate of the x and y variables is with respect to the center point (M_{xc}, M_{yc}), so this point corresponds to the origin.

2. The lower lip is represented by a parabola. This parabola is attracted to the edge field. The equation of the parabola can be written as

$$y = (height_s + height_d) - \frac{4(height_s + height_d)}{length^2} \times x^2 \tag{4.30}$$

where x is given by (4.29).

3. The upper lip is represented by parts of two similar parabolas. This is also attracted to the edge field. The equation of the upper lip consists of two parts

$$y = \frac{4height_u(x + \frac{length}{2} - \frac{length_u}{2})^2}{length_u^2} - height_u, \quad x \in \left[-\frac{length}{2}, 0\right] \tag{4.31}$$

$$y = \frac{4height_u(x + \frac{length}{2} - \frac{length_u}{2})^2}{length_u^2} - height_u, \quad x \in \left[0, \frac{length}{2}\right] \tag{4.32}$$

The mouth template is illustrated in Fig. 4.16. It has 6 parameters $\vec{x}_m$, *length*, $length_u$, $height_u$, $height_s$, and $height_d$, which are allowed to vary during the iterations.

Energy Function for The Mouth Template

The energy function for mouth template is defined in a similar way to the eye template. That is, the algorithm is divided into different stages. The complete energy function can be written as

$$E = E_v + E_e + E_{prior} \tag{4.33}$$

The energy potential is the line integral over the parabola being considered with the energy field, for example, the valley potential of the region between

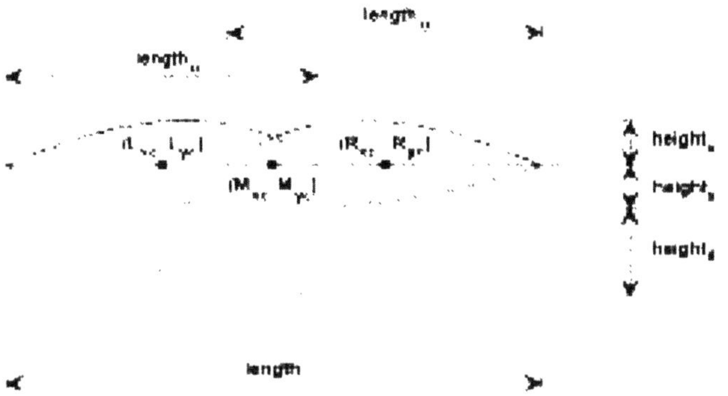

Figure 4.16: Deformable template for the mouth [41].

lips is given as

$$E_v = \int_{x=-\frac{length}{2}}^{\frac{length}{2}} \left(height_s - \frac{4height_s}{length^2} \times x^2 \right) dx \tag{4.34}$$

The prior potential is the energy term derived to make the thickness of the bottom lip to be twice the upper lip.

Implementation

Similar to the eye template, the mouth template uses search strategy to look for the minimum of the energy function. The algorithm is divided into stages with different variables and different energy field interacting at each stage. The parameter values are updated with method of steepest descent.

For the mouth template, 3 distinct epochs are defined. They are described as follows

1. Coefficients are high for the valley forces and zero for edge forces. Parameters varied in this step are mouth center $\vec{x}_m$, mouth length, and height of middle lip, $height_s$. This ensures the precise allocation of the middle lip. The position of the middle lip and hence the y-coordinate is considered to be quite fixed after this stage.

2. This epoch is similar to step 1 with the exception being that the mouth center is only moved in x-direction only to get a more precise location and shape of the middle lip.

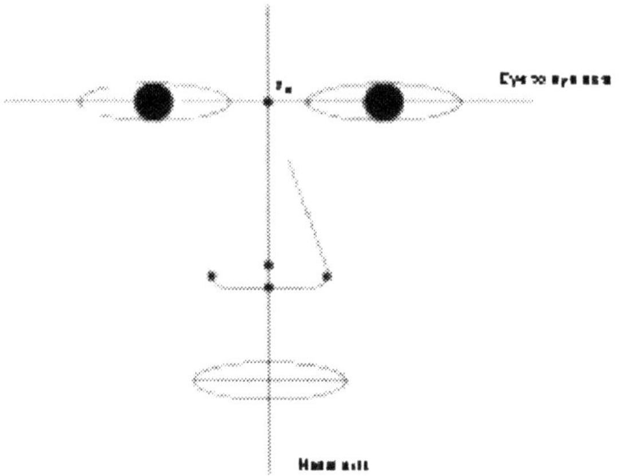

Figure 4.17: Illustration of nose control points extraction.

3. Edge field is now allowed to interact with zero for the coefficient of the valley forces. The upper and lower lips contours are extracted by adjusting the height of lower lip $height_d$, height of upper lip $height_u$, and length of upper lip $length_u$.

4.3.5 Nose Feature Points Extraction Using Geometrical Properties

The precise contour of the nose is hard to extract because it blends in with the side of the face. However, the nose control points shown in Fig. 4.6 are easy to extract. The point at the tip of the nose corresponds to a peak in image intensity, due to the illumination of light making it brighter while the point at the base of the nose corresponds to a valley in image intensity from the shadow between the base of the nose and the region above the upper lip (see Fig. 4.11) These feature points are extracted from the peak and valley images. The two points on the sides of the nose are then extracted from the edge image since the edges there are quite strong. The details of the nose control points extraction are as follows:

1. The point in the middle of the centers of the two eyes is defined as $\vec{x}_m$. An eye-to-eye axis is defined passing through the centers of the eyes. A nasal axis is also constructed passing through $\vec{x}_m$ and the center of the mouth. The two axes are illustrated in Fig. 4.17.

Figure 4.18: Facial contours extracted and nose feature points.

2. The control point at the base of the nose is derived by moving a cursor from the edge of the upper lip upward along the nasal axis and measuring the valley image intensity. When this point encounters a region of strong intensity values, then this is the desired point.

3. Similarly for the control point at the tip of the nose, it is derived by moving a cursor from the mid-eye point $\vec{x}_m$ downward along the nasal axis. The point corresponds to a region of strong intensity values in the peak image.

4. Control points on the sides of the nose are extracted by initially locating two points on the sides of the face with y-coordinate of the middle of the two points already derived and the x-coordinates at the left and right tips of the mouth. The starting points are derived assuming the mouth width is larger than the nose width. These points are then moved inwards towards the nose center to detect regions of strong edge intensities, which then correspond to the desired feature points.

Facial features contour extraction using the active contour models, deformable templates and nose control points derived in this section is illustrated in Fig. 4.18.

4.4 Automatic 3-D WFM Fitting and Adaptation to Facial Image

The 3-D wire-frame model is an important feature of 3-D model-based coding system as both analysis and synthesis are strongly dependent on it. Procedure for fitting and adaptation of the generic 3-D WFM to the facial image is outlined briefly in Section 4.2.1. Currently, this process is not yet fully automated, with the generic 3-D WFM manually adjusted through a user-interactive program.

The head and facial component models are adjusted to fit their respective features in the facial image. This is done by moving the control nodes of the WFM to feature points of the face. All the other nodes are interpolated according to the translations of these control nodes.

The generic 3-D WFM contains a detailed triangulated mesh of wireframes with a total number of 469 nodes. Each node is defined by its x, y and z coordinates. The x and y coordinates give the location of the node on the facial image and the z coordinate is used for the 3-D depth information. Each node is labeled by two numbers like (a, b), with the first number denoting the feature/location number with different values for different parts of the WFM or for different facial components, and the second number signifying the node number within the feature/location. Data for the WFM is stored in two files. The first one is a wire-frame datafile containing the locations for all the nodes, and the second file is a link datafile that stores the information of how the nodes of the WFM are inter-connected.

Work on automatic 3-D WFM fitting include T. Akimoto *et al.* [35] and M.J.T Reinders *et al.* [44] which adjust and scale the model to fit 2-D facial image. Fukuhara *et al.* [19] have developed 3-D WFM to include extraction of depth information from stereoscopic images.

In this section, the features of each component of the 3-D WFM and their automatic adjustment to fit the facial image are described in details.

4.4.1 Head Model Adjustment

The 3-D head model is used to describe the location of the head in the facial image. Although facial expressions do not adjust the head model significantly, it is important for synthesis of images containing translation and rotation of the head motion parameters. The detailed 3-D head model is given in Fig. 4.19.

Adjustment of the 3-D head model is divided into two stages. First initial adjustment requires five feature points (A,B,C,E,F) of the face as

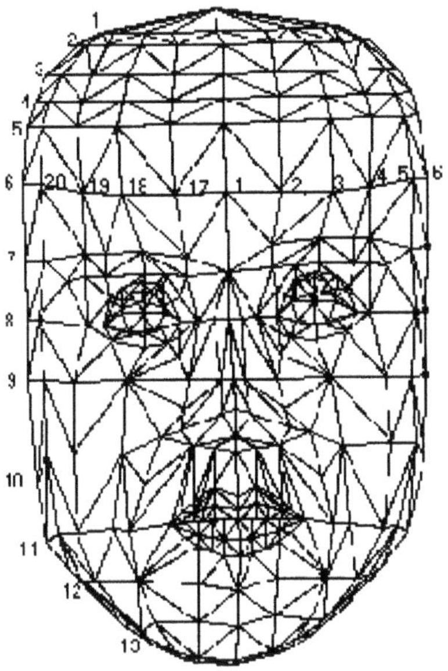

Figure 4.19: Generic 3-D head model. ©Univ. of Tokyo

shown in Fig. 4.3. The points E and F are derived from the final templates of the eyes as being the right tip of the left eye and left tip of the right eye respectively. Points B and C are points on the face profile contour that is on the eye-to-eye axis passing through the centers of the eyes. Similarly, point A is the lowest point on the facial profile contour that is also on the nasal axis passing through a midway point of E and F and the center of the mouth (see Fig. 4.17). The point D is calculated and the head model is adjusted to fit through the four points (A,B,C,D). The program 'Makeface' developed at University of Tokyo by Aizawa *et al.* is used for the 3-D WFM model adjustment. The five feature points entered in the program and the resulting adjusted 3-D head model are illustrated in Fig. 4.20 and Fig. 4.21, respectively.

The second stage of the 3-D head model adjustment is a refinement stage. This stage fits the 26 nodes of the WFM to the head boundary, with thirteen of the nodes shown in Fig. 4.19 labeled with numbers one to thirteen. These points are fitted to near points on facial profile contour

Figure 4.20: Five feature points (A,B,C,E,F) entered in the 'Makeface' program.

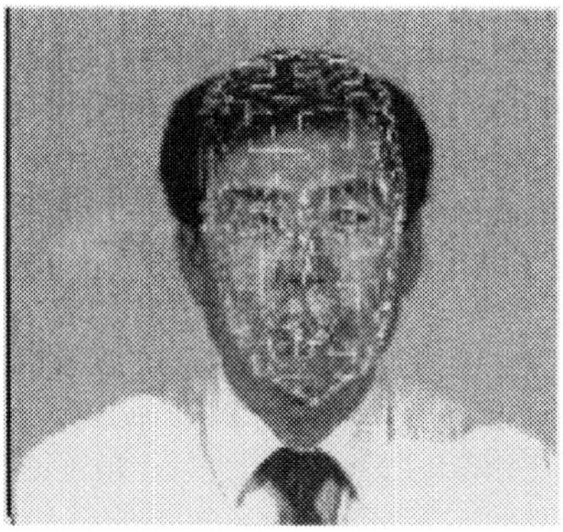

Figure 4.21: Adjusted 3-D WFM after first stage of 3-D head model fitting.

Figure 4.22: Feature points used in the second stage of 3-D head model adjustment.

extracted using active contours. Only ten of the points are adjusted since fourteen points around the upper head boundary are usually covered by the hair and are therefore not adjusted, and the other two points correspond to points B and C of Fig. 4.3 which need no adjustment. These points are entered in 'Makeface' program and the 3-D head model is adjusted as illustrated in Fig. 4.22 and Fig. 4.23, respectively.

4.4.2 Eye Model Adjustment

Eye expression is an important part of the face as it is often used during a conversation. Five feature points are defined on the eye to adjust the 3-D eye model. The facial feature points used to define the 3-D facial component models are illustrated in Fig. 4.6. The 3-D eye and eyebrow wire-frame models are given in Fig. 4.24.

The eye feature points are derived from the final eye template as described in Section 4.3.4.2. They correspond to the left-tip, right-tip, topmost, bottom-most point and center of the circle representing the iris of the template. These points are entered in the 'Makeface' program as illustrated in Fig. 4.25. The 3-D eye model is then adjusted to fit through these points.

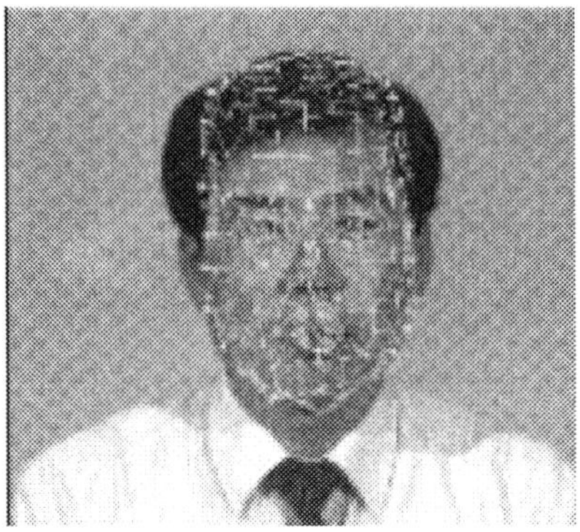

Figure 4.23: Adjusted 3-D head model after second stage of adjustment.

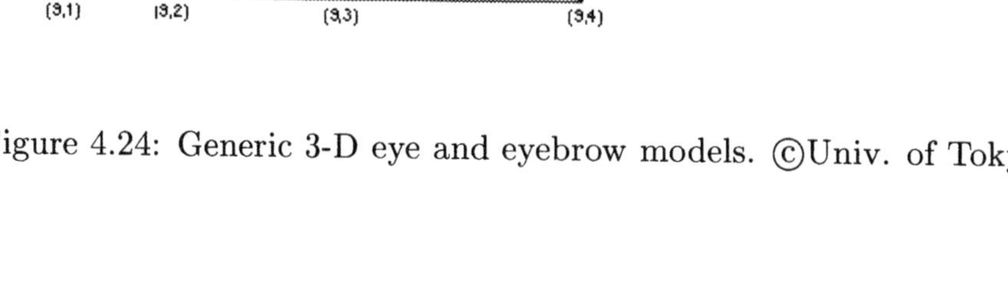

Figure 4.24: Generic 3-D eye and eyebrow models. ©Univ. of Tokyo

Figure 4.25: Feature points used in the 3-D eye model adjustment.

4.4.3 Eyebrow Model Adjustment

The shape of the eyebrow is an important feature of the face as it signifies the facial expressions on a person. Four feature points are used to defined the 3-D eyebrow model in Fig. 4.6. These points correspond to nodes (7,2), (7,3), (15,16) and (15,17) of the eyebrow model given in Fig. 4.24, with node (15,16) and (15,17) denoted by the numbers 16 and 17 respectively in the eyebrow model in the figure. These points are extracted from the eyebrow contours derived using active contours as described in Section 4.3.3.2. The left-most and right-most points on the contour give two of the feature points. The other two points are approximated at the middle of the left and right points at the lower and upper part of the contour. These points are entered in the 'Makeface' program as illustrated in Fig. 4.26, and the 3-D eyebrow model is adjusted to fit through them.

4.4.4 Mouth Model Adjustment

The mouth plays a vital part of the face as there is a continuous movement of the mouth throughout a conversation. The same assumption applies whereby the mouth is closed as in the derivation of mouth template. Five feature points are used to adjust the mouth model as depicted in Fig. 4.6. The 3-D wire-frame model of the mouth is given in Fig. 4.27.

The derivation of the feature points is quite straight-forward from the final template of the mouth, with the points corresponding to the center,

Figure 4.26: Feature points used in the 3-D eyebrow model adjustment.

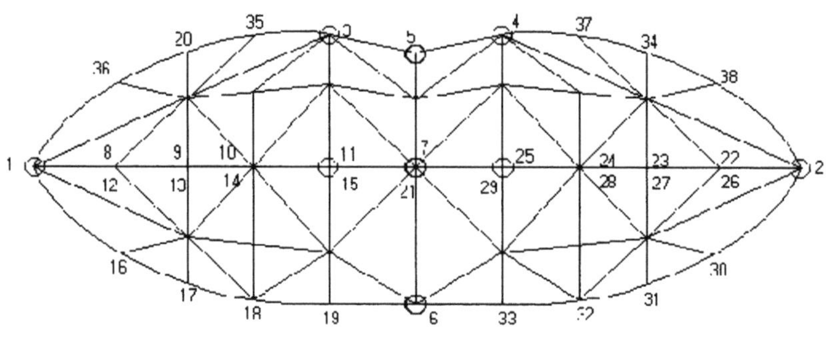

Figure 4.27: Generic 3-D mouth model. ©Univ. of Tokyo

Figure 4.28: Feature points used in the 3-D mouth model adjustment.

left-most, right-most, top and bottom tip of the template. The points are entered in the 'Makeface' program as illustrated in Fig. 4.28. The 3-D mouth model is then adjusted accordingly to fit these points.

4.4.4.1 Nose Model Adjustment

The feature points of the nose used for fitting the nose model to the facial image are shown in Fig. 4.6. They correspond to nodes (10,1), (10,2), (10,17) and (16,12), with (10,1) located in the middle of (10,2) and (10,17) as shown in Fig. 4.29. These points correspond to the nose feature points extracted in Section 4.3.5. They are entered in the 'Makeface' program and the 3-D nose model is then adjusted to fit through them. Figure 4.30 shows the feature points on the face and Fig. 4.31 gives the final adjusted 3-D WFM after the adjustment of all the facial component models.

4.5 Analysis of Facial Image Sequences

Analysis of image sequences represents a much more difficult problem compared to the synthesis of the images. The analysis part is strongly dependent on what is assumed as the model and what is synthesized as output images. Different aspects of analysis problems include segmentation of objects, estimation of global motion and estimation of local motion. In the context of 3-D model-based coding which is restricted to human facial images, the

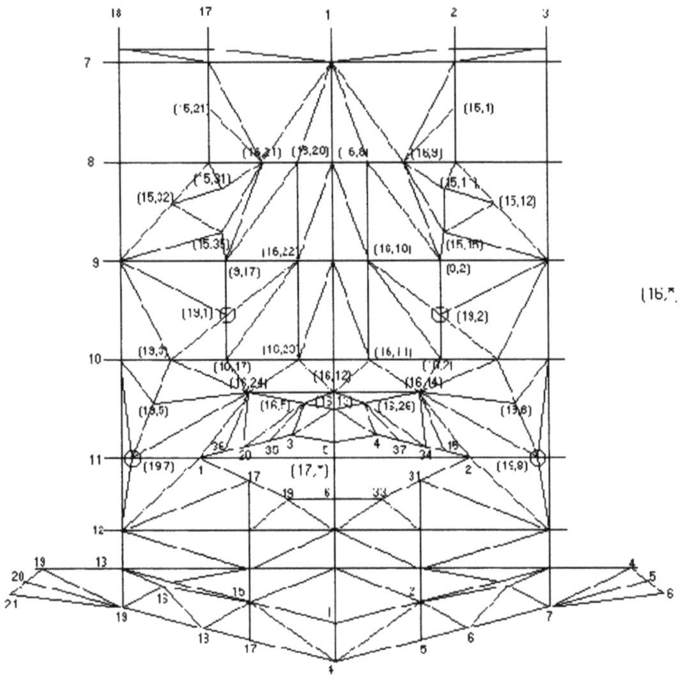

Figure 4.29: Generic 3-D nose model. ©Univ. of Tokyo

Figure 4.30: Feature points in the 3-D nose model adjustment.

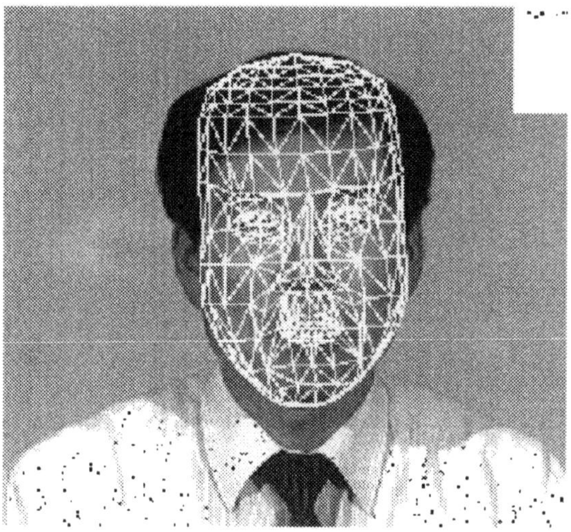

Figure 4.31: Final 3-D WFM with adjusted 3-D head and facial component models.

global motion refers to the head motion and local motion to the facial actions. Results of motion analysis have been reported under some restrictive situations such as the model is made in advance and the initial position of the face is known.

Different approaches of motion analysis include:

- feature points extraction by thresholding [51]
- motion estimation and tracking of marks plotted on the moving object [14]
- feature point extraction and tracking by using active contour model [52]
- estimation of facial expressions (Action Units) by using feature point displacements [53]
- direct estimation of head motion and facial expressions (Action Units) using spatio-temporal gradient [22]
- global motion estimation of the head using displacement vector field obtained by the block matching technique [54]
- facial muscle movement estimation based on optical flow [55]

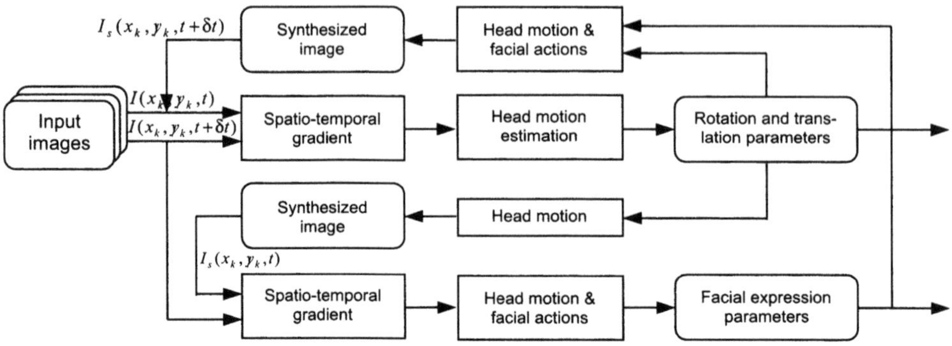

Figure 4.32: Estimation of head motion and facial expressions. ©IEEE 1994

- estimation of facial muscle movement by using the feature lines extracted by active contour model [56]

The direct estimation of facial motion using spatio-temporal gradient method is described in the following section. The block diagram of this method is given in Fig. 4.32. In this approach, the head motion is treated as rigid motion and facial expression is modelled as the Action Units (AUs) to be described in Section 3.1.

The analysis procedure can be divided into three steps: head motion estimation, facial expression estimation and iteration of these two steps for improving precision of analysis. The first step is the estimation of head motion parameters (HMP) between the current frame $I(x, y, t)$ and the next frame $I(x, y, t+\delta t)$ and synthesizes the image $I_s(x, y, t)$ using the estimated HMP. The next step estimates the facial expression parameters (FEP) between the synthesized image $I_s(x, y, t)$ and the next frame $I(x, y, t + \delta t)$ using both the estimated HMP and FEP. The next step repeatedly estimates HMP and FEP between $I_s(x, y, t+\delta t)$ and $I(x, y, t+\delta t)$ and corrects the HMP and FEP of the first iteration. These iterations continue until there is no more estimation error reduction.

The facial actions are assumed to be much smaller in magnitude compared to the head motion, so the face is treated as a non-rigid body which has small deviation from rigid body. The motion of the face can then be formulated as

$$V = \Omega \wedge (P + \delta P) + U \tag{4.35}$$

where $\wedge$ is the cross product operator for two vectors,
$V = (v_x, v_y, v_z)$ is the velocity of a point on the face,
$\Omega = (\omega_x, \omega_y, \omega_z)$ is the angular velocity of the point,
$U = (u_x, u_y, u_z)$ is the translational velocity of the point,
$P = (x, y, z)$ is a 3-D position vector of the point, and
$\delta P = (\delta x, \delta y, \delta z)$ is the non-rigid motion vector of the point.

The face motion can be separated as head motion and facial actions. That is,

$$V_h = \Omega \wedge P + U \tag{4.36}$$

$$V_f = \Omega \wedge \delta P \tag{4.37}$$

$$V = V_h + V_f \tag{4.38}$$

where V_h is the head motion, and V_f is the facial action.

4.5.1 Estimation of Head Motion Parameters

The first step of analysis involves the estimation of HMP between current frame $I(x, y, t)$ and next frame $I(x, y, t + \delta t)$. Following the optical flow approach, this leads to the following constraint between the velocity and the spatio-temporal gradient of the intensity between the two consecutive frames

$$I_x v_x + I_y v_y + I_t = 0 \tag{4.39}$$

where I_x, I_y and I_t are the partial derivatives of $I(x, y, t)$ with respect to x, y, and t, and v_x, v_y are the optical flow and the velocities for x- and y-direction.

The head motion is roughly estimated by firstly ignoring facial actions V_f. Substituting V_h into V in (4.38) gives

$$(-zI_y, zI_x, xI_y - yI_x, I_x, I_y)(\omega_x, \omega_y, \omega_z, u_x, u_y)^T = -I_t \tag{4.40}$$

The z-values of points other than the nodes of the WFM are obtained by linearly interpolating the z-values of the nodes. Let point P_i be the 3-D location of the point to be interpolated. Since the point P_i is assumed to lie on the surface of a triangle with nodes G_a, G_b, and G_c as its vertices (Fig. 4.33), the 3-D position vector of point P_i is then given as

$$P_i = s_i(G_b - G_a) + t_i(G_c - G_a) + G_a \tag{4.41}$$

The coefficients s_i and t_i are uniquely determined by the x and y components of G_a, G_b, G_c and P_i. The z-value (depth) of point P_i is obtained by using s_i, t_i and z-values of G_a, G_b and G_c.

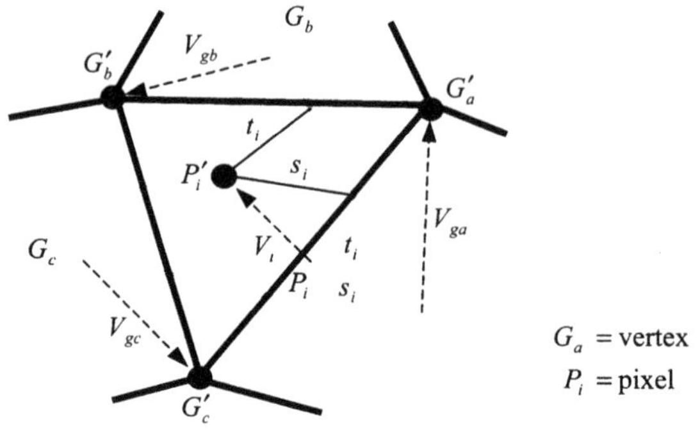

Figure 4.33: Linear interpolation of pixel points from vertices of the surface model. ©IEEE 1994

To obtain a system of linear equations, (4.40) is applied to all points on the face. This gives

$$HX = Y \tag{4.42}$$

where i-th row of H (1×5 matrix) is $(-zI_{yi}, zI_{xi}, xI_{yi} - yI_{xi}, I_{xi}, I_{yi})$, $X = (\omega_x, \omega_y, \omega_z, u_x, u_y)^T$, $Y = -(I_{t1}, I_{t2}, \ldots, Itp)$, p is the total number of pixels to be processed.

Equation (4.42) is solved using least mean squares method to minimize error $\|HX - Y\|^2$. This gives

$$X = [H^T H]^{-1} H^T Y \tag{4.43}$$

The solution gives only a rough estimation of the HMP, since facial actions V_f is ignored in the calculation, so the points affected by facial actions have to be removed. To improve the estimation, (4.40) is evaluated for each point by using the rough estimation of HMP. If the squared difference of the two sides of (4.42) for a point is larger than the mean error $\frac{\|HX-Y\|^2}{p}$, then this point is assumed to be included in the facial actions. This point is then removed for the next estimation. The HMP is calculated again using the same procedure. For the estimation of FEP in the next step, the image $I_s(x, y, t)$ is synthesized by moving the facial model using the estimated HMP.

4.5.2 Estimation of Facial Expression Parameters

The second step of analysis is the estimation of FEP between the synthesized image $I_s(x, y, t)$ and $I(x, y, t + \delta t)$. The FEP is formulated using the parameterization by action units as discussed previously. Similar to the constraint (4.40), we obtain the following equation

$$I_{sx}v_{fx} + I_{sy}v_{fy} + I_{st} = 0 \tag{4.44}$$

where I_{sx}, I_{sy} and I_{st} are the partial derivatives of $I(x, y, t)$ with respect to x, y,and t, and v_{fx}, v_{fy} are the optical flow and the velocities for x- and y-direction.

Utilizing the AU deformation rule which defines a set of displacement vectors of nodes of the WFM, the displacements of the nodes can be assumed as being the weighted sum of the action units. Therefore, the node displacement vector can be written as

$$\begin{aligned} g &= a(A_1, A_2, \dots, A_m) \\ &= a\bar{A} \end{aligned} \tag{4.45}$$

where $g = (g_{x1}, g_{y1}, g_{z1}, \dots, g_{zn})^T$ is a vector acting on the nodes of the WFM, A_i is the set of displacements which act on the nodes according to i-th AU of its unit intensity, $a = (a_1, a_2, \dots, a_m)$ are intensities of AUs, n is the number of nodes, and m is the number of AUs.

Suppose the triangle G_a, G_b and G_c is deformed to the triangle G'_a, G'_b and G'_c (see Fig. 4.33), and assume that s_i and t_i are not changed by the deformation. The relationship between the displacements of the pixels and those of the nodes can be written as

$$\delta P_i = s_i(g_b + g_a) + t_i(g_c + g_a) + g_a \tag{4.46}$$

where δP_i is the 3-D displacement vector of the point P_i in the triangle, and g_a, g_b and g_c are the 3-D displacement vectors of the nodes.

Applying (4.46) to all triangles in the WFM gives

$$\delta P = Sg \tag{4.47}$$

where $\delta P = (\delta P_1^T, \delta P_2^T, \dots, \delta P_n^T)^T$, S is a $3p \times 3n$ matrix describing the relationship the movements of the points and those of the nodes.

Substituting (4.45) into (4.47) gives

$$\begin{aligned} \delta P &= S\bar{A}a \\ &= \bar{A}_s a \end{aligned} \tag{4.48}$$

As defined previously in (4.39), the local facial actions can be written as

$$V_f = \Omega \delta P \tag{4.49}$$

Combining (4.49) and (4.48) results in linear equations in a, which can be solved by using least square estimation like in the previous section. Using the estimated FEP along with the previously estimated HMP, the image $I_s(x, y, t + \delta t)$ is synthesized.

4.5.3 High Precision Estimation by Iteration

Since only first order approximations are used to estimate the HMP and FEP, large displacements between frames will not give very accurate estimation. The following procedure is used to reduce the errors. The HMP is estimated between the current frame $I(x, y, t)$ and the next frame $I(x, y, t+\delta t)$, with the parameters used for synthesizing the image $I_s(x, y, t)$. The FEP is then estimated between $I_s(x, y, t)$ and $I(x, y, t+\delta t)$. The image $I_s(x, y, t+\delta t)$ is then synthesized using the HMP and FEP. The displacements between this synthesized image and the next frame will be smaller than the values between the two original frames. The second estimation is performed between $I_s(x, y, t + \delta t)$ and $I(x, y, t + \delta t)$ to give a more accurate estimation. This iteration continues until the error becomes smaller than some threshold.

4.6 Synthesis of Facial Image Sequences

The function of the decoder in the 3-D model-based coding system is to synthesize the output images using the 3-D WFM of the face using the analysis parameters. From the 3-D WFM, there is a number of ways to reproduce the facial movements. There is a hierarchy of levels for the parameterization for synthesizing facial actions.

1. Texture Reproduction Level: this method reproduces the facial movements by updating the texture. This includes clip-and-paste methods, template methods, model-based/waveform combined approaches and so on.

2. Node Control Level: this method synthesizes the images by controlling the nodes of the WFM of the face, both the intensities and node positions of the 3-D model templates are interpolated.

3. Shape Control Level: the images are synthesized by controlling the shape of the WFM of the face. Examples include shape parameterization [14] by making use of Facial Action Coding System (FACS) [57], heuristic control of shapes of facial components such as eyes and mouth and so on.

4. Muscle Control Level: synthesize facial images by controlling the muscles with the muscle models in the wire-frame model [56]

5. Abstract Level: the above parameters are used and controlled by incorporating them into more abstract parameters such as description of emotions.

In most of model-based coding proposals, texture reproduction level and shape control level are the two most commonly used. An example of texture reproduction level is the clip-and-paste synthesis method whilst the facial structure deformation synthesis method belongs to the shape control level.

The clip-and-paste method is performed by extracting specific region of the original image, transmitting them, and then pasting them in the corresponding region of the synthesized image. The facial structure deformation method deforms the structure of the 3-D WFM to simulate facial expressions. Facial expressions can be decomposed into 44 Action Units (AU), based on the FACS. The deformation of the WFM is done through the movement of vertices according to the deformation rules defined by the AUs. The deformation rules are defined using physical and anatomical knowledge.

The clip-and-paste method, even though is simpler to implement, it does not accurately render the facial expressions. In this next section, the facial structure deformation method is described in detail.

4.6.1 Facial Structure Deformation Method

This synthesis method deforms the structure of the 3-D facial model to describe the facial expressions. The facial actions are described using the Facial Action Coding System (FACS) [57] which was originally used in psychological studies. FACS describes a set of the minimal basic actions called Action Units (AUs) performable on a human face such as inner brow raise, outer brow raise and so forth. Each facial action can be decomposed as a combination of Action Units. There are 44 defined AUs in which 34 have been parameterized in the model-based coding system. Table 4.2 gives the list of Action Units for model-based coding system.

Table 4.2: List of Determined AU Deforming Rules. ©IEEE 1994

No.	AU Name	No.	AU Name
1	Inner Brow Raiser	20	Lip Stretcher
2	Outer Brow Raiser	23	Lip Tightener
4	Brow Lowerer	24	Lip Pressor
5	Upper Lid Raiser	25	Lips Part
6	Cheek Raiser	26	Jaw Drop
7	Lid Tightener	27	Mouth Stretch
8	Lips Toward	28	Lips Suck
9	Nose Wrinkler	29	Jaw Thrust
10	Upper Lip Raiser	30	Jaw Sideways
11	Nasolabial Furrow	32	Bite
	Deepener	35	Cheek Suck
12	Lip Corner Puller	41	Lid Droop
13	Sharp Lip Puller	42	Slit
14	Dimpler	43	Eyes Closed
15	Lip Corner Depressor	44	Squint
16	Lower Lip Depressor	45	Blink
17	Chin Raiser	46	Wink
18	Lip Pucker		

The upper face has AUs relevant to eyebrows and eyes. They are inner brow raiser (AU1), outer brow raiser (AU2) and brow lowerer (AU4). The lower face has many AUs defined such as the jaw rotation and lip actions. Figure 4.34 illustrates the deforming rules for AUs for the upper face.

The finely detailed 3-D WFM with the 3-D facial component models is used for this synthesis method as illustrated in Fig. 4.2.

The procedure for deforming the 3-D facial model is as follows:

1. Control points at the vertices of each facial component are set at locations determined by AUs deformation rule. The control points of the facial component are shown in Fig. 4.6. These points are moved to simulate the facial muscular actions.

2. The secondary control points are located according to the AUs and the original control points.

3. For all the other points, they are interpolated using either a linear or a quadratic curve depending on the facial components. These points

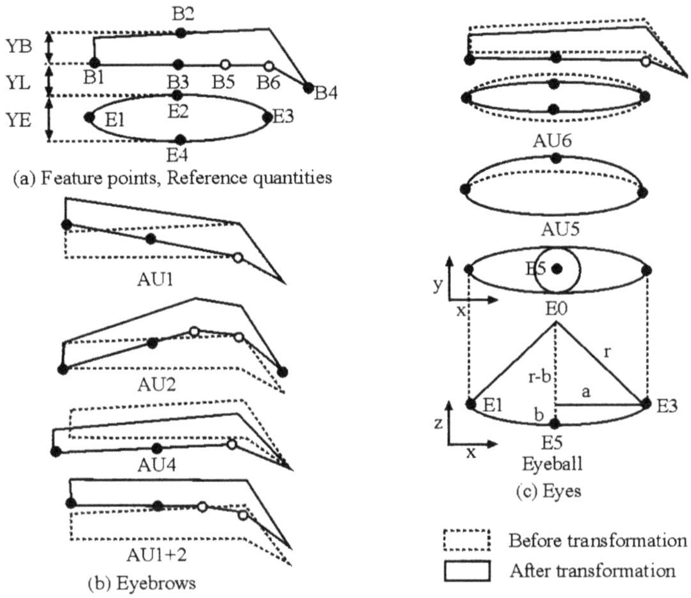

Figure 4.34: Deforming rules for AUs for the upper face. ©Univ. of Tokyo

connect the original control points and the secondary control points. Figure 4.35 shows an example of lip shape deformation.

4. The mapped data is then transformed at every triangular element according to the deformed wire-frame model.

Figure 4.36 illustrates the block diagram of facial structure deformation synthesis method. Figure 4.37 shows the synthesized images using this method for some AU combinations. The synthesized images appear natural as they are produced with the original texture of a full face and the updated texture. The textures of the furrows and the teeth are updated from other images for the person to be described in the next section.

4.7 Update of 3-D Facial Model and Expected Transmission Bit Rates

The 3-D facial model constructed using the synthesis technique described in the previous section consists of texture information from image in the first frame. This is insufficient for reconstructing all of the subsequent images

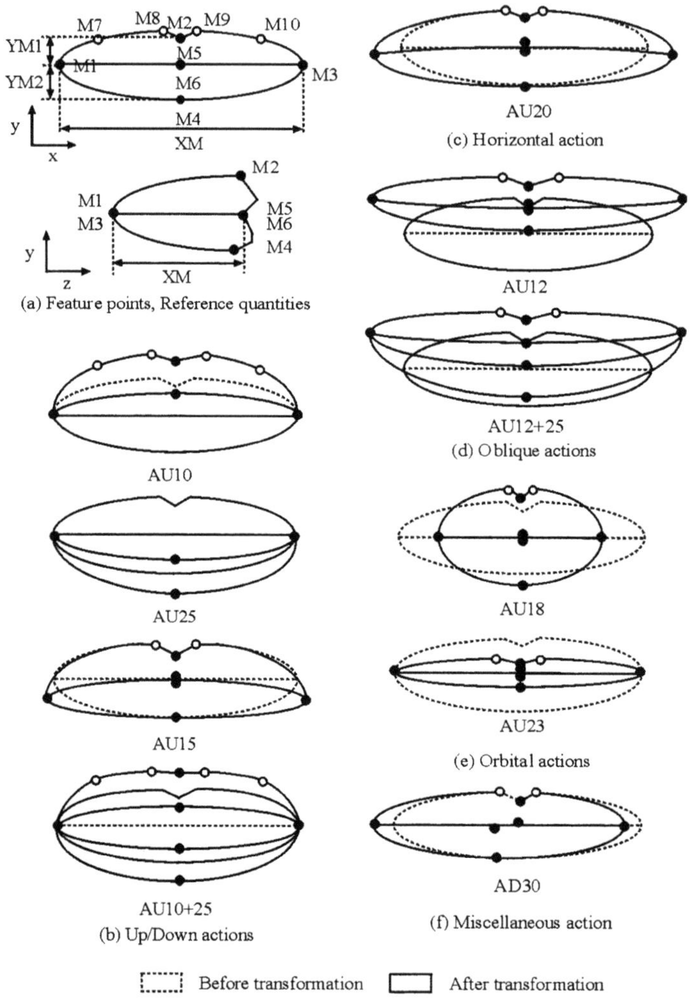

Figure 4.35: Deforming rules for AUs for lips. Black dots are the original control points and the white circles are the secondary control points. Dashed lines show the neutral shape and the solid line shows the deformed shape. ©Univ. of Tokyo

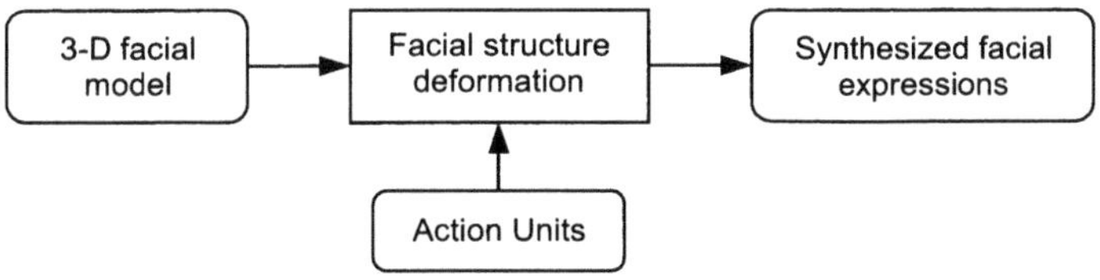

Figure 4.36: Block Diagram of Facial Structure Deformation Synthesis Method [14].

as sometimes new texture data is required. For example, when the mouth opens, the texture of teeth is not available from the first frame.

Depth information of the 3-D facial model is also approximated, so there is a need for updating this information. This section discusses the update methods for texture and depth information and estimates the transmission bit rates of the 3-D model-based coding system.

4.7.1 Update of Texture Information

4.7.1.1 Method 1

This method updates the texture on the basis of the estimated facial action parameters. For example, when the lip corners are pulled upward (AU12), the nasolabial furrows become deep and clear, which is not visible in the initial facial image with a neutral face. Let $I(x, y, 0)$ and $I(x, y, t_f)$ be the first neutral face image and the image with the furrow texture. Let $I_m(x, y, t_f)$ be the facial texture after normalization, in which the input facial texture $I(x, y, t_f)$ is warped so that the facial shape becomes the same as the initial face model. The furrow texture $I_f(x, y, t_f)$ in the region F_j can be defined as

$$I_f(x, y, t_f) = W(x, y)[I_m(x, y, t_f) - I(x, y, 0)], \quad (x, y) \in F_j \tag{4.50}$$

where $W(x, y)$ is a function to smooth the boundaries of the region.

The textures need to be updated only when the AUs are clearly observed. The texture update is required for areas around nasolabial furrows, glabella, chin, root of the nose and teeth. The new texture information is used in the synthesis process as follows

$$I_s(x, y, t_s) = I(x, y, 0) + \frac{a_{is}}{a_{if}} I_f(x, y, t_f) \tag{4.51}$$

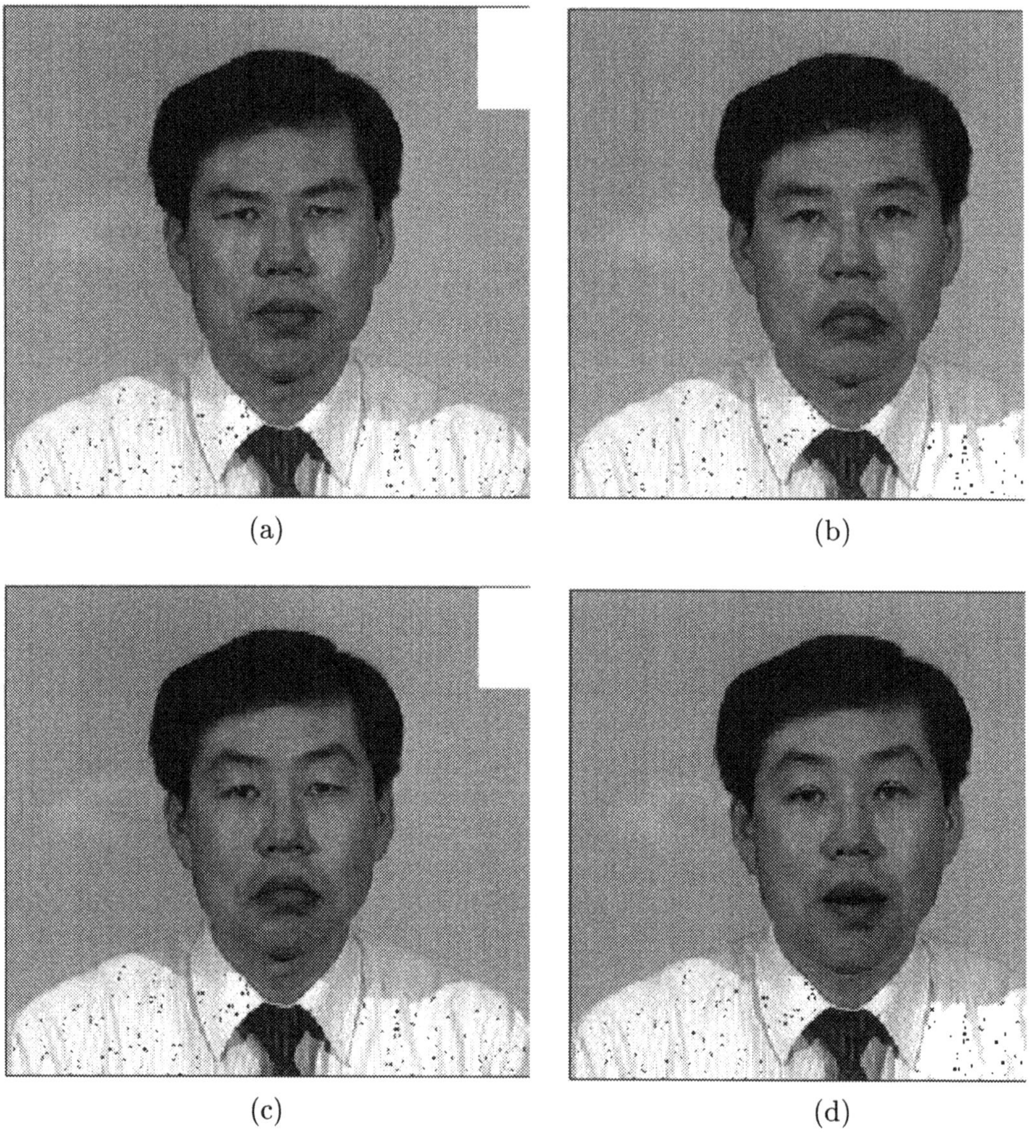

Figure 4.37: Synthesized images of AU combination: (a) AU4 (Anger), (b) AU15 (Frown), (c) AU1+4AU+AU15 (Sad), (d) AU1+AU2+AU5+AU26 (Surprise).

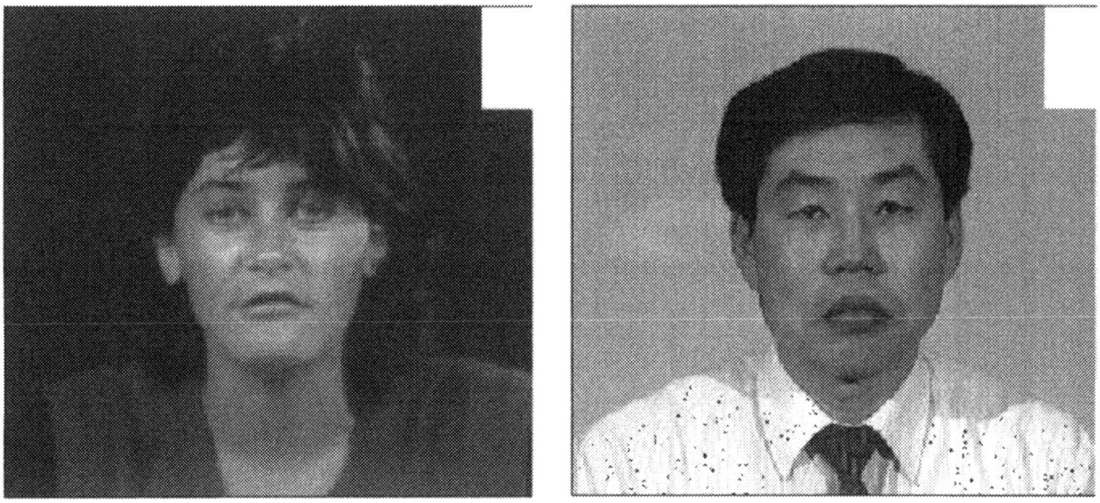

Figure 4.38: Synthesized images with Texture Update Method I.

where a_{is} has AU intensities of the synthesized image,
a_{if} has AU intensities of the image used for texture updating.

Figure 4.38 shows the synthesized images using this texture update method.

4.7.1.2 Method 2

This method updates texture by generating texture basis from the input images by Gram-Schmidt orthonormalization and describing textures with the texture basis. The steps for generating the texture basis are as follows:

1. The facial texture is normalized by deforming and mapping the input texture onto the initial WFM. This is defined as vector I_i components of which are the pixel brightness of the normalized texture.

2. The first texture basis is

$$B_1 = \frac{I_1}{\|I_1\|} \tag{4.52}$$

3. The second texture basis is derived as follows. The normalized second image frame I_2 is represented with B_1. The remaining vector is $R_2 = I_2 - (B_1, I_2)B_1$, this vector R_2 is orthogonal to B_1. If the magnitude

of this vector $\|R_2\| > Threshold$, $B_2 = \frac{R_2}{\|R_2\|}$ is adopted as the second texture basis, otherwise R_2 is neglected. This process is continued to produce a set of orthogonal vectors which represents the facial image sequence efficiently.

Parameters which represents an image texture are inner products of the image and the texture basis. The texture is reproduced with the texture basis and those inner product values, that is, texture at jth frame is $\sum_{i=1}^{q}(B_i, I_j)B_i$, where q is the number of texture bases. The texture information are transmitted continuously as side information when coding sequences. However, it may not be necessary to send additional information after a while.

4.7.2 Update of Depth Information

The depth information is approximated during the 3-D WFM fitting and adaptation. The depth information can be estimated by making use of the estimated motion parameters. The following is the procedure for estimation of depth information.

Let $P' = P + \delta P$ and $P = (x', y', z')^T$. Substituting (4.36) into (4.40) gives

$$(I_x\omega_y - I_y\omega_z)z' = -(x'I_y\omega_z - y'I_x\omega_z + I_x u_x + I_y u_y + I_t) \tag{4.53}$$

The depth z' can be obtained directly from the above equation. Applying (4.53) to all points gives

$$H_z Z' = Y_z \tag{4.54}$$

where $Z' = (z'_1, z'_2, \dots, z'_n)^T$ and components of H_z and Y_z are the left and right hand side of (4.53). The relationship between z-values of the nodes and the points in triangles is given by (4.41). Therefore,

$$Z' = S_z Z'_g, \tag{4.55}$$

where $Z'_g = (z_{g1}, z_{g2}, \dots, z_{gn})^T$ is a vector representing the depth of the nodes, and S_z is a $1 \times n$ matrix determined by using (4.41).

Similar to the estimation of HMP and FEP, by substituting (4.55) into (4.54) gives an equation similar to (4.42) which can be solved by least square method. The motion estimation and depth correction can be done alternately for convergence of results.

Table 4.3: Estimates of transmission bit rates according to texture update methods. [22]

Texture Update Methods	Transmission Bit Rates	Image Quality
No update	1.4 Kbits/s	Good
Method I	3.5 Kbits/s	High
Method II	10.5 Kbits/s	Very high

4.7.3 Transmission Bit Rates

The transmission bit rates of the 3-D model-based coding system presented in this chapter may vary according to the texture update method used. The transmission bit rate estimates are given in Table 4.3 for comparison. The results tabulated are using 360 × 238 frame size, with frame rate of 10 frames/second.

4.7.3.1 No update of texture information

The transmitted information includes the HMP and the FEP. The FEP usually consists of a combination of 10 AUs/frame. The HMP consists of 6 values/frame (3 rotations, 3 translations). Both the HMP and FEP are quantized to 32 levels. Since there are 44 AUs, the AU number is quantized to 64 levels. The transmission bit rate is therefore

$$[6HMP \times 5bits + 10AUs \times \{6bits(AUnumbers) + 5bits(AUintensities)\}] \times 10frame/s = 1.4Kbits/s$$

4.7.3.2 Method 1

The furrow, teeth, eyeball parts occupy some 10,000 pixels, and their textures are updated five times. The furrow textures are quantized to 32 levels. The HMP and FEP are transmitted along with the extra texture information. The mean transmission bit rate for a call of three minutes is

$$\begin{aligned} 1.4Kbits/s + \{&10,000\ pixels\ (updated\ parts) \times 5bits \times 1.5(luminance \\ &+ chrominance) \times 5times\}/(3minutes \times 60seconds) \\ &\approx 3.5Kbits/s \end{aligned}$$

4.7.3.3 Method 2

In this case, the texture bases (B_i) and the texture description parameters (TDP) which are the coefficients of B_i are transmitted in addition to the HMP and FEP. The face is presumed to occupy some 20,000 pixels. Assuming that ten texture bases are transmitted, and the other texture bases, (excluding the first basis) are quantized to 32 levels, and the TDP are quantized to 256 levels, the mean transmission bit rate for a call of three minutes is

$$\begin{aligned}
&1.4Kbits/s + (20,000\ pixels\ (face\ part) \times 5bits \times 1.5(luminance \\
&\qquad + chrominance) \times 10\ texture\ bases/(3\ minutes \times 60seconds) \\
&\qquad + 10TDP \times 8bits \times 10frames/s \approx 10.5Kbits/s
\end{aligned}$$

There is temporal redundancy in the parameters and spatial redundancy in the updated textures in all the three cases. By using the conventional waveform coding technique to reduce the redundancy, this will achieve a further reduction of the transmission bit rates.

References

[1] K. Aizawa and T.S.Huang, "Model-based image coding: Advanced video coding techniques for very low bit-rate applications," *Proceedings of the IEEE*, vol. 83, pp. 259–271, Feb. 1995.

[2] M. Kunt, A. Ikonomopoulos, and M. Kocher, "Second-generation image-coding techniques," *Proceedings of the IEEE*, vol. 73, no. 4, pp. 549–574, Apr. 1985.

[3] S. Carlsson, "Sketch based coding of gray level images," *Signal Processing*, vol. 15, pp. 57–83, 1998.

[4] M. Gilge, T. Englehardt, and R. Mehlan, "Coding of arbitrarily shaped segments based on a generalized orthogonal transform," *Signal Processing: Image Communication*, vol. 1, no. 2, pp. 153–180, 1989.

[5] M. Hötter and J. Ostermann, "Analysis synthesis coding based on planar rigid moving objects," in *Int. Workshop on 64 Kbps Coding of Moving Video, Hannover, Germany*, 1988.

[6] Y. Nakaya and H. Harashima, "Motion compensation based on spatial transformation," *IEEE Trans. Circuits Syst. for Video Technol.*, vol. 4, pp. 339–356, June 1994.

[7] H.G. Musmann, M. Hötter, and J. Ostermann, "Object-oriented analysis-synthesis coding of moving images," *Signal Processing: Image Communication*, vol. 1, no. 2, pp. 117–138, Oct. 1989.

[8] N. Diehl, "Object-oriented motion estimation and segmentation in image sequences," *Signal Processing: Image Communication*, vol. 3, no. 1, pp. 23–56, Feb. 1991.

[9] J. Ostermann, "Modelling of 3-d moving objects for an analysis-synthesis coder," in *Proc. SPIE Sensing and Reconstruction of Three-dimensional Objects and Scenes*, 1990, vol. 1260, pp. 240–249.

[10] H. Morikawa and H. Harashima, "3-d structure extraction coding of image sequences," *Journal of Visual Communication and Image Representation*, vol. 2, no. 4, pp. 332–344, 1991.

[11] R. Koch, "Dynamic 3-d scene analysis through synthesis feedback control," *IEEE Trans. Pattern Analysis and Machine Intelligence*, vol. 15, pp. 556–568, June 1993.

[12] R. Fochheimer, O. Fahlander, and T. Kronander, "Low bit-rate coding through animation," in *Int. Picture Coding Symposium, PCS'83, Davis, CA, USA*, Mar. 1983, pp. 113–114.

[13] R. Fochheimer and T. Kronander, "Image coding - from waveforms to animation," *IEEE Trans. Acoustics, Speech, and Signal Processing*, vol. 37, pp. 2008–2023, Dec. 1989.

[14] K. Aizawa, H. Harashima, and T. Saito, "Model-based analysis synthesis image coding for a person," *Signal Processing: Image Communication*, vol. 1, no. 2, pp. 139–152, 1989.

[15] K. Aizawa, H. Harashima, and T. Saito, "Model-based synthetic image coding system," in *Int. Picture Coding Symposium, PCS'87, Stockholm, Sweden*, 1987.

[16] W.J. Welsh, "Model-based coding of moving images at very low bit rate," in *Int. Picture Coding Symposium, PCS'87, Stockholm, Sweden*, 1987.

[17] W.J. Welsh, A.D. Simons, A.D. Hutchinson, and R.A. Searby, "Synthetic face generation for enhancing a user interface," in *Proc. Image Comm. '90, France*, 1990, pp. 177–182.

[18] Y. Nakaya, K. Aizawa, and H. Harashima, "Texture updating methods in model-based coding of facial images," in *Int. Picture Coding Symposium, PCS'90, Boston, MA, USA*, 1990.

[19] T. Fukuhara, K. Asai, and T. Murakami, "Model-based image coding using stereoscopic images and hierarchical structuring of new 3-d wireframe model," in *Int. Picture Coding Symposium, PCS'90, Tokyo, Japan*, 1991.

[20] T. Minami, I. So, T. Mizuno, and O. Nakamura, "Knowledge-based coding of facial images," in *Int. Picture Coding Symposium, PCS'90, Boston, MA, USA*, 1990.

[21] C.S. Choi, T. Okazaki, H. Harashina, and T. Takebe, "A system of analyzing and synthesizing facial images," in *IEEE Int. Symposium on Circuits and Systems, ISCAS'91, Singapore*, 1991, pp. 2665–2668.

[22] C.S. Choi, K. Aizawa, H. Harashina, and T. Takebe, "Analysis and synthesis of facial image sequences in model-based image coding," *IEEE Trans. Circuits Syst. for Video Technol.*, vol. 4, no. 3, pp. 257–275, June 1994.

[23] H. Li, P. Povivainen, and R. Fochheimer, "3-d motion estimation in model-based facial image coding," *IEEE Trans. Pattern Analysis and Machine Intelligence*, vol. 15, pp. 545–555, June 1993.

[24] G. Xu, H. Agawa, Y. Nagashima, F. Kishino, and Y. Kobayashi, "Three-dimensional face modelling for virtual space teleconferencing systems," *IEICE Transactions*, vol. E73, no. 10, 1990.

[25] M. Hayashi et al. Dec. 1992, ITEJ Tech. Rep.

[26] H. Holtzman, "3-d video modelling," in *CHI Conf. on Human Factors in Comp. Syst., CA*, May 1992.

[27] H.D. Lin and D.G. Messerschmitt, "Video composition methods and their semantics," in *IEEE Int. Conf. on Acoustics, Speech, and Signal Processing, ICASSP'91, Toronto, Canada*, 1991, pp. 2833–2836.

[28] S. Morishima, K. Aizawa, and H. Harashima, "An intelligent facial image coding driven by speech and phoneme," in *IEEE Int. Conf. on Acoustics, Speech, and Signal Processing, ICASSP'89, Glasgow, UK*, 1989.

[29] S. Luo, *Speech-enhanced model-based video facial image coding*, Ph.D. thesis, University of Sydney, Australia, 1995.

[30] F.I. Parke, *A paremetric model of human faces*, Ph.D. thesis, University of Utah, 1994.

[31] H. Yamada, H. Chiba, K. Tsuda, and K. Maiya, "New approach to the research on facial expressions: Model-based facial action synthesizing computer (masc) system," *Int. J. Psychol.*, vol. 27, pp. 47, 1992.

[32] F. Hara and H. Kobayashi, "Computer graphics for expressing robot-artificial emotions," in *IEEE Workshop on Robot and Human Communication*, 1992, pp. 155–160.

[33] M. Tanimoto and S. Nakashima, "Basic experiment of 2d-3d conversion for a new 3d visual communication," in *Int. Picture Coding Symposium, PCS'90, Boston, MA, USA*, 1990.

[34] F.I. Parke, "Parametrized models for facial animation," *IEEE Comp. Graphics and Applications*, vol. 12, pp. 61–68, Nov. 1982.

[35] T. Akimoto and Y. Suenaga, "3d facial model creation using genetic model and front and side views of faces," *IEICE Trans. Inf. & Sys.*, vol. 75-D, no. 2, pp. 191–197, 1992.

[36] J.Y. Zheng, Y. Nagashima, and Kishino, "3-d modelling from continuous aspect views," in *Int. Picture Coding Symposium, PCS'90, Tokyo, Japan*, 1991.

[37] K. Water and D. Terzopoulos, "Modelling and animating faces using scanned data," *The Jour. of Visualization and Computer Animation*, vol. 2, pp. 129–131, 1991.

[38] L.D. Harmon, M.K. Khan, R. Lasch, and P.F. Ramig, "Machine identification of human faces," *Pattern Recogn.*, vol. 13, pp. 97–110, 1981.

[39] Jr. G.J. Kaufmann and K.J. Breeding, "The automatic recognition of human faces from profile silhouttes," *IEEE Trans. Systems, Man, and Cybernetics*, vol. 6, pp. 113–120, 1976.

[40] O. Nakamura, S. Mathur, and T. Minami, "Identification of human faces based on isodensity maps," *Pattern Recogn.*, vol. 24, pp. 263–272, 1991.

[41] A.L. Yuille, P.W. Hallinan, and D.S. Cohen, "Feature extraction from faces using deformable templates," *Int. Journal of Computer Vision*, pp. 99–111, 1992.

[42] M. Kass, A. Witkin, and D. Terzopoulos, "Snakes: Active contour models," *Int. Journal of Computer Vision*, pp. 321–331, 1988.

[43] T. Kanade, *Comptuer recognition in human faces*, Birkhauser Verlag, Basel, Switzerland, 1977.

[44] M.J.T. Reinders, P.J.L. van Beck, B. Sankur, and J.C.A. van der Lubbe, "Facial feature localization and adaptation of a generic face model for mobel-based coding," *Signal Processing: Image Communication*, vol. 7, pp. 57–74, 1995.

[45] L.C. De Silva, K. Aizawa, and M. Hatori, "Detection and tracking of facial features," *Int. Journal of Computer Vision*, vol. 2501, pp. 1161–1172, 1995.

[46] C.L. Huang and C.W. Chen, "Human facial feature extraction for face interpretation and recognition," *Pattern Recogn.*, vol. 25, no. 12, pp. 1435–1444, 1992.

[47] T.Y. Cheng and C.L. Huang, "Color image segmentation using scale space filter and markov random field," in *SPIE Intelligent Robots and Computer Vision X, Boston, MA, USA*, Nov. 1991, vol. 1607, pp. 11–13.

[48] C.W. Chen, "Human face recognition using deformable template and active contours," M.S. thesis, National Tsing-Hua University, Hsin-chu, Taiwan, 1991.

[49] G.A. Baxes, *Digital Image Processing*, pp. 124–152, John Wiley & Sons, 1994.

[50] D.J. Williams and M. Shah, "A fast algorithm for active contours and curvature estimation," *Computer Vision, Graphics, and Image Processing*, vol. 55, no. 1, pp. 14–26, Jan. 1992.

[51] M. Kaneko, A. Koike, and Y. Hatori, "Real-time analysis and synthesis of moving facial images applied to model-based coding," in *Int. Picture Coding Symposium, PCS'90, Tokyo, Japan*, 1991.

[52] S. Reddy and K. Aizawa, "Human facial motion modelling, analysis and synthesis for video compression," in *SPIE Visual Communications and Image Processing, VCIP'91*, 1991, pp. 234–241.

[53] C.S. Choi and T. Takebe, "Analysis and synthesis of facial expressions in model-based image coding," in *Int. Picture Coding Symposium, PCS'90, Boston, MA, USA*, 1990.

[54] A. Koike, M. Kaneko, and Y. Hatori, "Model-based image coding with 3-d motion estimation and shape change detection," in *Int. Picture Coding Symposium, PCS'90, Boston, MA, USA*, 1990.

[55] K. Mase, "An application of optical flow - extraction of facial expressions," in *Proc. MVA'90, Tokyo, Japan*, 1990.

[56] K. Water and D. Terzopoulos, "Analysis of facial images using physical and anatomical models," in *Int. Conf. on Computer Vision, ICCV'90, Osaka, Japan*, 1990.

[57] T. Kanade, *Facial Action Coding System*, Consulting Psychologists Press, 1977.

Chapter 5

Video Object Plane Extraction and Tracking

5.1 Video Object Plane Extraction Techniques

The video and motion segmentation algorithms described in Section 1.6 are primarily focusing on coding. They segment video sequences into primitive regions that are homogeneous with respect to motion and possibly color or luminance (see Fig. 1.13). A small prediction error after motion-compensation is the indicator for a good segmentation. To support the content-based functionalities in MPEG-4, on the other hand, a decomposition into objects that are semantically meaningful to the human observer is required.

As was mentioned earlier, partitioning a video sequence into video object planes by means of automatic or semi-automatic segmentation is a very difficult task, and comparatively little research has been undertaken in this field. In fact, we are at the moment not aware of any algorithm that can automatically perform this segmentation accurately and reliably for generic video sequences. Semi-automatic techniques that get some input from humans, for example by tuning a few parameters, can significantly improve the segmentation result. Currently, this appears to be the most promising approach unless a very constrained situation is present.

An intrinsic problem of VOP generation is that objects of interest are not homogeneous with respect to low-level features such as color, intensity, or optical flow. Hence, conventional segmentation techniques, including the motion segmentation algorithms described in Section 1.6, will fail to obtain meaningful partitions.

Segmentation algorithms that specifically address video object plane generation have been proposed, many of them just recently with the development of the new video coding standard MPEG-4. Although VOP segmentation needs not to be standardized, it constitutes a crucial factor in the future success of MPEG-4 and has consequently been examined in the so-called *N2 Core Experiment on Automatic Segmentation Techniques* of ISO MPEG-4. There is also one section devoted to segmentation for VOP generation in Annex F of the committee draft of the MPEG-4 standard [1].

The most important cue exploited by a majority of VOP extraction techniques is motion, and therefore these algorithms must overcome the same problems associated with motion estimation as motion segmentation techniques. Consequently, the use of color, luminance, or spatial edges is strongly recommended to achieve high boundary accuracy. In fact, to extract an object from one sequence and to place it into another one requires a virtually perfect boundary location, because a human observer would immediately recognize any inaccuracies in the segmentation. So-called change detection masks or difference images and estimated motion fields are the most common forms of motion information incorporated into the segmentation process.

Temporal linking of VOPs in successive frames is another important issue that must be addressed. In an interactive multimedia application a user typically selects an object only once and would expect the application to recognize that object in subsequent frames.

Wang and Adelson [2, 3] proposed a layered representation of image sequences that corresponds to the VOP technique used by MPEG-4. They assume that regions undergoing a common affine motion (1.52) are part of the same physical object in the scene. Consequently, they group pixels that can be described by an affine transformation to layers or VOPs. The objective is to derive a single representative image for each layer that together with the affine motion parameters is sufficient to reconstruct the corresponding video object throughout the sequence. The algorithm starts by estimating an optical flow field, and then subdivides the frame into square blocks. The affine motion parameters are computed for each block by linear regression to get an initial set of motion models or hypotheses. The pixels are then grouped by an iterative adaptive K-means clustering algorithm. Pixel $\mathbf{x}$ is assigned to hypothesis or layer i if the difference between the optical flow at $\mathbf{x}$ and the flow vector synthesized from the affine parameters of layer i is smaller than for any other hypothesis. Obviously, this does not enforce spatial continuity of the label field. To construct the layers, information of a longer sequence is necessary. The frames are warped according to the affine motion of the layers such that coherently moving objects are aligned. A temporal median filter is then applied to the aligned sequence to obtain a single representative image for each object. This proposal has several shortcomings. For instance, it is not possible to represent an object which is rotating about it's axis so that different views are shown by a single image that is warped from frame to frame. Furthermore, the affine transformation (1.52) cannot describe the motion of a layer undergoing strongly non-rigid motion (for example a walking person). The algorithm also depends completely on the accuracy of the optical flow estimates since no color or intensity information is used. Finally, the layer construction process makes real-time execution impossible because a long sequence of frames is required.

A technique that closely follows [2, 3] was described by Torres *et al.* [4]. The main differences are the algorithm to estimate the dense motion field and the way similar motion models are merged.

Both color and motion are used by Zhong and Chang [5]. Primitive regions (see Fig. 1.13) that are coherent in color are obtained by an intra-frame segmentation technique. For that, adjoining regions for which the color distance is below a threshold are merged and small regions removed by a morphological opening-closing operator. These primitive regions can then be combined into VOPs based on spatial connectivity and motion similarity. Thus, motion is used to group low-level regions to semantically meaningful higher-level objects. Temporal continuity of the segmentation is achieved by initializing the intra-frame segmentation with the projected low-level partitions of the previous frame, whereby each region is assumed to follow the affine motion model (1.52). The model parameters are estimated from a dense correspondence vector field, which is obtained by hierarchical block matching [6]. Unfortunately, it is not further specified how motion similarity is defined and exploited in order to group primitive regions into VOPs. Nevertheless, this approach is interesting due to its clear distinction between different levels of segmentation.

The double partition approach based on morphology suggested by Marqués and Molina [7] also distinguishes between different levels of segmentation. The first partition contains primitive regions, while the second partition describes video objects as illustrated in Fig. 1.13. Initially, objects of interest have to be selected interactively leading to a partition at object level that corresponds to a decomposition into video object planes. These objects are normally not homogeneous in color or motion and are resegmented to obtain a fine partition that is spatially homogeneous. After estimating a dense motion field by block matching, the fine partition is projected onto the next frame using motion compensation. These projected regions are the starting point to extract the markers for the next frame, which is then segmented by the watershed algorithm based on luminance. To improve temporal stability, the segmentation process is guided by change detection masks that prevent markers of static areas from overgrowing into moving areas and vice versa. Finally, the new object level partition is computed from the projected and segmented fine partition. For that, the algorithm must keep track of the labels of each region to know the correspondence between fine regions and objects. This algorithm performs the high-level grouping of primitive regions in the fine partition to semantically meaningful objects with the help of user input. We will see later that such semi-automatic approaches are increasingly being investigated due to the difficulty of fully automatic segmentation techniques.

Garrido *et al.* [8] combine several features in a hierarchical way to obtain four different partitions or layers. The gray-level partition is obtained by merging flat zones of the original frame. Recall from Section 1.3.1 that a flat zone is the largest connected component in an image where the intensity is constant. The resulting regions are then merged based on an estimated motion field to arrive at the motion layer. Regions of the motion partition are grouped to obtain the depth layer by analyzing overlapping zones that appear during the motion-compensation step. The difficult step to the semantic layer, which is the highest level of the hierarchy, relies on the assumption that the image contains a face as the object to segment. Temporal continuity is accomplished by motion-compensating the partitions of the previous frame.

VOP extraction has not yet received the same attention as motion segmentation, which is reflected in the smaller number of proposals. Arguably the biggest contribution comes from the so-called ISO MPEG-4 *N2 Core Experiment on Automatic Segmentation Techniques.* In particular, three groups have been very active in the development of these segmentation methods: the University of Hannover (UH) in Germany [9, 10, 11, 12, 13], the Fondazione Ugo Bordoni (FUB) in Italy [14, 15, 16, 17, 18], and the Electronics and Telecommunications Research Institute (ETRI) in Korea [19, 20, 21, 22].

These techniques tackle VOP segmentation by formulating it as the problem of separating moving foreground objects from the background. For each area in the frame that is moving differently from the background one object is created. No assumption about the motion model of these objects is made. A foreground object is simply expected to exhibit high values for the frame difference FD (1.68). Obviously, parts of an object that are not moving or changing will be assigned to the background. Moreover, change detection masks mark uncovered background as changed, while the interior of objects remains unchanged unless it contains sufficient texture. The resulting holes inside objects must somehow be filled and uncovered background eliminated.

Mech and Wollborn [11] generate the video object planes or object masks from an estimated change detection mask (CDM). Initially, a change detection mask is generated by taking the difference between two successive frames using a global threshold. The boundaries of this CDM are then refined by an iterative relaxation technique that uses a locally adaptive threshold. Temporal stability is increased by incorporating a memory such that each pixel is labeled as changed if it belonged to an object in the previous frame and was at least once marked as changed in the last L change detection masks. A simplification step including a morphological closing removes small regions to obtain the final CDM. The object mask is calculated from the CDM by eliminating uncovered background. To this end, a correspondence vector field is calculated using hierarchical block-matching [6]. Pixels that belong to changed areas and for which the estimated displacement vector points outside the CDM are assigned to the uncovered background and are removed from the object mask. To improve the accuracy, boundaries in the object mask are adapted to gray-level edges. An extended version of this technique contains an additional global motion estimation and compensation step based on the eight-parameter model (1.54), a scene cut detector, and the memory length L has been made adaptive [9]. This algorithm is also part of the ISO MPEG-4 N2 Core Experiment [10].

Automatic segmentation is formulated by Neri *et al.* [14, 16] as the problem of separating moving objects from a static background. In the case of moving background, the frames must first be aligned by motion-compensation. Potential foreground regions are detected in a preliminary stage by applying a higher-order statistics (HOS) test to a group of inter-frame differences. The non-zero values in these difference frames are either due to noise or moving objects, with the noise being assumed to be Gaussian in contrast to the moving objects, which are highly structured. For all difference frames, the zero-lag fourth-order moments are calculated because of their ability to suppress Gaussian noise. These moments are then thresholded, resulting in a preliminary segmentation map containing moving objects and uncovered background. To identify uncovered background, the motion analysis stage calculates the displacement of pixels that are marked as changed. The displacement is estimated at different lags from the fourth-order moment maps by block matching. If the displacement of a pixel is zero for all lags, it is classified as background and otherwise as foreground. Finally, a regularization phase applies morphological opening and closing operators to achieve spatial continuity and to remove small holes inside moving objects of the segmentation map. The resulting segmented foreground objects are slightly too large, because the boundary location is not directly determined from the gray-level or edge image. A version of this technique is under investigation in the ISO MPEG-4 N2 Core Experiment on Automatic Segmentation Techniques [15].

Note that if the background is moving, the use of frame differences or change detection masks can cause some problems, because a perfect registration of two frames is rarely possible. Spatial edges in the background will then lead to highly structured non-Gaussian components of large values in the corresponding frame difference image.

While the two proposals [10, 15] to the ISO MPEG-4 N2 Core Experiment perform segmentation mainly based on temporal information, Choi *et al.* [19] presented a spatial morphological segmentation technique. It starts with a global motion estimation and compensation step. The global affine motion parameters (1.52) are calculated by linear regression from the correspondence field, which is obtained by a block matching algorithm. After examining the presence of a scene cut, the actual segmentation commences by simplifying the frame with a morphological opening-closing by reconstruction filter. The thresholded morphological gradient image, calculated from the luminance and chrominance components of the frame, serves as input for the watershed algorithm, which detects the location of the object boundaries. To avoid oversegmentation, regions smaller than a threshold are merged with their neighbors. Finally, a foreground/background decision is made to create the video object planes. Every region for which more than half of its pixels are marked as changed in a change detection mask is assigned to the foreground. To enforce temporal continuity, the segmentation is aligned with that of the previous frame and those regions for which a majority of pixels belonged to the foreground before are added to the foreground, too. This allows tracking an object even when it stops moving for an arbitrary time. In contrast, the techniques [14] and [9, 11] will lose track after a certain number of frames depending on the size of the group of frames and memory length, respectively.

Investigations into a combination of the spatial segmentation method [19] with the two temporal segmentation techniques [10, 15] to form one algorithm were initiated in [12]. Three proposals for such a combined scheme were then presented in [13, 20]. A straightforward approach selects those regions obtained by the spatial segmentation for which a majority of pixels correspond to the foreground mask computed using one of the temporal segmentation algorithms.

Many researchers have recognized that a fully automatic VOP segmentation is probably premature. The main difficulty is to formulate semantic concepts in a form suitable for a segmentation algorithm. In Fig. 1.13, this corresponds to the step from primitive regions to semantically meaningful video objects. User-assisted semi-automatic segmentation could provide a viable alternative to fully automatic versions [14, 17, 18, 21, 22, 23, 24].

In [21], a user initially has to select objects in the scene by manual segmentation. These objects are then tracked throughout the sequence. To this end, the partition of frame $(n-1)$ is first projected onto frame n. The motion of each video object is modeled by the affine transformation (1.52) with the parameters being calculated in the least squares sense from a displacement field. The latter is estimated by block matching. These projected regions serve as markers for a morphological segmentation, whereby pixels near boundaries as well as uncovered and overlapping zones are defined as uncertainty areas. The boundaries in uncertainty areas are then located using the watershed algorithm. A more realistic motion model was used in [22]. Every VOP now consists of several regions with an independent set of affine motion parameters each. However, these regions must also be initially selected by the user.

Toklu *et al.* represent each video object by a deformable 2-D triangular mesh [24]. A user must manually segment an object of interest, which is then fitted to a mesh. This mesh is tracked and updated in subsequent frames. As for the techniques in [21, 22], this method relies on a two-stage approach. Semantics is incorporated by means of a user-assisted initialization of the segmentation. The algorithm then knows what the semantically meaningful object looks like and is able to track it in the following frames.

An interesting semi-automatic technique was proposed by Gu and Lee [23]. It divides the task into user-assisted manual segmentation to initialize the partition and automatic tracking of the VOPs in the remaining frames similar to [21, 22, 24]. Thus, the concept of semantics is elegantly incorporated into the segmentation process. A user first selects interactively an object of interest. The specified object boundary is then dilated to make it thicker, and the correct boundary location within the resulting uncertainty band is determined using the morphological watershed procedure. Hence, the need for a precise VOP selection by the user is avoided. The obtained VOP is automatically tracked and updated in successive frames. For that, the VOP boundary is predicted from the previous partition using motion-compensation. The mapping of VOPs is modeled by the eight-parameter model (1.54), and the parameters are estimated by linear regression from an estimated dense motion field. To accommodate variations of the contour, the predicted VOP boundary is dilated again and the correct location is determined by the watershed algorithm. This method works well as long as the VOPs are limited to fairly rigid motion that can be described by only one set of motion parameters.

The usefulness of user interaction to incorporate high-level information that is intrinsic to VOPs has also been reported in [17]. The performance of the segmentation algorithm is improved by letting a user tune a few crucial parameters on a frame by frame basis. In [18], the user is in addition able to select an area containing the object of interest. This allows the algorithm to estimate critical parameters only on the region with the object instead of the whole image that might consist of several regions with different characteristics.

5.2 Outline of VOP Segmentation and Tracking Algorithm

The main point to be addressed by a VOP segmentation algorithm is how to detect semantically meaningful objects, which corresponds to the introduction of semantics into the segmentation process (see also Fig. 1.13). Motion is by far the most widely adopted cue to identify such physical objects. Therefore, another important issue regarding VOP segmentation is how to deal with the intrinsic inaccuracies of motion estimation.

In the remainder of this chapter, a description will be given of the VOP segmentation algorithm that was published in [25, 26, 27, 28]. The algorithm performs a separation of foreground objects from the background based on motion information. The main hypothesis underlying this approach is the existence of a dominant global motion that can be assigned to the background. Areas in the frame that do not follow this background motion then indicate the presence of independently moving physical objects. Note that in almost all practical cases the background is either still or its motion is smooth and can be described by a simple model such as the affine transformation (1.52).

The estimated flow field in Fig. 5.1 demonstrates that foreground objects can be characterized by a motion that is different from the dominant global background motion. However, it also becomes clear that color, intensity, or spatial edges must be included to accurately locate the object boundaries. Furthermore, Fig. 5.1 highlights the difficulty of grouping flow vectors to objects based on their similarity. Consequently, it is sometimes better to group flow vectors which show a significant deviation from the global motion

(a) (b)

Figure 5.1: (a) Original frame 39 of sequence *Alexis* and (b) estimated optical flow field using the Horn-Schunck method [29].

and assign them to foreground objects. In other words, departure from global motion is often the better characteristic to group flow vectors to VOPs than their similarity.

The key idea of the described VOP segmentation algorithm is to derive a two-dimensional binary model for the object of interest and to track it throughout the video sequence as illustrated in Figs. 5.2 and 5.3. The model points correspond to edge pixels that were detected by the Canny operator [30]. Notice that model points are not restricted to contours and can also include interior points of the objects.

No prior information about the objects in the scene is needed. The only inputs are the video signal and possibly some user assistance by tuning a few parameters. However, this technique expects, like most VOP segmentation algorithms, that physical objects move differently from the background.

Temporal linkage is established by an object tracker that matches the binary models against subsequent frames. The matching criterion employed is the Hausdorff distance [31, 32, 33], which is computationally efficient and very robust to noise and changes in shape. Tracking is crucial for VOP extraction because motion is the main cue for segmentation, and the algorithm must still be able to keep track of objects when they stop moving. Moreover, the user of an interactive multimedia application normally selects an object only once. It is then the task of the application to identify this object in subsequent frames.

While the best match found by the Hausdorff object tracker indicates

Figure 5.2: Two-Dimensional binary model sequence for the person in the sequence *Hall monitor*. These models, which consist of edge pixels detected by the Canny operator [30], will guide the VOP extraction.

the translation the object has undergone, a two-stage model update method accommodates rotations and changes in shape of the object. This updating technique also ensures continuity in time of the resulting VOPs. Finally, the binary models guide the actual VOP extraction. Since these model points correspond to spatial edges detected by the Canny operator, very accurate object boundaries can be obtained.

There exist two versions of this VOP segmentation algorithm, with both of them having proven to be efficient at extracting physical objects. They are very similar, but depending on the situation one or the other version performs better. The main difference is the object motion detection part, i.e., how semantically meaningful objects are detected. The version incorporating the morphological motion filter (Fig. 5.4) is more suitable for sequences involving little and fairly rigid motion, and the change detection mask based version (Fig. 5.32) is stronger at dealing with fast moving non-rigid objects.

5.3 Version I: Morphological Motion Filtering

The flowchart of the algorithm depicted in Fig. 5.4 contains five major functional blocks. These are global motion estimation, object motion detection, model initialization, model update, and VOP extraction.

Since physical objects are identified by locating areas that are moving differently from the background, the first step is to compute the background motion. This is the objective of the first block. For that purpose, the global motion is modeled by the six-parameter affine transformation (1.52)

Figure 5.3: A sequence of two-dimensional binary models for *Alexis*.

with the parameters being obtained using a robust least median of squares algorithm, which is applied to a dense motion field estimated by a standard technique [29, 6].

The morphological motion filter aims to detect motion which is distinct from that of the background, and thereby locates independently moving physical objects in the scene. Its output forms an integral part of both the model initialization stage and the model update.

The core of this VOP extraction technique is the object tracker that establishes temporal correspondence of objects throughout the video sequence. When an object is detected for the first time, its binary model must be initialized by selecting an appropriate set of edge pixels. Afterwards, the model has to be tracked by minimizing the Hausdorff distance and updated by a two-stage method. The latter takes account of both slowly and rapidly changing parts of the object. Again, the updated model consists of edge pixels that were detected by the Canny operator [30]. The actual VOP extraction is then guided by the binary model, whereby a boundary post-processor ensures high boundary accuracy.

There is an optional filter to remove stationary background. It will be described in Section 5.4.4 because it is more useful in sequences with fast moving objects.

5.3.1 Global Motion Estimation

5.3.1.1 Least Median of Squares (LMS) Estimation

For each frame, the VOP segmentation algorithm first computes a dense optical flow field and estimates the global motion parameters. In principal,

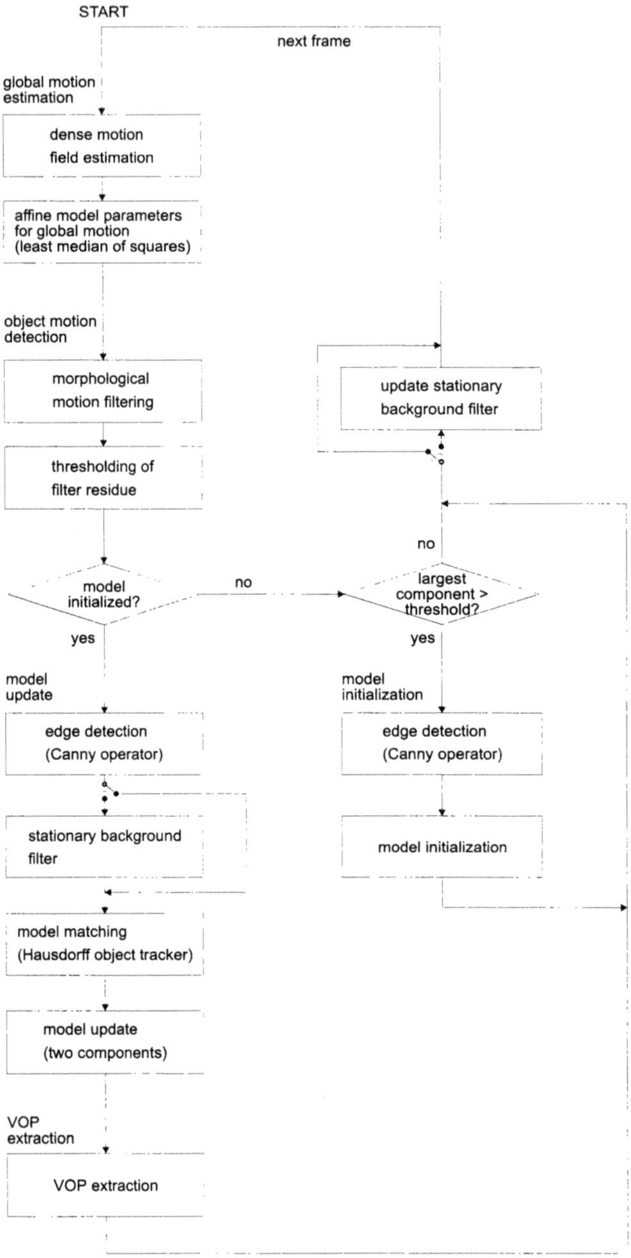

Figure 5.4: Flowchart of the VOP segmentation algorithm based on morphological motion filtering.

any of the non-parametric techniques in Section 1.5 serves the purpose, but the Horn-Schunck method [29] and hierarchical block matching [6] have proven to be particularly effective. The estimated dense motion field is then the starting point for calculating the global motion parameters.

In many cases, global motion is very simple and consists only of a pan and possibly zoom. Therefore, the six-parameter affine transformation (1.52) is normally sufficient to describe the global motion. The relation (1.52) is separable so that the parameter triples $A_x = (a_1, a_2, a_3)^T$ and $A_y = (a_4, a_5, a_6)^T$ can be found separately by regression. The following discussion will concentrate on the estimation of A_x, however, the same procedure also applies to A_y.

Each independent vector in the dense motion field provides one observation to obtain an estimate

$$\hat{A}_x = \begin{bmatrix} \hat{a}_1 \\ \hat{a}_2 \\ \hat{a}_3 \end{bmatrix} \tag{5.1}$$

of the unknown parameter vector $A_x = (a_1, a_2, a_3)^T$. Let x_i' be the dependent variable and x_i and y_i the independent variables of the ith observation. Note that given an optical flow vector (u, v) at pixel (x_i, y_i), x_i' is obtained by $x_i' = x_i + u$. The predicted value $\hat{x}_i'$ corresponding to the affine model is then given by

$$\hat{x}_i' = \hat{a}_1 x_i + \hat{a}_2 y_i + \hat{a}_3. \tag{5.2}$$

Further, the residual or error

$$e_i = x_i' - \hat{x}_i' \tag{5.3}$$

is defined as the difference between the observed and the predicted value.

Traditionally, the least squares (LS) method has been the most widely adopted technique to solve for the unknown parameters. It fits the model by minimizing the sum of the squared residuals

$$\hat{A}_x = \arg\{\min_{A_x} \sum_i e_i^2\}. \tag{5.4}$$

The lack of robustness against outliers is a major drawback of the LS method. Moreover, for global motion estimation in the presence of independently moving foreground objects we know that many motion vectors will not belong to the background. All of these factors will introduce errors into the resulting estimate $\hat{A}_x$; these errors will increase as the area covered by foreground objects increases. For instance, the optical flow vectors of the person in the foreground of Fig. 5.1 (b) are non-zero in contrast to those of the still background.

The least median of squares (LMS) method [34], on the other hand, does not suffer from these shortcomings. Its estimator is given by

$$\hat{A}_x = \arg\{\min_{A_x} \operatorname*{median}_i e_i^2\}. \tag{5.5}$$

While the least squares estimator (5.4) minimizes the sum of all residues, least median of squares only minimizes the median value of the residues. Therefore, the observations belonging to foreground objects do not affect the estimate $\hat{A}_x$ even for arbitrarily large errors e_i as long as they constitute less than 50% of the pixels. This makes least median of squares regression very suitable for global motion estimation.

5.3.1.2 Finding the LMS Estimate

The enormous popularity of the least squares (LS) method over the last two hundred years can partially be explained by its ease of computation. Unfortunately, no such simple solution is known for the LMS estimator. The approach described in [34] repeatedly draws subsamples of three observations. Each subsample leads to a system of three linear equations with three unknowns that is sufficient to obtain an estimate $\hat{A}_x$ using Gauss-Jordan elimination or LU decomposition [35]. With (5.2) and (5.3) it is then easy to calculate the value

$$\operatorname*{median}_i e_i^2 = \operatorname*{median}_i (x_i' - \hat{a}_1 x_i - \hat{a}_2 y_i - \hat{a}_3)^2. \tag{5.6}$$

The estimate $\hat{A}_x$ among all subsamples that yields the lowest value for the median (5.6) is our LMS estimate.

If n independent motion vectors are available, then there exist $\frac{n!}{(n-3)!3!}$ different subsamples of three observations. With one independent motion vector per pixel, this becomes $\frac{101376!}{(101376-3)!3!} \approx 1.7 \times 10^{14}$ for a CIF size image of 352×288 pixels. Instead of evaluating all these subsamples, which is computationally infeasible, only a small subset is considered. To this end, 1500 out of all possible subsamples are selected at random, as described in [34].

The actual LMS estimation is computed on standardized data. This is a common procedure to avoid numerical inaccuracies caused by different units of measurement. The standardization is carried out by transforming the observations according to [34]

$$x_{i,std} = \frac{x_i - \operatorname*{median}_k x_k}{1.4826 \cdot \operatorname*{median}_l |x_l - \operatorname*{median}_k x_k|} \tag{5.7}$$

$$y_{i,std} = \frac{y_i - \operatorname*{median}_k y_k}{1.4826 \cdot \operatorname*{median}_l |y_l - \operatorname*{median}_k y_k|} \tag{5.8}$$

$$x'_{i,std} = \frac{x'_i - \operatorname*{median}_k x'_k}{1.4826 \cdot \operatorname*{median}_l |x'_l - \operatorname*{median}_k x'_k|} \tag{5.9}$$

where $x_{i,std}$, $y_{i,std}$, and $x'_{i,std}$ are the respective standardized values of x_i, y_i, and x'_i.

The LMS estimator applied to the standardized data returns a parameter vector $\hat{A}_{x,std}$ for which

$$\hat{x}'_{i,std} = \hat{a}_{1,std} \cdot x_{i,std} + \hat{a}_{2,std} \cdot y_{i,std} + \hat{a}_{3,std}. \tag{5.10}$$

To obtain $\hat{A}_x$ from $\hat{A}_{x,std}$, an inverse transformation must be performed. Let x_{med}, y_{med}, and x'_{med} denote the median values for x, y, and x', respectively, calculated over all observations. By substituting (5.7), (5.8), and (5.9) into (5.10) and comparing the coefficients with (5.2) we finally arrive at

$$\begin{aligned}
\hat{a}_1 &= \hat{a}_{1,std} \cdot \frac{\operatorname*{median}_l |x'_l - x'_{med}|}{\operatorname*{median}_l |x_l - x_{med}|} \\
\hat{a}_2 &= \hat{a}_{2,std} \cdot \frac{\operatorname*{median}_l |x'_l - x'_{med}|}{\operatorname*{median}_l |y_l - y_{med}|} \\
\hat{a}_3 &= x'_{med} + 1.4826 \cdot \operatorname*{median}_l |x'_l - x'_{med}| \cdot \hat{a}_{3,std} - \hat{a}_1 \cdot x_{med} - \hat{a}_2 \cdot y_{med}
\end{aligned} \tag{5.11}$$

5.3.2 Object Motion Detection Using Morphological Motion Filtering

After calculating the global motion, the object motion detection block illustrated in Fig. 5.4 identifies objects that are moving differently from the background. The major work of this block is performed by the morphological motion filter, which removes components that do not follow the dominant global motion while perfectly preserving other parts of the image.

In fact, the filtering process has to be carried out twice. In the first run, dark components are removed and in the second run bright components are removed. Each run consists of three steps: representation of the image by an appropriate tree structure, filtering of the image by pruning the tree, and transformation of the pruned tree back into an image. The resulting filter achieves comparatively accurate object boundary locations because of the incorporated gray-level information.

5.3.2.1 Connected Operators

The morphological motion filter belongs to a class of morphological operators called connected operators. Recall from Section 1.3.1 that a gray-level connected operator Ψ is an operator such that the partition of flat zones of an image I is finer than the partition of flat zones of $\Psi(I)$.

Generally speaking, connected operators merge flat zones according to a specified criterion, and so they do not create any new contours. The merging process is controlled by a filtering criterion that in our case determines how well a flat zone follows the global motion. Such motion-oriented filters were originally proposed in [36, 37, 38].

5.3.2.2 Max-Tree Representation

As mentioned above, motion filtering is performed by pruning a tree representing the image. The information contained in this tree is equivalent to that of the image and would be sufficient to reconstruct the image. However, the tree will not be transformed back into a gray-level image until it has been pruned according to the specified motion criterion. In the following we will describe the construction of the so-called Max-Tree, which allows the elimination of bright components moving differently from the global motion. The dual Min-Tree for removing dark components can be created in the same way, as will be shown later.

The Max-Tree is recursively generated by considering thresholded versions of the image at all gray-levels. The three-gray-level image of size 8×5 in Fig. 5.5 (a) consists of nine flat zones $Z_1, \ldots, Z_9$ as illustrated in Fig. 5.5 (b). Firstly, all flat zones at the lowest level 0 are assigned to the root, in this example $C_0^1 = \{Z_2, Z_6\}$. Following the notation in [37, 38], C_h^k refers to tree node k at level h. Each connected component of flat zones with gray-level higher than 0 forms one child node of the root in the tree. From Fig. 5.5 (c) it follows that there are two such components leading to the child nodes $\{Z_1, Z_3, Z_4, Z_5, Z_7\}$ and $\{Z_8, Z_9\}$ shown in Fig. 5.5 (d).

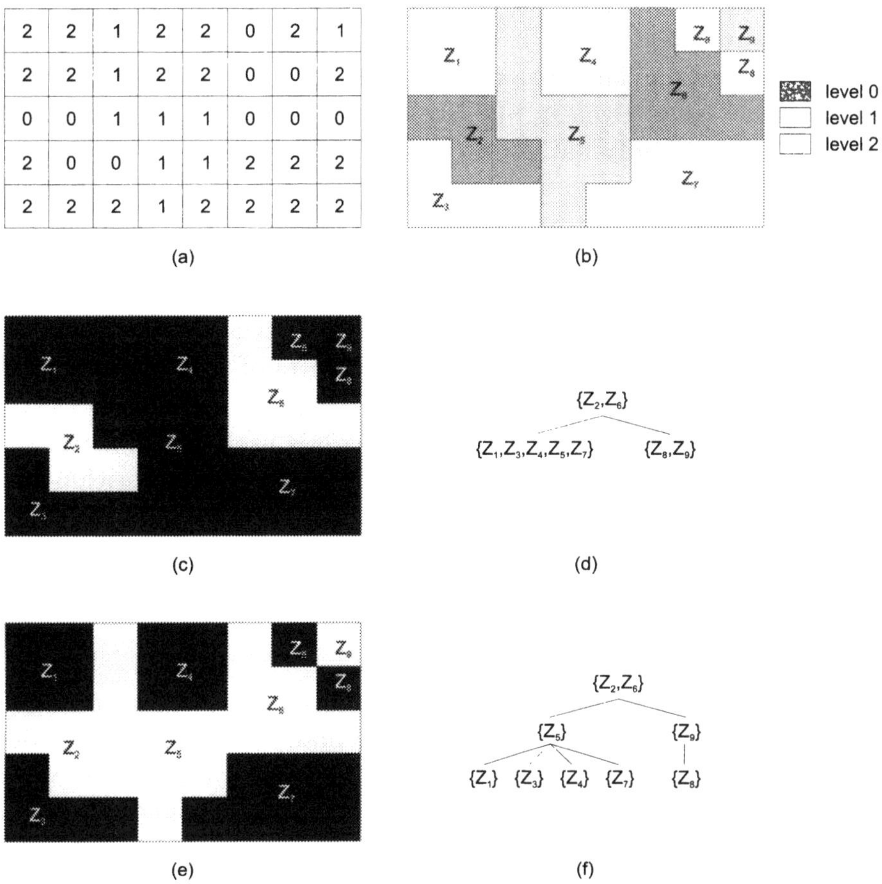

Figure 5.5: Creation of Max-Tree. (a) Original 8×5 image consisting of the three gray-levels 0, 1, and 2. (b) Corresponding partition of flat zones, resulting in nine components or zones. (c) The two components Z_2 and Z_6 (white) at the lowest level 0 are assigned to the root, whereas the other flat zones (black) form two connected components. These are assigned to two separate child nodes of the root in (d). (e) shows the thresholded partition of flat zones at the next higher level and (f) contains the final Max-Tree representing the image of (a).

At the next higher gray-level 1 there are five connected components left (see Fig. 5.5 (e)), for which new nodes are created. These are $C_2^1 = \{Z_1\}$, $C_2^2 = \{Z_3\}$, $C_2^3 = \{Z_4\}$, $C_2^4 = \{Z_7\}$, and $C_2^5 = \{Z_8\}$. The parent node of the new nodes C_2^1, C_2^2, C_2^3, and C_2^4 is $C_1^1 = \{Z_5\}$, because Z_1, Z_3, Z_4, and Z_7 belonged to that node at the previous level in Fig. 5.5 (d). For the same reason the parent node of C_2^5 is $C_1^2 = \{Z_9\}$.

Since there are no flat zones with gray-level higher than the next level 2, the final Max-Tree is given in Fig. 5.5 (f). Note that in the final Max-Tree each node contains only flat zones having the same gray-level. Moreover, the level in the tree represents the corresponding gray value and is sufficient to transform the tree back into an image. The name Max-Tree stems from the fact that the gray-level is increasing as we move from the root towards the leaves with the maxima being in the leaf nodes. There exists a dual Min-Tree with the leaves containing the minima. It is generated in exactly the same way by using $-I(x, y)$ for the gray-level of pixel (x, y) instead of $I(x, y)$.

The construction procedure described here is useful for illustrating the properties of the Max-Tree. However, the tree creation algorithm in [38], which relies on FIFO queues, is more efficient in practical applications and does not need explicit thresholding of the image.

5.3.2.3 Filter Criterion

Once an image is represented by its Max-Tree, the pruning process can begin. To this end, a criterion $M(C_h^k)$ for node C_h^k must be specified to decide whether C_h^k has to be removed or preserved. In the case where it is removed, all pixels of the node C_h^k and all its descendant nodes will be assigned to C_h^k's parent node.

Consider, for instance, the partition of flat zones in Fig. 5.6 (a) and its Max-Tree representation. Assume that according to some criterion the tree must be pruned as marked by the crosses (x). The flat zones Z_8 and Z_9 will then be merged with the root node, whereas Z_7 will join the node containing Z_5 as shown in Fig. 5.6 (b). To transform the pruned tree back into an image, we have to assign each flat zone the gray-level corresponding to the level in the tree. As a result, Z_8 and Z_9 have the new gray value 0 of the root and Z_7 takes on 1 like Z_5.

The remaining task is to find a suitable criterion that describes the deviation from the global motion. The average value for the DFD (1.61) was proposed in [36, 37, 38]. Objects or parts thereof that are well compensated by the global motion are expected to have smaller values for the DFD than

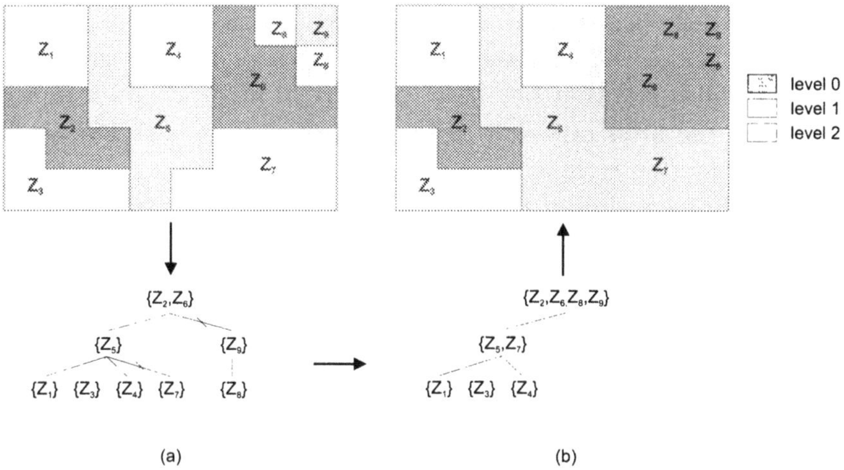

Figure 5.6: Filtering by pruning the Max-Tree. (a) Original partition of flat zones and corresponding Max-Tree. The crosses (x) mark where the tree has to be pruned. (b) Filtered image after pruning. To obtain the filtered image, each pixel was assigned the gray-level h of the node C_h^k it belongs to.

those that move differently. The pruning process then terminates when all nodes are sufficiently well motion-compensated by the global motion.

Here, we will employ a different criterion that takes the difference between synthesized global motion and estimated local motion. As part of the prior global motion estimation step both the dense motion field and the affine parameters of the global motion were estimated. Let $(p(x,y), q(x,y))$ be the estimated local displacement vector at pixel (x,y) in the dense field. Further, $(\hat{p}(x,y), \hat{q}(x,y))$ denotes the displacement vector at (x,y) synthesized according to the affine global motion model

$$\begin{aligned} \hat{p}(x,y) &= \hat{x}' - x = (\hat{a}_1 - 1)x + \hat{a}_2 y + \hat{a}_3 \\ \hat{q}(x,y) &= \hat{y}' - y = \hat{a}_4 x + (\hat{a}_5 - 1)y + \hat{a}_6, \end{aligned} \tag{5.12}$$

whereby $\hat{a}_i$ $(1 \leq i \leq 6)$ are the parameters estimated in the global motion estimation stage (see Section 5.3.1).

The motion criterion for the morphological motion filter to measure the deviation of the estimated local motion from the synthesized global motion is then given by

$$M(x,y) = (p(x,y) - \hat{p}(x,y))^2 + (q(x,y) - \hat{q}(x,y))^2. \tag{5.13}$$

$M(x, y)$ is low for background pixels that conform with the global motion and high for pixels belonging to independently moving objects. The morphological filter is based on a tree structure and requires a criterion for nodes. Therefore, $M(C_h^k)$ for the tree node C_h^k is defined as the average of $M(x, y)$ over all pixels that belong to C_h^k and all its descendant nodes.

Note that the filter criterion (5.13) is fairly robust with respect to the quality of the motion estimation, because pixels within the same object are not required to have similar motion vectors. The flow vectors only have to be different from the global motion.

An important issue regarding the selection of a filter criterion is increasingness. Most classical criteria are increasing, which means that if $C_{h_1}^{k_1}$ is a child node of $C_{h_2}^{k_2}$, then $M(C_{h_1}^{k_1}) \leq M(C_{h_2}^{k_2})$. The biggest advantage of increasing criteria is the well defined location where the tree must be pruned. Consider, for instance, the criterion defined as the number of pixels belonging to node C_h^k and all its descendant nodes. When we move from a leaf node towards the root, the criterion steadily increases until the specified threshold for pruning is reached. This position is easily found, because the value of the criterion would only be further increased by moving even closer to the root.

Motion criteria like (5.13) or the ones reported in [36, 37, 38], on the other hand, are non-increasing. This makes it much harder to decide where to prune the tree. The criterion can both increase and decrease along the path from a leaf node to the root. As a result, the value for the criterion might fluctuate around the specified threshold. In [36] it was suggested to apply a median filter to the criterion sequence to reduce these fluctuations. A more elegant solution to this problem is the Viterbi algorithm proposed in [37, 38].

5.3.2.4 Viterbi Algorithm

The basic idea of using the Viterbi algorithm [39] is to assign a cost to each possible decision for a node. The goal is then to find the paths of lowest cost running from the leaves to the root. Fig. 5.7 shows part of a single branch of the Max-Tree with the corresponding trellis. For a particular node C_h^k there exist two choices: preserve or remove. A branch that is pruned at node C_h^k will have all pixels belonging to C_h^k and all its descendant nodes assigned to the parent node of C_h^k. This is the same with real trees where you cannot prune a branch while keeping the leaves. Consequently, there is no transition from a preserve state to a remove state in Fig. 5.7.

The costs assigned to preserving and removing C_h^k are $M(C_h^k) - \lambda$ and

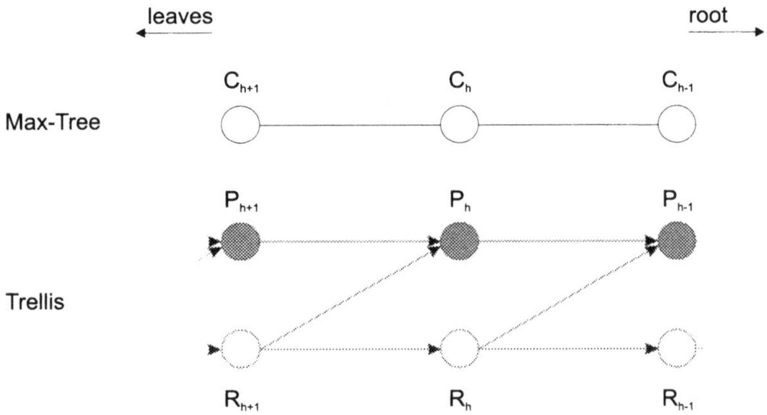

Figure 5.7: Trellis for a single branch of the Max-Tree. Note that there is no transition from the preserve state to the remove state.

$\lambda - M(C_h^k)$, respectively, where λ is a specified threshold. More specifically, the former cost applies to transitions going to a preserve node and the latter to transitions going to a remove node. Assume that we wish to remove node C_h^k if $M(C_h^k) > \lambda$. If $M(C_h^k) > \lambda$, we have a positive cost $M(C_h^k) - \lambda$ for preserving and a negative cost $\lambda - M(C_h^k)$ for removing. This obviously favors removal, which is exactly what we want.

A strength of the Viterbi algorithm is that all decisions can be made locally. Suppose we know the paths of lowest cost ending at P_{h+1} and R_{h+1}, denoted by $Path_{h+1}^P$ and $Path_{h+1}^R$ (see Fig. 5.7). The optimum paths ending at P_h and R_h are then given by the following simple rule (Note that the cost of going to the preserve node P_h is the same for transitions originating from P_{h+1} and R_{h+1}.)

optimum path ending at P_h:
`If` $\quad Cost(Path_{h+1}^P) \leq Cost(Path_{h+1}^R)$
`then` $\quad Path_h^P = (Path_{h+1}^P) \cup \{P_{h+1} \rightarrow P_h\}$
`else` $\quad Path_h^P = (Path_{h+1}^R) \cup \{R_{h+1} \rightarrow P_h\}$
optimum path ending at R_h:
$Path_h^R = (Path_{h+1}^R) \cup \{R_{h+1} \rightarrow R_h\}$

The corresponding cost functions $Cost(Path_h^P)$ and $Cost(Path_h^R)$ are updated according to

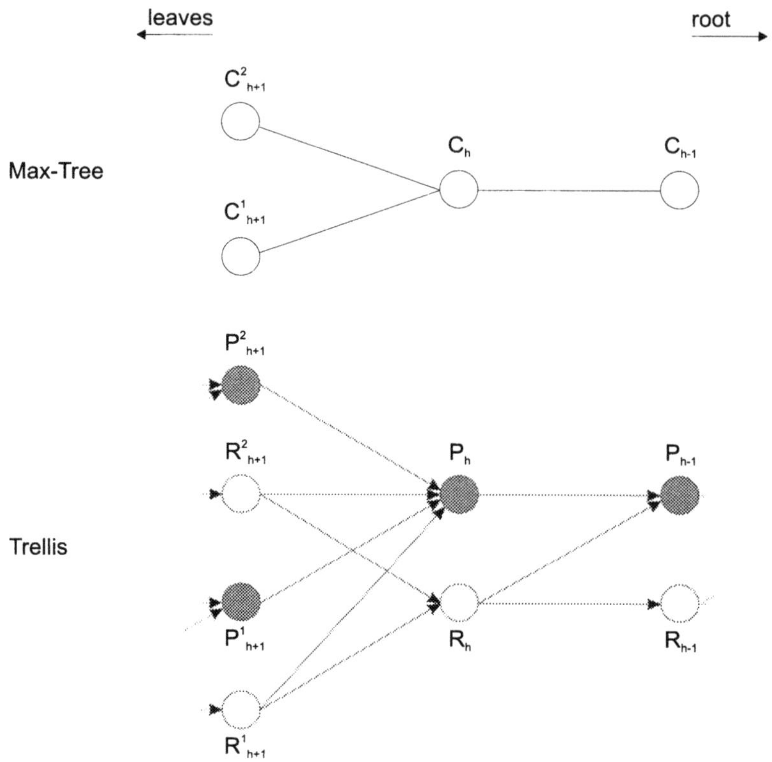

Figure 5.8: Trellis for a junction of the Max-Tree.

cost of path ending at P_h:
$Cost(Path_h^P) = \min\{Cost(Path_{h+1}^P), Cost(Path_{h+1}^R)\} + M(C_h^k) - \lambda$
cost of path ending at R_h:
$Cost(Path_h^R) = Cost(Path_{h+1}^R) + \lambda - M(C_h^k)$.

Along the paths from leaf nodes to the root there will normally be some junctions as illustrated in Fig. 5.8. These junctions only require a slight modification of the rules above due to the independence of the subbranches that are joined. The modified rules are

optimum path ending at P_h (junction):
`If` $Cost(Path_{h+1}^{P1}) \leq Cost(Path_{h+1}^{R^1})$
`then` $TempPath_h^P = (Path_{h+1}^{P^1}) \cup \{P_{h+1}^1 \rightarrow P_h\}$
`else` $TempPath_h^P = (Path_{h+1}^{R^1}) \cup \{R_{h+1}^1 \rightarrow P_h\}$
`If` $Cost(Path_{h+1}^{P^2}) \leq Cost(Path_{h+1}^{R^2})$

$$\begin{array}{l}
\texttt{then}\quad Path_h^P = TempPath_h^P \cup (Path_{h+1}^{P^2}) \cup \{P_{h+1}^2 \rightarrow P_h\} \\
\texttt{else}\quad Path_h^P = TempPath_h^P \cup (Path_{h+1}^{R^2}) \cup \{R_{h+1}^2 \rightarrow P_h\} \\
\textbf{optimum path ending at } R_h \textbf{ (junction):} \\
Path_h^R = (Path_{h+1}^{R^1}) \cup \{R_{h+1}^1 \rightarrow R_h\} \cup (Path_{h+1}^{R^2}) \cup \{R_{h+1}^2 \rightarrow R_h\}
\end{array}$$

and the cost functions must be updated by

$$\begin{array}{l}
\textbf{cost of path ending at } P_h \textbf{ (junction):} \\
Cost(Path_h^P) = \min\{Cost(Path_{h+1}^{P^1}), Cost(Path_{h+1}^{R^1})\} \\
\qquad\qquad + \min\{Cost(Path_{h+1}^{P^2}), Cost(Path_{h+1}^{R^2})\} + M(C_h^k) - \lambda \\
\textbf{cost of path ending at } R_h \textbf{ (junction):} \\
Cost(Path_h^R) = Cost(Path_{h+1}^{R^1}) + Cost(Path_{h+1}^{R^2}) + \lambda - M(C_h^k).
\end{array}$$

As can be seen, the subbranches are independent and their costs can simply be added up. Hence, an extension to junctions with more than two branches is straightforward.

These rules are applied to all nodes in the Max-Tree until reaching the root, which only allows a preserve decision to ensure that the whole tree is not removed. The paths of lowest cost computed by this procedure then uniquely define where the Max-Tree must be pruned. After pruning the tree, the corresponding filtered image is obtained by assigning to each pixel (x, y) the gray value h, whereby C_h^k is the node in the pruned tree to which (x, y) belongs to (see Fig. 5.6).

5.3.2.5 Detection of Moving Objects

To summarize, the morphological motion filter described so far consists of three major steps. These are Max-Tree creation, pruning using the motion criterion (5.13), and transformation of the tree back into an image. This filter or operator $\Psi(I)$ accomplishes a simplification of the image I by removing bright components that move differently from the background. Naturally, there exists a dual operator $\Psi^*(I) = -\Psi(-I)$ acting on the so-called Min-Tree representation with the same filtering effect but for dark components. Thresholding the residue or difference between original and filtered image, $I - \Psi^*(\Psi(I))$, exposes *moving components* that do not follow the background motion. These moving components will later be used for the model initialization and updates.

Fig. 5.9 shows the output of the morphological motion filter with the criterion (5.13), the filter residue, and the moving component for a frame of the *Coastguard* sequence. The camera is following the boat so that the background appears to be moving while the boat is still. This is reflected

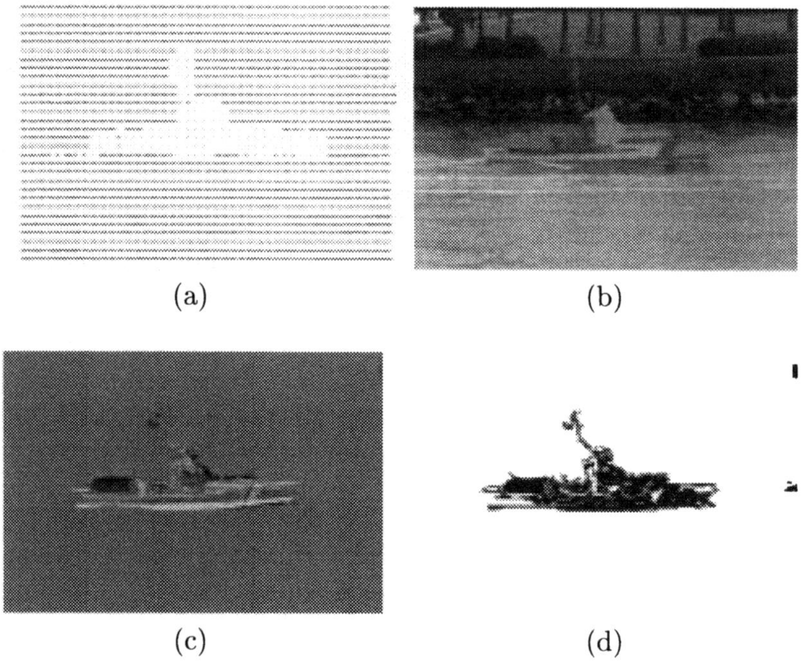

Figure 5.9: (a) Correspondence vector field estimated by block matching for frame 233 of the sequence *Coastguard*, (b) output of the morphological motion filter with $\lambda = 1.5$, (c) residue $I - \Psi^*(\Psi(I))$, and (d) corresponding moving component derived by thresholding the residue ($T = 10$)).

in the correspondence vector field in Fig. 5.9 (a), which was estimated by full-search block matching using blocks of 12×12 pixels and a 17×17 search window allowing for a maximum displacement of ± 5 pixels. Since a majority of pixels are undergoing horizontal translation, the affine global model describes that motion. As a result, it is the boat that has been correctly detected as "moving" object.

Alexis (Fig. 5.10) is a typical head-and-shoulder sequence characterized by still background and only very little motion of the person. To obtain reasonable results from the motion estimation, frames had to be skipped so that frames n and $n-5$ served as inputs for the Horn-Schunck method [29]. This was also necessary for the sequence *Akiyo* in Fig. 5.11. Note that grouping the flow vectors in Fig. 5.10 (a) into regions based on their similarity would be a very difficult task. In addition, the motion boundaries exhibited by

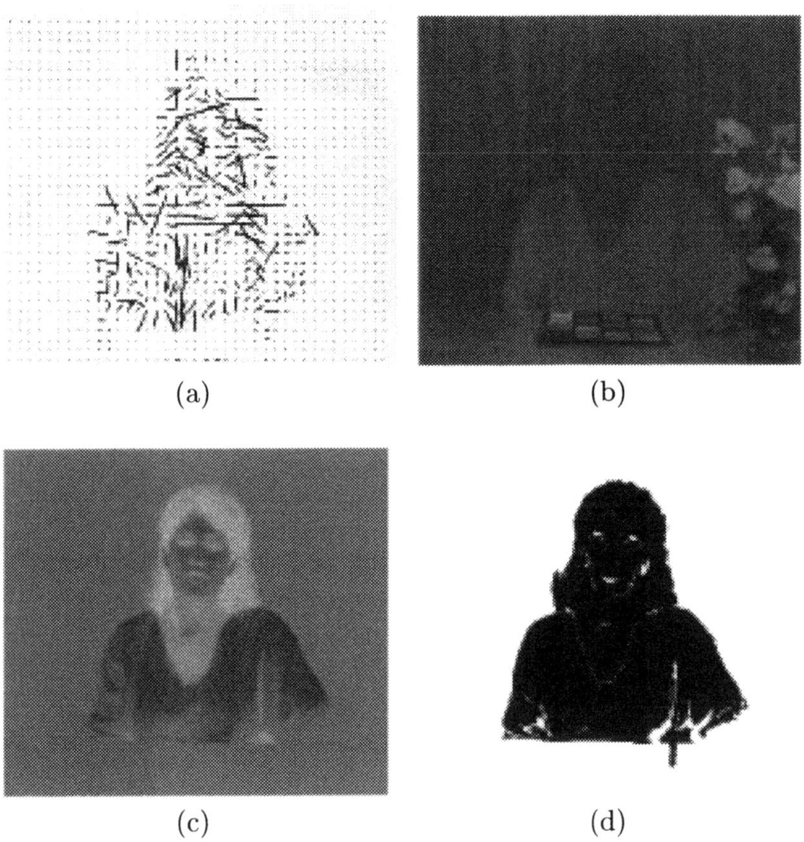

Figure 5.10: (a) Optical flow field estimated by method [29] for frame 40 of the sequence *Alexis*, (b) output of the morphological motion filter with $\lambda = 1.6$, (c) residue $I - \Psi^*(\Psi(I))$, and (d) corresponding moving component $(T = 1)$.

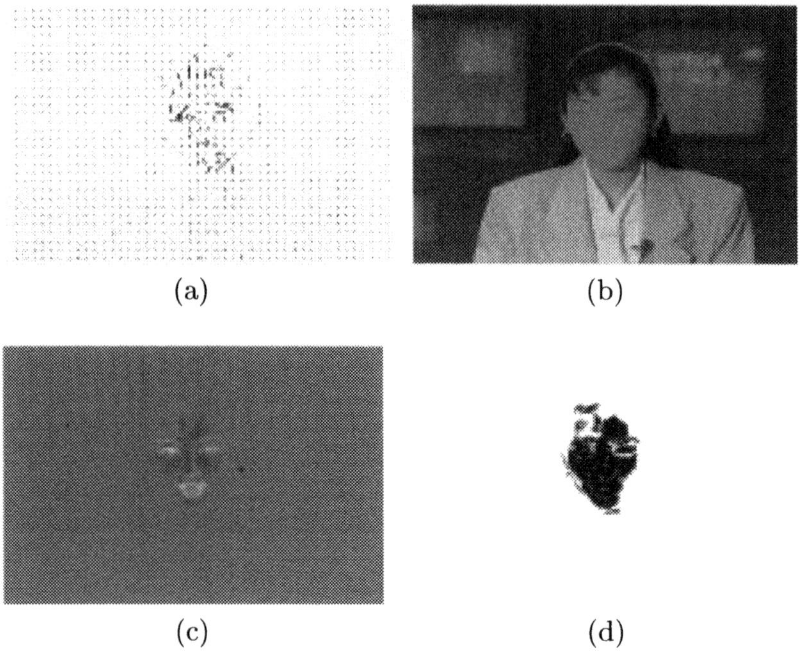

Figure 5.11: (a) Optical flow field estimated using method [29] for frame 18 of the sequence *Akiyo*, (b) output of the morphological motion filter with $\lambda = 0.4$, (c) difference between filtered and original frame $I - \Psi^*(\Psi(I))$, and (d) corresponding moving component ($T = 1$).

the flow field are far less accurate than those in Fig. 5.10 (d). This demonstrates the strength of the morphological motion filter at extracting moving objects.

Sometimes the motion of objects is not strong enough to locate them. In head-and-shoulder sequences, for example, there is often not sufficient motion of the person's body to classify it as moving. Despite the lower value $\lambda = 0.4$ for the threshold in Fig. 5.11 compared to $\lambda = 1.5$ and $\lambda = 1.6$ for *Coastguard* and *Alexis*, only the face of *Akiyo* could be detected. This highlights the need for temporal integration of moving components. If parts of an object only move occasionally, it is crucial that the VOP segmentation algorithm is capable of identifying these parts in successive frames to enforce temporal continuity. This will be achieved by deriving binary models for the objects to track, as explained shortly.

All in all, the morphological motion filter with the criterion (5.13) has proven to be most effective in the case of rather slow object motion. In Section 5.4, a different method to detect moving objects more suitable for faster motion will be presented.

5.3.3 Model Initialization

Initially, the position of objects is unknown and no models exist. With the assumption that physical objects are characterized by a different motion from that of the background, the moving components obtained by the morphological motion filter can be used for the model initialization. The models consist of an appropriate set of edge pixels detected by the Canny operator [30], and therefore edge detection constitutes the first step of the model initialization as illustrated in Fig. 5.4.

The binary model of a moving object is then initialized by selecting all edge pixels that belong to the moving component. However, sufficient evidence for a moving object is necessary to increase robustness and to avoid tracking noise. This is accomplished by requiring the moving components to have a specified minimum size (for example 2000 pixels for 352×288 CIF size video sequences).

Fig. 5.12 and Fig. 5.13 demonstrate how the initial binary model is extracted from the edge image using the moving component. The model for the boat in Fig. 5.12 is already a good representative in contrast to the model in Fig. 5.13. The morphological motion filter can only detect components which are undergoing some motion relative to the background. Parts of an object that do not move will be assigned to the background. Accordingly, it might take several frames to pick up the whole object. This is achieved by the model update described in Section 5.3.5 and can be considered as a temporal integration enforcing temporal continuity.

5.3.4 Object Tracking Using the Hausdorff Distance

Temporal continuity and linking are two crucial aspects of VOP segmentation. The former enforces the shape of the extracted VOPs to be a smooth function of time, whereas the latter ensures that objects do not get lost even when they stop moving and enables applications to identify objects in subsequent frames. These functions are provided by the model update block (see Fig. 5.4) and in particular by the model matching block. Matching is performed on edge images because it is computationally efficient and fairly insensitive to changes in illumination. It is also a natural choice given that objects are represented by binary models.

(a) (b) (c)

Figure 5.12: (a) Edge image computed by Canny operator [30], (b) moving component obtained by morphological motion filter, and (c) initial model for the sequence *Coastguard*.

(a) (b) (c)

Figure 5.13: (a) Output of Canny edge detector [30], (b) moving component obtained by morphological motion filter, and (c) initial model for the sequence *Akiyo*.

The binary model of the tracked object is matched against subsequent frames by using the Hausdorff distance [31, 32, 33] as matching criterion. The Hausdorff distance is a simple but powerful measure for comparing binary images and can be efficiently implemented using a distance transformation [40]. It is also very robust when objects are partially occluded, rotated or change their shape.

5.3.4.1 Hausdorff Distance

The Hausdorff distance was proposed in [32] as a measure to compare binary images or portions thereof. Let $O_q = \{o_1, \ldots, o_m\}$ denote the set of binary model points of the object to track for frame q where m is the number of model points. Similarly, we define $E_{q+1} = \{e_1, \ldots, e_n\}$ as the set of all edge pixels detected by the Canny operator in the whole image of frame $q+1$. Note that O_q is the representative of the object for frame q, and our

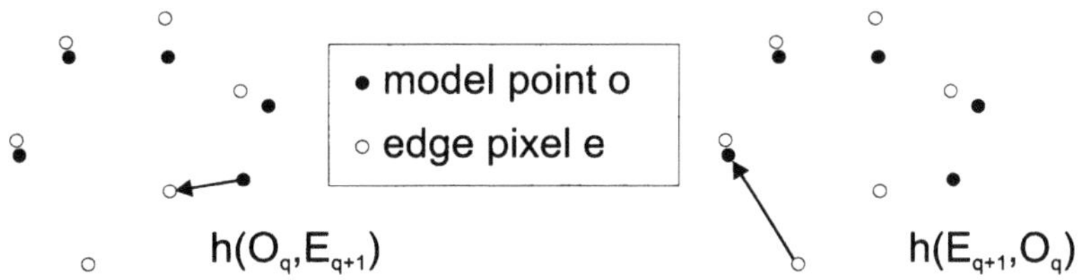

Figure 5.14: Definition of Hausdorff distance: (a) $h(O_q, E_{q+1})$ measures the maximum distance of a model point to the nearest edge pixel, and (b) $h(E_{q+1}, O_q)$ measures the maximum distance of an edge pixel to the nearest model point. In this example, $h(E_{q+1}, O_q)$ is larger than $h(O_q, E_{q+1})$ and therefore the Hausdorff distance is equal to $h(E_{q+1}, O_q)$.

goal is to find its position in the next frame $q+1$. The Hausdorff distance $H(O_q, E_{q+1})$ between O_q and E_{q+1} is then given by

$$H(O_q, E_{q+1}) = \max\{h(O_q, E_{q+1}), h(E_{q+1}, O_q)\} \tag{5.14}$$

with

$$h(O_q, E_{q+1}) = \max_{o \in O_q} \min_{e \in E_{q+1}} \|o - e\| \tag{5.15}$$

and

$$h(E_{q+1}, O_q) = \max_{e \in E_{q+1}} \min_{o \in O_q} \|e - o\|, \tag{5.16}$$

which is illustrated in Fig. 5.14.

Thus, for every model point $o \in O_q$ the distance to the nearest edge pixel $e \in E_{q+1}$ is calculated, and the maximum value is assigned to $h(O_q, E_{q+1})$. Then, for each edge pixel $e \in E_{q+1}$ the distance to the nearest model point $o \in O_q$ is computed, and $h(E_{q+1}, O_q)$ is set to the maximum distance. Finally, the Hausdorff distance is the larger of the two maxima. It is easy to see that for $h(O_q, E_{q+1}) = d$ every model point must be within distance d of some point in E_{q+1}.

A shortcoming of the definitions in (5.15) and (5.16) is the large impact that outliers have, because one outlying model point or edge pixel will lead to a large Hausdorff distance even if all other points perfectly match. Therefore, it is preferable to use the generalized Hausdorff distance [31, 32] as it does not suffer from this problem. Instead of using the maximum value in (5.15), the distances are sorted in ascending order and the k-th value is chosen:

$$h_k(O_q, E_{q+1}) = \underset{o \in O_q}{k\text{th}} \min_{e \in E_{q+1}} \|o - e\|. \tag{5.17}$$

Equation (5.17) is equivalent to (5.15) for $k = m$. In the case of $k < m$, there may be $m - k$ outlying points without influencing the Hausdorff distance. This is also a very useful property when dealing with objects that are partially occluded or rapidly changing their shape where a good match for all points is not possible and not needed. Similarly, $h_l(E_{q+1}, O_q)$ is defined as the l-th value of the ordered distances

$$h_l(E_{q+1}, O_q) = \underset{e \in E_{q+1}}{l\text{th}} \min_{o \in O_q} \|e - o\|. \tag{5.18}$$

With the parameters k and l we can essentially choose how many model points have to be near edge pixels and vice versa. A reasonable choice for a wide range of sequences is $k \approx 0.6m$ and $l \approx 0.6n$, where m and n are the number of model points in O_q and the number of edge pixels in E_{q+1}, respectively. Unlike some other matching techniques, the Hausdorff distance does not require point correspondences between model points and edge pixels because it automatically selects the k (or l) best matching points.

The best match is now found by minimizing the Hausdorff distance between the edge image E_{q+1} and the object model O_q for all translations of the model relative to the edge image. The smallest distance indicates the new position of the tracked object in the next frame. Fig. 5.15 shows the best match according to the Hausdorff distance with the model of the previous frame shifted to the position of the best match and superimposed on the current frame. Despite the change in shape due to the differently moving right leg, the match is very accurate.

Finally, the computation of the Hausdorff distance can be efficiently implemented using the Chamfer distance transformation [40] and early scan termination [31, 32], as we will see now.

5.3.4.2 The Chamfer Distance Transformation

As can be seen from equations (5.17) and (5.18), to calculate the Hausdorff distance it is necessary to know for each model point the distance to the nearest edge pixel and vice versa. Here, an efficient solution to this problem based on the Chamfer distance transformation [31, 32, 40] is outlined. Only the calculation of the distance to the nearest edge pixel is considered,

Figure 5.15: The model of the person for frame 40 (black) of the sequence *Hall monitor* is shifted to the position of best match in frame 41. The match is accurate except for the right leg, which was moving differently from the rest of the person.

although the same method can be used to obtain the smallest distance to the nearest model point.

The distance to the nearest edge pixel is obviously 0 for edge pixels, while their horizontal and vertical neighbors, if not edge pixels themselves, have a distance of 1. For diagonal neighbors the corresponding distance is $\sqrt{2}$ unless they or their horizontal or vertical neighbors are edge pixels.

Unfortunately, computing these distances is a global operation and computationally prohibitive. An algorithm that operates locally and approximates the Euclidean distance well enough is described in [40]. This so called *distance transformation (DT)* defines small masks containing integer approximations of distances in a small neighborhood. There are two such masks suggested, Chamfer 3-4 and Chamfer 5-7-11 (see Fig. 5.16). The horizontal and vertical distances for Chamfer 3-4 are 3 and the diagonal one is 4. This gives a ratio of 1.333 compared to 1.414 for Euclidean distances.

The DT is initialized by assigning zero to edge pixels and infinity or a suitably large number to non-edge pixels. In two iterations the distances are calculated by centering the mask at each pixel in turn and updating the distance of this pixel. A binary image and its distance transform using

4	3	4
3	0	3
4	3	4

(a)

	11		11	
11	7	5	7	11
	5	0	5	
11	7	5	7	11
	11		11	

(b)

Figure 5.16: (a) Mask Chamfer 3-4 and (b) Chamfer 5-7-11 suggested as integer approximation of Euclidean distances for the distance transformation [40].

Chamfer 3-4 are given in Fig. 5.17. Note that the distances are about three times higher than the corresponding Euclidean distances because of the approximation chosen by Chamfer 3-4.

For the remainder of this chapter we will use the metric Chamfer 5-7-11 because of its higher accuracy. The edge image obtained by the Canny operator and the corresponding distance transform for a frame of the sequence *Grandma* are shown in Fig. 5.18. Dark values for the distance transform represent small distances to the nearest edge pixel, while bright values indicate large distances.

The term $\min_{e \in E_{q+1}} \|o - e\|$ for the distance $h_k(O_q, E_{q+1})$ in (5.17) can now be directly read from the distance transform for any model point o. Let (x, y) be the coordinates of the model point o. The value of the distance transform at (x, y) is then the distance of o to the nearest edge pixel, i.e., $\min_{e \in E_{q+1}} \|o - e\|$. This is repeated for all model points $o \in O_q$ and the resulting distances are sorted in ascending order so that the kth value of the sorted distances is equal to $h_k(O_q, E_{q+1})$.

5.3.4.3 Implementation of Hausdorff Distance

Several suggestions for efficient implementation are presented in [31, 32]. The main idea is to assume that the Hausdorff distance is smaller than a threshold τ so that bad matches can be ruled out early. Obviously, matches can only be found if the Hausdorff distance is indeed smaller than τ.

Here, early scan termination will be described. Assume that we want to find only those translations of the object O_q relative to the edge image E_{q+1} for which $h_k(O_q, E_{q+1}) \leq \tau$ for some specified constant τ. It can be seen from the definition (5.17) that this is satisfied if and only if no more

0	0	0	0	0	0	0	0
0	0	0	0	0	0	0	1
0	1	0	0	0	0	1	1
0	0	1	1	1	1	1	1
0	0	0	0	0	1	0	0
0	0	0	0	0	1	0	0
0	0	0	0	0	1	0	0
0	0	0	0	0	0	0	0

(a)

7	6	7	8	8	7	4	3
4	3	4	6	6	4	3	0
3	0	3	3	3	3	0	0
4	3	0	0	0	0	0	0
7	4	3	3	3	0	3	3
8	7	6	6	3	0	3	6
11	10	9	6	3	0	3	6
14	13	10	7	4	3	4	7

(b)

Figure 5.17: (a) Binary image with edge pixels having value of 1 and non-edge pixels being 0. (b) The corresponding distance transform using Chamfer 3-4 indicates for each pixel the distance to the nearest edge pixel.

than $m - k$ model points $o \in O_q$ have a distance to the nearest edge pixel $e \in E_{q+1}$ larger than τ. Therefore, as soon as $m - k + 1$ model points have been found with a distance larger than τ, the corresponding translation can be ruled out and does not need any further investigation. In the case where no match is found, this procedure is repeated for a higher value of τ, although this should not be required with a conservative choice of τ.

The Hausdorff distance is now computed as follows. First, the Chamfer 5-7-11 distance transform is calculated for the edge image so that for each model point the distance to the nearest edge pixel is known. Then, for all translations $t = (t_x, t_y)$ we calculate $h_{k,t}(O_q, E_{q+1})$, whereby the index t indicates that $h_k(O_q, E_{q+1})$ depends on the translation t. The object O_q is translated by the vector t and the distance transform at the location of model points o directly gives the distance between o and the nearest edge pixel. These distances are then sorted in ascending order and the kth value is selected to get $h_{k,t}(O_q, E_{q+1})$. Because of early scan termination $h_{k,t}(O_q, E_{q+1})$ is only found for translations with $h_{k,t}(O_q, E_{q+1}) \leq \tau$. For these translations, $h_{l,t}(E_{q+1}, O_q)$ is calculated in the same way to finally obtain $H_t(O_q, E_{q+1})$. The smallest Hausdorff distance $H_t(O_q, E_{q+1})$ indicates the new position of the object, i.e., the translation t the model has undergone.

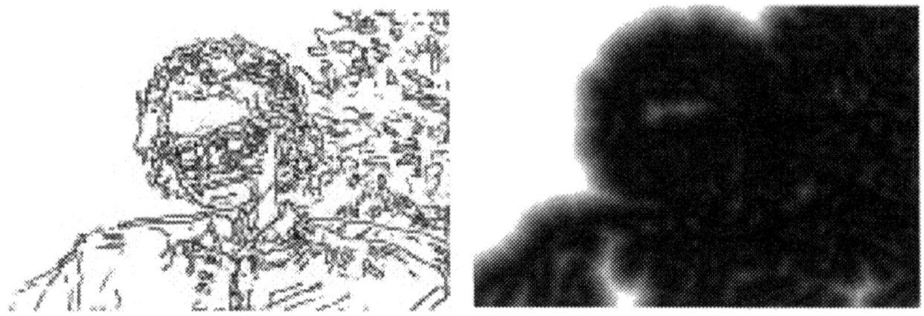

(a) (b)

Figure 5.18: (a) Edge image computed using Canny operator [30] and (b) resulting distance transform using the mask Chamfer 5-7-11 [40] for a frame of the *Grandma* sequence. Darker pixels indicate a smaller distance to the nearest edge pixel.

5.3.5 Model Update

The tracked object might rotate or change its shape as it moves through the video sequence, and consequently the corresponding model must be updated every frame by the model update stage in Fig. 5.4. More precisely, the model is actually not updated but a new model is derived by selecting an appropriate set of edge pixels from the edge image of the current frame. However, the object model of the previous frame is an important cue for choosing the set of edge pixels forming the new model.

In many situations there are parts of an object that are moving differently from the overall motion of the object. The legs and arms of a walking person, for example, normally move faster than the body. A two-stage updating technique that accommodates both rigid and non-rigid moving parts of an object, referred to as slowly changing and rapidly changing components, respectively, was suggested in [25, 26, 27, 28].

5.3.5.1 Slowly Changing Component

The slowly changing component is essentially the same as the updating mechanism reported in [31, 33]. It takes care of quasi-rigid parts of the object that exhibit only small changes in successive frames. Since these parts are by definition not expected to change significantly, they can be updated based on the old model O_q of frame q. To this end, the old model is shifted to the position of best match in frame $q + 1$ and all edge pixels within a specified small distance T_{slow} of the shifted old model are added to the new model O_{q+1}.

Let O_{q+1}^{slow} denote the slowly changing component of the updated model and $O_q \oplus t = \{o+t | o \in O_q\}$ be the old model O_q shifted by the translation t to the position of best match in frame $q+1$, whereby t is determined using the Hausdorff distance. Then, we have

$$O_{q+1}^{slow} = \{e \in E_{q+1} | \min_{x \in O_q \oplus t} \|e - x\| \leq T_{slow}\}, \tag{5.19}$$

with T_{slow} typically being one or two pixels.

The component O_{q+1}^{slow} is obtained by computing the distance transform of the shifted old model $O_q \oplus t$ and finding all edge pixels $e \in E_{q+1}$ with a value for this distance transform that is smaller than or equal to T_{slow}. The larger T_{slow} is chosen, the more likely the whole object will be included into the new model. However, this also increases the possibility of picking up cluttered background, which can at least partly be avoided by filtering stationary background beforehand if necessary (see Section 5.4.4).

Fig. 5.19 (a) and 5.20 (a) show the slowly changing component for two different frames of the sequence *Alexis*. Although they have adapted to minor changes of the object, they look in both cases very similar to the corresponding model of the previous frame in Fig. 5.19 (c) and 5.20 (c), respectively. In fact, temporal continuity is exactly what the slowly changing component must achieve. Notice that the value for the parameter T_{slow} has been scaled with respect to the mask Chamfer 5-7-11 in Fig. 5.16 (b) so that the value of 5 is actually a Euclidean distance of 1.

Obviously, the slowly changing component is not capable of picking up new parts of an object that become visible when they start moving. While this does not matter in Fig. 5.20 where the whole object has already been identified, it is a problem in Fig. 5.19, and an updating mechanism consisting only of the component (5.19) is insufficient. This highlights the need for an additional component that is able to include newly appearing parts of an object.

5.3.5.2 Rapidly Changing Component

The aim of the rapidly changing component is to incorporate non-rigid moving and newly appearing parts of an object into the model update. This ensures that non-rigid components of an object, which cannot be compensated by a pure translation t, still remain part of the model. In addition, it enables the object tracker to pick up those parts which were initially assigned to the background.

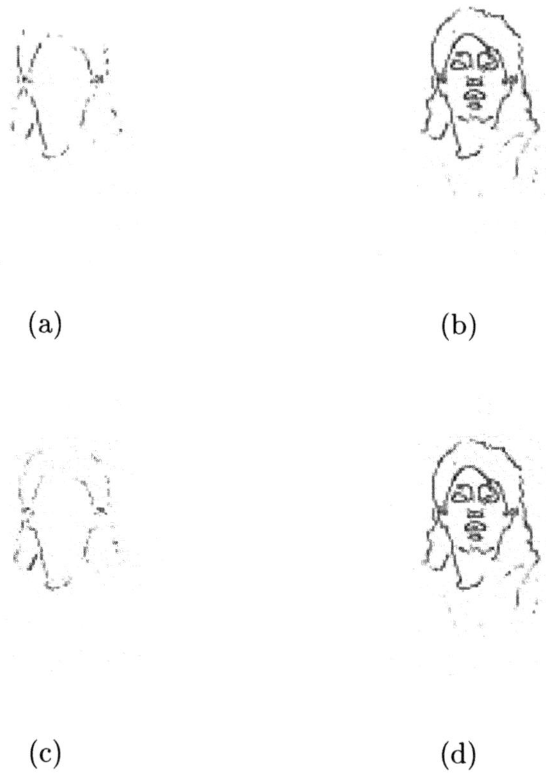

Figure 5.19: Model update: (a) slowly changing component ($T_{slow} = 5$), (b) rapidly changing component ($T_{rapid} = 0$), (c) old model O_{33} for frame 33, and (d) updated model O_{34} obtained by combining the slowly changing and rapidly changing component for frame 34 of the sequence *Alexis*.

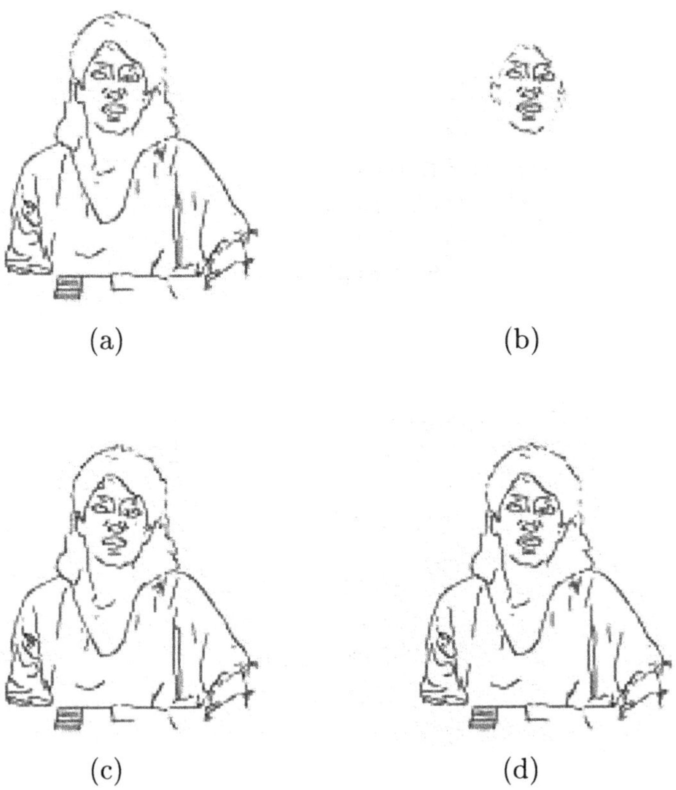

Figure 5.20: (a) slowly changing component ($T_{slow} = 5$), (b) rapidly changing component ($T_{rapid} = 0$), (c) old model O_{73} for frame 73, and (d) updated model O_{74} for frame 74 of the *Alexis* sequence.

Rapidly changing components are obtained by identifying moving components using morphological motion filtering in the same way as for the model initialization (see Section 5.3.3). Moving components that overlap the shifted old model $O_q \oplus t$ are assumed to be induced by the corresponding object, and all edge pixels within a small distance T_{rapid} of these moving components are added to the new model O_{q+1}.

Formally, the rapidly changing component O_{q+1}^{rapid} is defined as

$$O_{q+1}^{rapid} = \{e \in E_{q+1} | \min_{x \in MC} \|e - x\| \leq T_{rapid}\}, \tag{5.20}$$

where MC stands for the set of all pixels belonging to those moving components that overlap the shifted old model $O_q \oplus t$.

Fig. 5.19 demonstrates the importance of this second component for the update process. The slowly changing component in Fig. 5.19 (a) alone was clearly not able to include the facial features and the improved contour of *Alexis*' head in contrast to the rapidly changing component (Fig. 5.19 (b)), which could detect them thanks to a slight motion of the head. On the other hand, in Fig. 5.20 (b) the rapidly changing component is less important because no new parts of the object have appeared.

Finally, the updated model O_{q+1} is given by combining the two components:

$$O_{q+1} = O_{q+1}^{slow} \cup O_{q+1}^{rapid}. \tag{5.21}$$

The updating technique (5.21) yields a robust mechanism which can handle both objects that change quickly and objects that stop moving. Depending on the motion characteristics of the sequence and the objects, which can change from frame to frame, either O_{q+1}^{slow} (Fig. 5.20) or O_{q+1}^{rapid} (Fig. 5.19) can dominate the update process. This co-operation is a kind of temporal integration of the model. O_{q+1}^{rapid} detects and includes newly appearing parts of the object, whereas O_{q+1}^{slow} is the "memory" that stores the accumulated model components.

5.3.6 VOP Extraction

The output of the model update stage (see Fig. 5.4) is a sequence of binary edge images modeling the tracked object. The remaining step is to extract the corresponding object from the video sequence, i.e., the VOP for the object of interest has to be created. Unfortunately, the VOP extraction is not straightforward because the binary model points normally do not form a closed contour as illustrated in Fig. 5.23 (a).

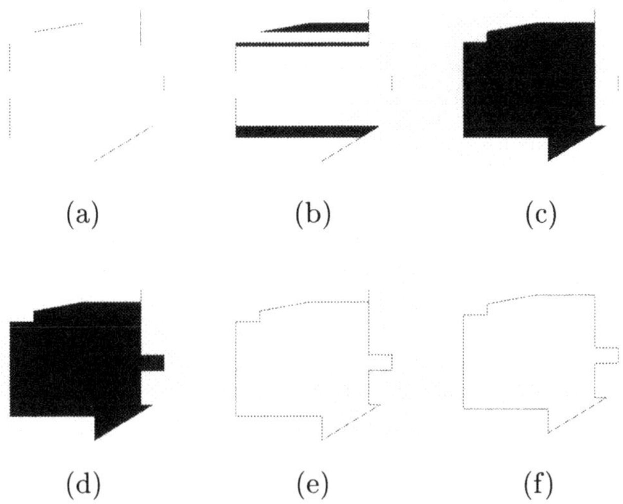

Figure 5.21: Filling-in technique: (a) Binary model with non-closed contour, (b) VOP (black) after first scan of rows, (c) VOP after first scan of columns, (d) VOP after second scan of rows, and (e) resulting contour by filling-in method. (f) VOP contour after removing the short branch in the top right corner of the boundary in (e).

The object boundary is determined by a two-step method. Firstly, the closed VOP boundary is approximated by a simple filling-in technique. This starts by assigning for each row the pixels between the first and last model point to the VOP. Fig. 5.21 (b) shows the VOP for the model in Fig. 5.21 (a) after the first row scan with pixels belonging to the VOP marked in black. This procedure is then repeated for each column (Fig. 5.21 (c)) and once more for each row, leading to the VOP illustrated in Fig. 5.21 (d) and the corresponding contour in Fig. 5.21 (e).

Before applying the boundary post-processor, which is the second step of the VOP extraction method, it must be ensured that the boundary forms a closed loop without any protruding branches or thin lines such as the one in the top right corner of Fig. 5.21 (e). This task can be accomplished by eroding the contour alternately from North, South, West, and East by the thinning algorithm described in [41], which has the important property that it does not alter connectedness of components.

Since the decision whether to remove a boundary pixel (x, y) is made by examining a 3×3 neighborhood centered at (x, y), the thinning algo-

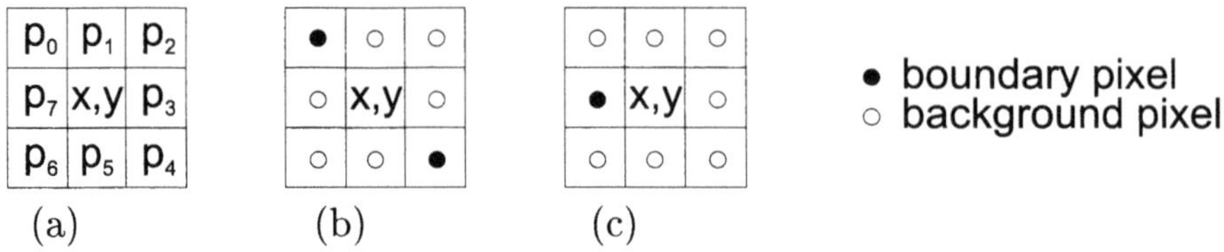

Figure 5.22: (a) Labeling of neighbor pixels for adjacency code, (b) an example where the pixel in question (x, y) must not be removed, and (c) an example where (x, y) is removed.

rithm can be efficiently implemented using an adjacency code (AC) combined with a look-up table. The adjacency code at pixel (x, y) is defined as (see Fig. 5.22 (a))

$$AC(x, y) = \sum_{i=0}^{7} p_i \cdot 2^i \tag{5.22}$$

for an eight-neighbor system with p_i being one if neighbor i is a contour pixel and zero otherwise. For each of the 256 possible combinations it can be determined in advance whether the pixel (x, y) should be removed or not, resulting in an efficient look-up table (see Table 5.1).

In Fig. 5.22 (b), we have $AC(x, y) = 2^0 + 2^4 = 17$ and the pixel may not be removed because otherwise the connectedness of the top left and bottom right pixel would be broken. Thus, the look-up table entry for AC=17 is "do not remove". On the other hand, the pixel (x, y) in Fig. 5.22 (c) must be removed because it is the end of a branch or line and is not part of a closed loop. Therefore, the entry in Table 5.1 for $AC(x, y) = 2^7 = 128$ is "remove" (x).

Fig. 5.21 (f) shows the final VOP boundary obtained by the filling-in technique after applying the thinning algorithm. Notice that the thin line in the top right hand corner has been successfully removed.

The contour of *Claire* in Fig. 5.23 (b) exhibits the typical jagged look of the boundary caused by the simple filling-in technique. This is corrected in the second step by the boundary post-processor. Most boundary pixels coincide with model points; these are assumed to be correct and so do not require any further processing. However, those parts of the contour that do not correspond to model points must have been created by the filling-in technique and need correction. Fig. 5.23 (c) shows these wrong boundary segments for frame 25 of the sequence *Claire*.

Table 5.1: Look-up table for adjacency code (x=remove boundary pixel).

$AC(x,y)$	0	1	2	3	4	5	6	7	8	9	10	11	12	13	14	15
remove		x	x	x	x		x	x	x		x	x	x		x	x
$AC(x,y)$	16	17	18	19	20	21	22	23	24	25	26	27	28	29	30	31
remove	x								x		x	x	x		x	x
$AC(x,y)$	32	33	34	35	36	37	38	39	40	41	42	43	44	45	46	47
remove	x								x		x	x	x		x	x
$AC(x,y)$	48	49	50	51	52	53	54	55	56	57	58	59	60	61	62	63
remove	x								x		x	x	x		x	x
$AC(x,y)$	64	65	66	67	68	69	70	71	72	73	74	75	76	77	78	79
remove	x															
$AC(x,y)$	80	81	82	83	84	85	86	87	88	89	90	91	92	93	94	95
remove																
$AC(x,y)$	96	97	98	99	100	101	102	103	104	105	106	107	108	109	110	111
remove	x								x		x	x	x		x	x
$AC(x,y)$	112	113	114	115	116	117	118	119	120	121	122	123	124	125	126	127
remove	x								x		x	x	x		x	x
$AC(x,y)$	128	129	130	131	132	133	134	135	136	137	138	139	140	141	142	143
remove	x	x	x	x			x	x			x	x			x	x
$AC(x,y)$	144	145	146	147	148	149	150	151	152	153	154	155	156	157	158	159
remove											x	x			x	x
$AC(x,y)$	160	161	162	163	164	165	166	167	168	169	170	171	172	173	174	175
remove	x	x	x	x			x	x	x	x	x	x	x	x	x	x
$AC(x,y)$	176	177	178	179	180	181	182	183	184	185	186	187	188	189	190	191
remove	x	x	x	x			x	x	x	x	x	x	x	x	x	x
$AC(x,y)$	192	193	194	195	196	197	198	199	200	201	202	203	204	205	206	207
remove	x	x	x	x			x	x			x	x			x	x
$AC(x,y)$	208	209	210	211	212	213	214	215	216	217	218	219	220	221	222	223
remove											x	x			x	x
$AC(x,y)$	224	225	226	227	228	229	230	231	232	233	234	235	236	237	238	239
remove	x	x	x	x			x	x	x	x	x	x	x	x	x	x
$AC(x,y)$	240	241	242	243	244	245	246	247	248	249	250	251	252	253	254	255
remove	x	x	x	x			x	x	x	x	x	x	x	x	x	x

Each wrong boundary segment is corrected separately. After removing one of these, there will be a gap in the otherwise closed contour, as illustrated in Fig. 5.23 (d). This is also depicted in Fig. 5.24 (a), where a magnified portion is shown. The two end points A and B of the gap can easily be found, and the correct boundary between these two points is determined by Dijkstra's shortest path algorithm [42]. To this end, a weight or distance is assigned to the different types of pixels, since we are not interested in minimizing the Euclidean distance.

Naturally, the contour between A and B in Fig. 5.24 (a) should coincide with binary model points, which are shown in Fig. 5.24 (b). Hence, a small weight or distance d_0 is assigned to these model points. The weight is also set to d_0 for all pixels along the four frame boundaries, because in many situations they form part of the VOP contour. In head-and-shoulder sequences, for instance, the bottom boundary of the frame is also a boundary of the VOP representing the person. All remaining pixels are less likely to belong to the VOP boundary and a weight d_1 with $d_1 > d_0$ is assigned to them. Furthermore, our goal is to close the gap between A and B with the corrected boundary segment. Therefore, all pixels that already belong to

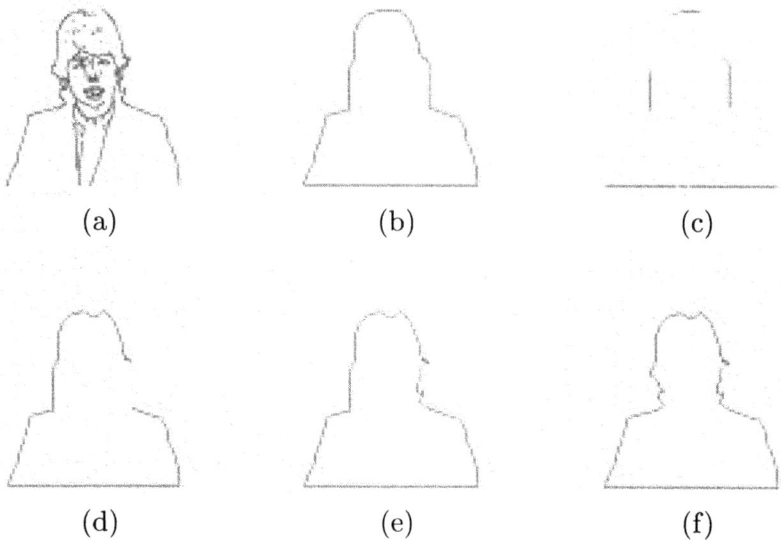

Figure 5.23: (a) Binary model for frame 25 of the sequence *Claire*, (b) VOP boundary obtained by simple filling-in technique, and (c) wrong VOP boundary segments that do not correspond to model points. (d) One wrong boundary segment has been removed and (e) replaced by the shortest path according to Dijkstra's algorithm [42]. (f) The VOP boundary after correcting all wrong segments is far more accurate than that in (b).

the VOP contour after removing the wrong boundary segment (these pixels are marked in black in Fig. 5.24 (a)) must be excluded. This is accomplished by setting the corresponding weights to infinity.

To summarize, pixels that already belong to the VOP boundary must be excluded from the search by assigning infinity or a suitably large distance to them. The distance is set to d_0 both for object model points that are not yet VOP boundary pixels and for pixels along the four frame boundaries. All other pixels have a weight of d_1. To enforce the VOP boundary to coincide with model points wherever possible, we choose $d_1 > d_0$. Good results for various types of sequences can be achieved by choosing $d_0 = 1$ and $d_1 = 10$.

Fig. 5.24 (c) visualizes the weights assigned to the pixels in our example. Notice that all model points of Fig. 5.24 (b) were assigned the small weight d_0, except for those pixels that are marked in black in Fig. 5.24 (a). The

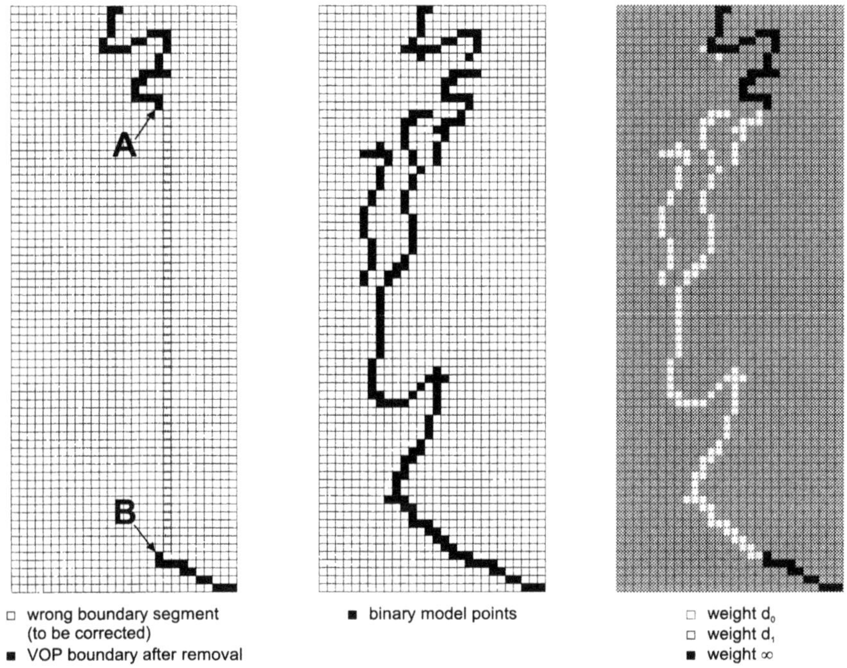

(a) (b) (c)

Figure 5.24: Correction of a boundary segment for frame 25 of *Claire*: (a) enlarged portion of the wrong boundary segment removed in Fig. 5.23 (d); (b) corresponding binary model points, and (c) weights assigned for Dijkstra's algorithm.

weight for these pixels was set to ∞, because they already belong to the VOP boundary. As can be seen from Fig. 5.24 (c), the shortest path between A and B follows the model points as desired, because d_0 is much smaller than d_1.

Dijkstra's algorithm can now be applied to find the shortest path between A and B with respect to these defined distances. The algorithm starts by temporarily assigning the shortest distance 0 to the starting point A and ∞ to all other pixels. It then finds the pixel X with the shortest temporary distance and labels it as permanent, because this must be the shortest possible distance from X to A. For the first pixel, X will obviously be equal to A. Then, the shortest distance is updated for all neighbor pixels Y of X that are still labeled as temporary. It is possible that the shortest path from Y to A becomes smaller when using the path via X. In that case, the shortest distance of Y is replaced by the sum of the shortest distance of X plus the weight assigned to Y.

This procedure is then repeated by finding the next pixel X with the shortest distance among the remaining pixels still labeled as temporary. The algorithm terminates as soon as B has been labeled as permanent. The gap between A and B is then closed by the shortest path as shown in Fig. 5.23 (e). Should there be multiple shortest paths of equal length, the shortest path with respect to the Euclidean distance is chosen.

The significant improvement of the VOP boundary after post-processing compared to that in Fig. 5.23 (b) is demonstrated in Fig. 5.23 (f). Note that in the case of a closed model contour, the proposed post-processor is, for an appropriate choice of d_0 and d_1, capable of finding the correct boundary, in contrast to the filling-in technique. This is due to the fact that the weight assigned to model points is smaller than that for other pixels.

5.3.7 Results

Let us now have a look at the results obtained by the described VOP segmentation algorithm. The dense motion field, which is required as input to the morphological motion filter, was estimated by the Horn-Schunck method [29] for *Claire*, *Alexis*, and *Grandma*, whereas block matching with full search was used for *Coastguard*.

Ten iterations of the Horn-Schunck algorithm with $\alpha^2 = 30$ (see (1.57)) proved to be a good choice for these sequences. Block matching, on the other hand, was carried out using 12×12 blocks with a maximum displacement of 5 pixels in horizontal and vertical directions.

The critical parameter for the morphological motion filter is λ (see Section 5.3.2). Ideally, λ is chosen based on the frame with the fastest motion of the tracked object so that as much of this object as possible is filtered without filtering the background.

Claire is a sequence with uncluttered, stationary background. The original frame, the binary model, and the resulting VOP for frame 50 are shown in Fig. 5.25. As can be seen, the VOP boundary is virtually perfect for this simple sequence.

The background in the sequence *Alexis* is also stationary, but cluttered in contrast to *Claire*. Due to a lack of sufficient motion, it took until frame 33 to initialize the object tracker and the initial model in Fig. 5.26 still did not yet represent the whole object. However, more parts of the person were included in the subsequent frames 34 and 35 as illustrated in Fig. 5.27 and

(a) (b) (c)

Figure 5.25: (a) Original frame 50 of sequence *Claire*, (b) binary model, and (c) extracted video object plane.

(a) (b) (c)

Figure 5.26: Initial model: (a) Original frame 33 of sequence *Alexis*, (b) binary model, and (c) resulting video object plane.

Fig. 5.28, respectively. The extracted VOP after 99 frames is depicted in Fig. 5.29; the model update stage picked up small parts of the object in front of *Alexis* because of reflections on the desk when *Alexis* was moving. The resulting variations in illuminance were then interpreted as motion.

The camera in the sequence *Coastguard* is following the boat so that the background appears to be moving. In Fig. 5.30, the extracted VOP for the boat is shown for frame 250 of this sequence. The boat, which is moving relative to the background, was properly identified and extracted. However, as expected, the boundary location for this more complicated sequence is not as accurate as for the previous head-and-shoulder sequences. Most problems were caused by the clutter and stones in the background and the waves below the boat. Because of their vicinity to the tracked object and the temporal variations of the waves, the tracker included parts of them

(a) (b) (c)

Figure 5.27: (a) Original frame 34 of sequence *Alexis*, (b) binary model, and (c) extracted VOP.

(a) (b) (c)

Figure 5.28: (a) Original frame 35 of sequence *Alexis*, (b) binary model, and (c) extracted VOP.

(a) (b) (c)

Figure 5.29: (a) Original frame 99 of sequence *Alexis*, (b) binary model, and (c) extracted video object plane.

(a) (b) (c)

Figure 5.30: (a) Original frame 250 of sequence *Coastguard*, (b) binary model for the boat, and (c) extracted video object plane.

(a) (b) (c)

Figure 5.31: *Grandma*: (a) Original frame 99, (b) binary model, and (c) extracted VOP.

into the model.

Finally, the results on *Grandma* after 99 frames in Fig. 5.31 demonstrate that only those parts of an object can be extracted that are moving or changing. While the head was very well segmented, there was not enough motion of the body to be included into the VOP. This is a problem of most if not all VOP segmentation algorithms that identify physical objects based on motion.

5.4 Version II: Change Detection Masks

The second version of the VOP segmentation algorithm relies on the same concept as the first version. A two-dimensional binary model for the object of interest is tracked and updated throughout the sequence. The major differences are the object motion detection stage and the global motion estimation stage, which is reflected in the flowcharts of Fig. 5.32 and Fig. 5.4. This version of the algorithm is restricted to sequences with stationary background because the global motion estimation has been omitted, as we will explain shortly.

Consequently, there are only four major functional blocks left: object motion detection, model initialization, model update, and VOP extraction. In addition, the optional stationary background filter will be described.

Change detection masks obtained by taking the difference between successive frames are a simple and fast way to detect moving objects. Fig. 5.33 (a) demonstrates that the moving person can be identified in the difference image. The fact that mainly the occlusion regions are marked as changed is no drawback, since we are interested in deriving a binary model from the object contour, which is located in or at least near occlusion areas.

Unfortunately, the use of change detection masks can become problematic when the background is moving, because it is often not possible to achieve a perfect registration of the background. If the background is highly textured or contains clutter such as in *Coastguard* (see Fig. 5.30 (a)), there will be connected components in the change detection mask which cannot be distinguished from moving objects. Moreover, these components cannot be eliminated by means of noise filtering. Therefore, the second version of the VOP segmentation algorithm is restricted to sequences with stationary background.

Again, the model update consists of two stages, although change detection masks are employed to update rapidly changing components instead of the morphological motion filter. The model matching and VOP extraction blocks, on the other hand, are exactly the same as described in Sections 5.3.4 and 5.3.6, respectively.

The second version of the VOP extraction algorithm has proven to be most effective for objects with fast and non-rigid motion. For head-and-shoulder or videophone sequences with only little motion, the occlusion regions identified by the change detection mask are often too small to initialize the model.

5.4.1 Object Motion Detection Using Change Detection Masks

High values for the absolute difference of gray-levels between two consecutive frames indicate objects that are moving or changing their shape (see Fig. 5.33 (a)). There might also be noise present in these so-called change detection masks (CDMs). This noise can easily be suppressed based on the size of connected components in the CDM, because pixels belonging to moving objects are connected, while noisy pixels form isolated clusters.

The connected component labeling algorithm in [43] is a simple tool to

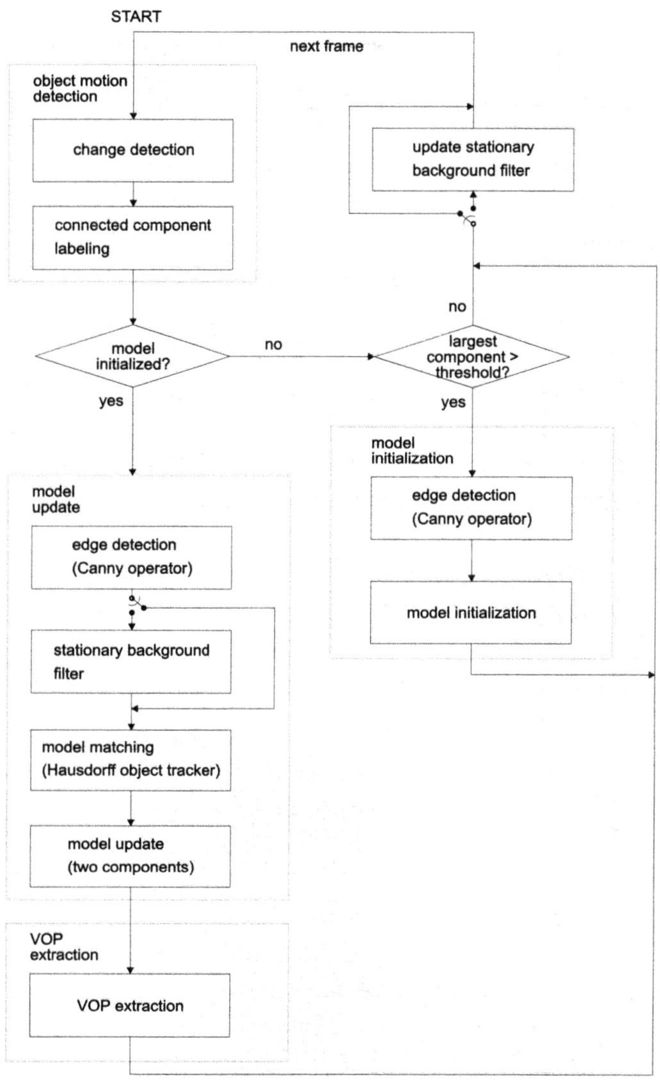

Figure 5.32: Flowchart of the VOP segmentation algorithm using change detection masks.

(a) (b) (c)

Figure 5.33: (a) Gray-level difference image for frame 31 of the sequence *Hall monitor* after thresholding (change detection mask), (b) moving connected component, and (c) rapidly changing model component obtained by selecting all edge pixels near the corresponding moving connected component.

find connected components in binary images. If the size of a component exceeds a threshold, it is assumed to belong to a moving object and is referred to as a *moving connected component.* Thus, moving objects can be detected by finding moving connected components in the CDM. An example is shown in Fig. 5.33 (b) for frame 31 of the *Hall monitor* sequence. It is the largest connected component in Fig. 5.33 (a) consisting of 1577 pixels, with the next larger component containing only 47 pixels.

5.4.2 Model Initialization

As we mentioned above, difference images or change detection masks normally exhibit high values in occlusion areas. Since the correct object boundaries are in these occlusion zones or are at least in their vicinity, the boundaries of moving objects are detected by simply selecting edge pixels near moving connected components.

Thus, the object tracker is initialized based on the first detected moving connected component. Let MCC denote the set of all pixels belonging to that moving connected component and as usual let E_q be the set of edge pixels in the corresponding frame. Then, the initial model O_q is given by selecting all edge pixels within a small distance T_{rapid} of MCC, i.e.,

$$O_q = \{e \in E_q | \min_{x \in MCC} \|e - x\| \leq T_{rapid}\}, \tag{5.23}$$

which can be implemented using the distance transformation [40].

Notice that the rapidly changing component of the model update described below will use the same procedure as (5.23). Hence, the initial model is essentially the first rapidly changing component detected in the sequence. Fig. 5.33 (c) shows such a rapidly changing component corresponding to the moving connected component in Fig. 5.33 (b).

5.4.3 Model Update

The model update mechanism is very similar to the one described in Section 5.3.5 and also consists of two stages. The slowly changing component accounts for rigid moving parts and the rapidly changing component accommodates non-rigid moving parts as well as parts that were assigned to the background in the model initialization.

The slowly changing component O_{q+1}^{slow} is defined by (5.19). Thus, the model O_q of the previous frame q is shifted by the translation t to the position of best match determined by the Hausdorff distance. Edge pixels within a small distance T_{slow} of the shifted old model $O_q \oplus t$ are then assigned to O_{q+1}^{slow}.

The rapidly changing component O_{q+1}^{rapid} requires the detection of moving components like in (5.20). However, these are identified from change detection masks as described in Section 5.4.1, instead of using morphological motion filtering. Let MCC be the set of all pixels belonging to moving connected components that overlap the shifted old model $O_q \oplus t$. O_{q+1}^{rapid} is then found by selecting edge pixels according to

$$O_{q+1}^{rapid} = \{e \in E_{q+1} | \min_{x \in MCC} \|e - x\| \leq T_{rapid}\}. \tag{5.24}$$

The same threshold T_{rapid} as for the model initialization in (5.23) is used. In fact, equation (5.24) is the same as (5.23) except that MCC consists here of moving connected components overlapping the shifted old model.

Fig. 5.34 demonstrates how the two updating components complement each other. The slowly changing component in Fig. 5.34 (a) is capable of updating most of the model. Only the left leg, which was moving differently from the rest of the person, could not be identified. The rapidly changing component shown in Fig. 5.34 (b), on the other hand, managed to detect the left leg but not the right foot, which was not moving. The combination of these two components according to (5.21) then yields an updated model that comprises the whole tracked object as illustrated in Fig. 5.34 (c).

5.4.4 Background Filter

Object tracking would be fairly easy if all pixels in the edge image belonged to objects, but unfortunately many sequences contain cluttered background

(a) (b) (c)

Figure 5.34: (a) Slowly changing component, (b) rapidly changing component, and (c) updated model (combination of slowly and changing component) for frame 48 of the sequence *Hall monitor.*

(see for example Fig. 5.35 (a)). Normally, it is desirable to remove cluttered background prior to model matching and updating for two reasons. Filtering background edges will drastically reduce the number of pixels in the set E_{q+1} and therefore the computation time for $h_l(E_{q+1}, O_q)$ in (5.18). However, it will not affect $h_k(O_q, E_{q+1})$ because the calculation of the distance transform for E_{q+1} is independent of the number of pixels in the set. Notice that background filtering generally does not affect the quality of matching as the Hausdorff object tracker can handle cluttered background very well. It is only a matter of reducing the computation time.

More important is the background filter for the model update stage. There is a possibility that the model update stage picks up some edge pixels belonging to the cluttered background if they are close enough to the model. This can be a problem, especially in noisy environments such as the sequence *Hall monitor* in Fig. 5.35. It is therefore often preferable to filter stationary background beforehand.

Let us in the following assume that the background is stationary. The usefulness of eliminating stationary background was also reported in [31, 33]. To perform the task, they suggested simple differencing where all edge pixels are removed that were already classified as edge in the previous frame. Unfortunately, this approach also filters those parts of an object that stop moving. For example, in Fig. 5.35 (b) not only the stationary background

(a) (b) (c)

Figure 5.35: (a) Binary edge image of frame 40 of sequence *Hall monitor* obtained by Canny operator, (b) simple binary difference image, and (c) binary edge image obtained by proposed background filter.

has been removed but also the person's left leg, which did not move between two frames.

The filtering technique in [25, 26, 27, 28] counts for each pixel (x, y) how often it has been classified as edge. If the ratio of this counter to the number of frames exceeds a threshold T_{bg} (e.g. 75%), then (x, y) is assumed to be part of the background and removed from the set of edge pixels. Further, it is recommended that the counter be updated only for those pixels that are not occluded by an object. This is accomplished by updating the counter after processing a frame when the position of all objects is known. Hence, information is collected only on pixels that are really classified as background. This avoids increasing the counter at the location of objects, and therefore objects or parts thereof will not be removed even when they stop for an arbitrarily long time. The flowcharts in Fig. 5.4 and Fig. 5.32 show that the actual filtering is applied before model matching and updating, while the counter is updated as the last step for each frame.

For the first few frames, the counter cannot give reliable results and simple differencing is applied until enough data has been collected. After a few frames the filter will come into effect for most pixels. Fig. 5.35 (c) demonstrates that edge pixels belonging to objects are much better preserved than by simple differencing in Fig. 5.35 (b).

The assumption of stationary background is valid for many applications but not for all. An extension to filtering moving background is unfortunately not straightforward, since global motion estimation and compensation cannot perfectly align edge images. However, there is normally less danger of picking up moving background and including it into the model, because such a background is constantly changing.

(a) (b) (c)

Figure 5.36: Initial model for sequence *Bowing*: (a) Original frame 48, (b) binary model, and (c) extracted video object plane.

5.4.5 Results

In this section, results for the second version of the VOP segmentation algorithm are shown. *Bowing* is a sequence with stationary background, containing fast motion and large changes in shape of the moving person. As soon as the person enters the scene from the right, the object tracker gets initialized. While the initial model in Fig. 5.36 is very small, the model update incorporates newly appearing parts of the person immediately as illustrated in Fig. 5.37. In Fig. 5.38, Dijkstra's shortest path algorithm wrongly followed from the head to the shoulder along the right frame boundary, because the four frame boundaries also have the smaller weight d_0 assigned (see Section 5.3.6). Only one frame later, as the person has moved away from the right frame boundary, the VOP boundary follows the correct object contour again and the corresponding VOP depicted in Fig. 5.39 was virtually perfectly extracted.

An interesting sequence of VOPs comprising the frames 129, 132, 135, 138, 145, and 150 is given in Fig. 5.40. The tracked person is bowing, leading to fast non-rigid motion. Nevertheless, the segmentation algorithm adapted very well to the enormous changes in shape and could still provide VOPs of high accuracy.

In the sequence *Hall monitor*, there is a high level of noise present, and the stationary background is very cluttered. The frames 45, 60, and 90 in Fig. 5.41, Fig. 5.42, and Fig. 5.43, respectively, demonstrate how the VOP segmentation algorithm adapted to changes in shape as the person turns to his left to walk down the floor, then stops and turns again.

The results in Sections 5.3.7 and 5.4.5 demonstrate that the described segmentation algorithm can successfully extract VOPs of moving objects for

(a) (b) (c)

Figure 5.37: (a) Original frame 53 of sequence *Bowing*, (b) binary model, and (c) resulting VOP.

(a) (b) (c)

Figure 5.38: Sequence *Bowing*: (a) Original frame 58, (b) binary model for person, and (c) resulting video object plane.

(a) (b) (c)

Figure 5.39: (a) Original frame 59 of sequence *Bowing*, (b) binary object model, and (c) VOP obtained by our algorithm.

(a) (b) (c)

(d) (e) (f)

Figure 5.40: VOPs for a phase of fast motion in the sequence *Bowing*: (a) frame 129, (b) frame 132, (c) frame 135, (d) frame 138, (e) frame 145, and (f) frame 150.

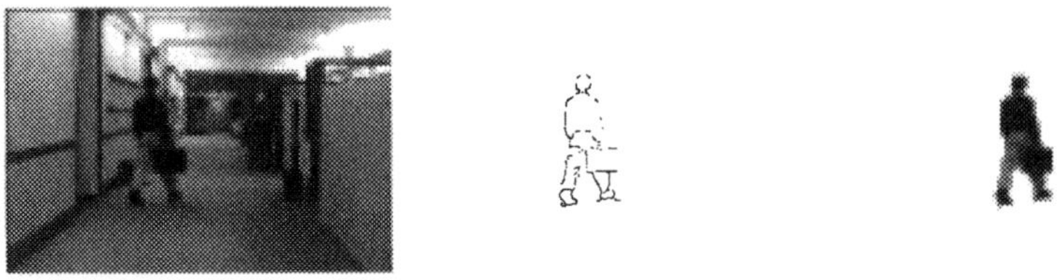

(a) (b) (c)

Figure 5.41: (a) Original frame 45 of sequence *Hall monitor*, (b) binary model, and (c) resulting VOP.

(a) (b) (c)

Figure 5.42: Sequence *Hall monitor*: (a) original frame 60, (b) binary object model, and (c) corresponding VOP.

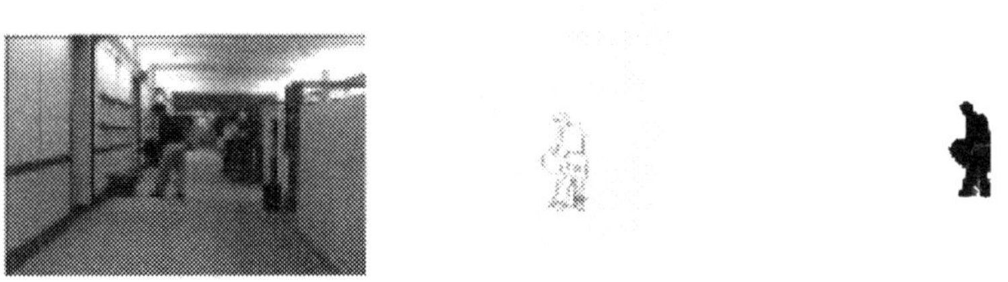

(a) (b) (c)

Figure 5.43: (a) Original frame 90 of *Hall monitor*, (b) binary model, and (c) corresponding video object plane.

different types of sequences. In particular, the resulting VOP boundaries are highly accurate. This is due to the fact that the location of object boundaries is guided by the binary models, which consist of edge pixels detected by the Canny operator [30].

The first version of the algorithm based on morphological motion filtering is more effective for sequences where the objects are moving only a little, like, for example, in head-and-shoulder or videophone sequences. If an object moves too fast or changes its shape significantly between successive frames, then there will be large occlusion zones where the dense motion field cannot be correct (see Section 1.4.5). This might cause the morphological motion filter to incorrectly assign the occlusion areas to a moving object. For that reason, the second version based on change detection masks is more suitable for objects with fast and non-rigid motion. Objects are expected to move reasonably fast or change their shape because large occlusion areas allow large moving connected components to be extracted from the change detection masks. For objects characterized by little motion, on the other hand, the corresponding occlusion regions would be to small to yield useful moving connected components for the model initialization or update.

The VOP extraction techniques described in this chapter also have some limitations. Since physical objects are identified based on motion, those parts of an object which are not moving or changing will be assigned to the background. This is intrinsic to the assumption that semantically meaningful objects are defined as independently moving, connected areas and is common to other VOP segmentation techniques. Hence, it is not always possible to immediately derive a binary model comprising the whole object of interest. Instead, it might take several frames to update the initial model until the complete object has been identified.

In some sequences, however, there is insufficient motion to extract the whole object of interest (for example, see *Grandma* in Fig. 5.31). Furthermore, apparent motion can also be induced by noise, which might be wrongly picked up as part of an object, especially in the case of clutter in the background. As a result, the obtained VOPs are not always suitable for all the content-based functionalities of MPEG-4. For instance, the floor that is visible between the legs of the person in Fig. 5.43 (c) does not allow that VOP to be copied into a scene with different background.

It is obvious that extracting and placing a physical object into another scene requires complete segmentation of the object with perfect boundary accuracy. Currently, segmentation for content-based representation is still premature, and none of the published techniques is able to do this for generic video sequences. However, the VOPs obtained by this algorithm could be used by an MPEG-4 coder to provide other content-based functionalities such as content-based scalability.

All in all, more research is needed to include higher-level information into the segmentation process. Currently, this VOP extraction algorithm expects a user to tune a few crucial parameters. Such semi-automatic techniques provide a viable way to incorporate some intelligence, but they fall short of a fully automatic segmentation. Therefore, future work should concentrate on how to associate semantic concepts with groups of low-level objects without user interaction. This might involve the use of domain knowledge, image understanding, and some kind of artificial intelligence.

References

[1] ISO/IEC 14496-2, "Information technology - coding of audio-visual objects: Visual (committee draft)," in *ISO/IEC JTC1/SC29/WG11 N1902, Fribourg, Switzerland*, Oct. 1997.

[2] J.Y.A. Wang and E.H. Adelson, "Layered representation for image sequence coding," in *IEEE Int. Conf. on Acoustics, Speech, and Signal Processing, ICASSP'93, Minneapolis, MN, USA*, Apr. 1993, vol. V, pp. 221–224.

[3] J.Y.A. Wang and E.H. Adelson, "Representing moving images with layers," *IEEE Trans. Image Processing*, vol. 3, no. 5, pp. 625–638, Sept. 1994.

[4] L. Torres, D. Garcia, and A. Mates, "A robust motion estimation and segmentation approach to represent moving images with layers," in *IEEE Int. Conf. on Acoustics, Speech, and Signal Processing, ICASSP'97, Munich, Germany*, Apr. 1997, vol. 4, pp. 2981–2984.

[5] D. Zhong and S.F. Chang, "Video object model and segmentation for content-based video indexing," in *IEEE Int. Symposium on Circuits and Systems, ISCAS'97, Hong Kong*, June 1997, vol. 2, pp. 1492–1495.

[6] M. Bierling, "Displacement estimation by hierarchical blockmatching," in *SPIE Visual Communications and Image Processing, VCIP'88, Cambridge, MA, USA*, Nov. 1988, vol. 1001, pp. 942–951.

[7] F. Marqués and C. Molina, "Object tracking for content-based functionalities," in *SPIE Visual Communications and Image Processing, VCIP'97, San Jose, CA, USA*, Feb. 1997, vol. 3024, pp. 190–199.

[8] L. Garrido, F. Marqués, M. Pardàs, P. Salembier, and V. Vilaplana, "A hierarchical technique for image sequence analysis," in *Workshop on Image Analysis for Multimedia Application Services, WIAMIS'97, Louvain-la-Neuve, Belgium*, June 1997, pp. 13–20.

[9] R. Mech and M. Wollborn, "A noise robust method for 2D shape estimation of moving objects in video sequences considering a moving camera," *Signal Processing*, vol. 66, no. 2, pp. 203–217, Apr. 1998.

[10] R. Mech and P. Gerken, "Automatic segmentation of moving objects (partial results of core experiment N2)," in *ISO/IEC JTC1/SC29/WG11 MPEG97/m1949, Bristol, UK*, Apr. 1997.

[11] R. Mech and M. Wollborn, "A noise robust method for segmentation of moving objects in video sequences," in *IEEE Int. Conf. on Acoustics, Speech, and Signal Processing, ICASSP'97, Munich, Germany*, Apr. 1997, vol. 4, pp. 2657–2660.

[12] P. Gerken, R. Mech, G. Russo, S. Colonnese, C. Ahn, and M.H. Lee, "Merging of temporal and spatial segmentation," in *ISO/IEC JTC1/SC29/WG11 MPEG97/m1948, Bristol, UK*, Apr. 1997.

[13] M. Wollborn, R. Mech, S. Colonnese, U. Mascia, G. Russo, P. Talone, J.G. Choi, M. Kim, M.H. Lee, and C. Ahn, "Description of automatic segmentation techniques developed and tested for MPEG-4 Version 1," in *ISO/IEC JTC1/SC29/WG11 MPEG97/m2702, Fribourg, Switzerland*, July 1997.

[14] A. Neri, S. Colonnese, G. Russo, and P. Talone, "Automatic moving object and background separation," *Signal Processing*, vol. 66, no. 2, pp. 219–232, Apr. 1998.

[15] S. Colonnese, U. Mascia, G. Russo, and C. Tabacco, "New FUB results on core experiment N2 on automatic segmentation techniques," in *ISO/IEC JTC1/SC29/WG11 MPEG97/m1633, Sevilla, Spain*, Feb. 1997.

[16] A. Neri, S. Colonnese, and G. Russo, "Automatic moving object and background segmentation by means of higher order statistics," in *SPIE Visual Communications and Image Processing, VCIP'97, San Jose, CA, USA*, Feb. 1997, vol. 3024, pp. 246–256.

[17] S. Colonnese and G. Russo, "Segmentation techniques: Towards a semi-automatic approach," in *ISO/IEC JTC1/SC29/WG11 MPEG98/m3093, San Jose, CA, USA*, Feb. 1998.

[18] S. Colonnese and G. Russo, "User interaction modes in semi-automatic segmentation: Development of a flexible graphical user interface in java," in *ISO/IEC JTC1/SC29/WG11 MPEG98/m3320, Tokyo, Japan*, Mar. 1998.

[19] J.G. Choi, M. Kim, M.H. Lee, and C. Ahn, "Automatic segmentation based on spatio-temporal information," in *ISO/IEC JTC1/SC29/WG11 MPEG97/m2091, Bristol, UK*, Apr. 1997.

[20] J.G. Choi, M. Kim, M.H. Lee, C. Ahn, S. Colonnese, U. Mascia, G. Russo, P. Talone, R. Mech, and M. Wollborn, "Combined algorithm

of ETRI, FUB and UH on core experiment N2 for automatic segmentation algorithm of moving objects," in *ISO/IEC JTC1/SC29/WG11 MPEG97/m2383, Stockholm, Sweden*, July 1997.

[21] J.G. Choi, M. Kim, M.H. Lee, and C. Ahn, "Partial experiments on a user-assisted segmentation technique for video object plane generation," in *ISO/IEC JTC1/SC29/WG11 MPEG98/m3147, San Jose, CA, USA*, Feb. 1998.

[22] J.G. Choi, M. Kim, J. Kwak, M.H. Lee, and C. Ahn, "User-assisted video object segmentation by multiple object tracking," in *ISO/IEC JTC1/SC29/WG11 MPEG98/m3349, Tokyo, Japan*, Mar. 1998.

[23] C. Gu and M.C. Lee, "Semantic video object segmentation and tracking using mathematical morphology and perspective motion model," in *IEEE Int. Conf. on Image Processing, ICIP'97, Santa Barbara, CA, USA*, Oct. 1997, vol. II, pp. 514–517.

[24] C. Toklu, A.M. Tekalp, and A.T. Erdem, "Simultaneous alpha map generation and 2-D mesh tracking for multimedia applications," in *IEEE Int. Conf. on Image Processing, ICIP'97, Santa Barbara, CA, USA*, Oct. 1997, vol. I, pp. 113–116.

[25] T. Meier and K.N. Ngan, "Automatic segmentation of moving objects for video object plane generation *(Invited paper)*," *IEEE Trans. Circuits Syst. for Video Technol.*, vol. 8, no. 5, pp. 525–538, Sept. 1998.

[26] T. Meier and K.N. Ngan, "Video object plane segmentation using a morphological motion filter and Hausdorff object tracking," in *IEEE Int. Conf. on Image Processing, ICIP'98, Chicago, IL, USA*, Oct. 1998, vol. TP 05.05.

[27] T. Meier and K.N. Ngan, "Video object plane extraction for content-based functionalities in MPEG-4," in *Int. Workshop on Very Low Bitrate Video Coding, VLBV'98, Urbana, IL, USA*, Oct. 1998, pp. 121–124.

[28] T. Meier and K.N. Ngan, "Extraction of moving objects for content-based video coding," in *SPIE Visual Communications and Image Processing, VCIP'99, San Jose, CA, USA*, Jan. 1999, vol. 3653, pp. 1178–1189.

[29] B.K.P. Horn and B.G. Schunck, "Determining optical flow," *Artificial Intelligence*, vol. 17, pp. 185–203, 1981.

[30] J. Canny, "A computational approach to edge detection," *IEEE Trans. Pattern Analysis and Machine Intelligence*, vol. PAMI-8, no. 6, pp. 679–698, Nov. 1986.

[31] W. Rucklidge, *Efficient Visual Recognition Using the Hausdorff Distance*, Springer-Verlag, Berlin, Germany, 1996.

[32] D.P. Huttenlocher, G.A. Klanderman, and W.J. Rucklidge, "Comparing images using the Hausdorff distance," *IEEE Trans. Pattern Analysis and Machine Intelligence*, vol. 15, no. 9, pp. 850–863, Sept. 1993.

[33] D.P. Huttenlocher, J.J. Noh, and W.J. Rucklidge, "Tracking non-rigid objects in complex scenes," in *Fourth Int. Conf. on Computer Vision, ICCV'93, Berlin, Germany*, May 1993, pp. 93–101.

[34] P.J. Rousseeuw and A.M. Leroy, *Robust Regression and Outlier Detection*, John Wiley & Sons, New York, NY, 1987.

[35] W.H. Press, S.A. Teukolsky, W.T. Vetterling, and B.P. Flannery, *Numerical Recipes in C - The Art of Scientific Computing*, Cambridge University Press, Cambridge, UK, 1992.

[36] P. Salembier, A. Oliveras, and L. Garrido, "Motion connected operators for image sequences," in *Eurasip EUSIPCO'96, Trieste, Italy*, Sept. 1996, number ME.5.

[37] L. Garrido, A. Oliveras, and P. Salembier, "Motion analysis of image sequences using connected operators," in *SPIE Visual Communications and Image Processing, VCIP'97, San Jose, CA, USA*, Feb. 1997, vol. 3024, pp. 546–557.

[38] P. Salembier, A. Oliveras, and L. Garrido, "Antiextensive connected operators for image and sequence processing," *IEEE Trans. Image Processing*, vol. 7, no. 4, pp. 555–570, Apr. 1998.

[39] A.J. Viterbi and J.K. Omura, *Priciples of Digital Communications and Coding*, McGraw-Hill, New York, NY, 1979.

[40] G. Borgefors, "Distance transformations in digital images," *Computer Vision, Graphics, and Image Processing*, vol. 34, pp. 344–371, 1986.

[41] E.R. Davies, *Machine Vision: Theory, Algorithms, Practicalities*, Academic Press, London, UK, 1990.

[42] E.W. Dijkstra, "A note on two problems in connexion with graphs," *Numerische Mathematik*, vol. 1, pp. 269–271, 1959.

[43] A. Rosenfeld and J.L. Pfaltz, "Sequential operations in digital picture processing," *Journal of the Association for Computing Machinery*, vol. 13, no. 4, pp. 471–494, Oct. 1966.

Chapter 6

MPEG-4 - Standard for Multimedia Applications

6.1 Introduction

After having completed the tasks of setting two highly acclaimed international standards for audiovisual applications, i.e.

- MPEG-1 (ISO/IEC 11172), a standard for storage and retrieval of moving pictures and audio on storage media
- MPEG-2 (ISO/IEC 13818), a standard for digital television.

MPEG is now working to produce MPEG-4, a standard for multimedia applications, scheduled for the status of International Standard in December 1998 with the ISO number 14496 [1]. This is prompted by a need for a new standard with the convergence of the three separate technologies in the fields of entertainment and communications (e.g., digital television), interactive graphics applications (e.g., animation) and interactive multimedia applications (e.g., World Wide Web). There are clearly commonalities in the technologies with regard to the production, distribution and access of content and MPEG-4 aims to provide the standardized elements enabling the integration of the three processes within a common framework.

6.2 MPEG-4 Development Process

MPEG-4 work started in July 1993. The first Call for Proposal was made two and a half years later after the definition of the scope has been achieved.

All interested parties, whether within or outside MPEG, are invited to participate so as to attract state-of-the-arts technology to be considered for incorporation into the new standard. After that first call, other calls were issued for other technology areas.

The proposals of technology received were assessed and, if found promising, incorporated in the so-called Verification Models (VMs). A VM describes, in text and some sort of programming language, the operation of encoder and decoder. VMs are used to carry out simulations with the aim to optimize the performance of the coding schemes.

It is envisaged that with the rapid advance of computer processing power, software platforms will gain increasing importance in the implementation of the standard. Therefore MPEG decided to maintain software implementation of the different parts of the standard which can be used for the purpose of testing in the development process and for commercial implementations.

When sufficient confidence has been achieved in the stability of the standard under development, a Working Draft (WD) is produced which may undergo several revisions. The WDs already had the structure and form of a standard but they were kept internal to MPEG for revision. The WD then becomes a Committee Draft (CD) which undergoes a formal ballot by National Bodies (NBs). Ballots by NBs are usually accompanied by technical comments. If the number of positive votes is more than 2/3 of the total, the CDs will become Final CDs or FCDs.

The FCDs will be sent again to the National Bodies for a second ballot, the outcome of which will be considered with a similar process as for the CD stage. After that, the FCD becomes a Final Draft International Standard (FDIS). It will then be sent to National Bodies for a final ballot where NBs are only allowed to cast a yes/no ballot without comments. If the number of positive votes is above 75%, the DIS will become International Standard (IS) and is sent to the ISO Central Secretariat for publication.

6.3 Features of the MPEG-4 Standard [2]

Unlike the previous video coding standards where the coding unit is frame or set of frames, MPEG-4 adopts a different content-based approach [2]. The shift in paradigm is prompted by the realisation that the traditional coding techniques do not take into account the semantic properties of the image content coded and therefore result in objectionable artefacts such as "blocking" artefacts. Moreover, the overriding need of a new standard to provide new functionalities renders the current standards inappropriate.

MPEG-4 addresses the following content-based functionalities:

- content-based interactivity
  - content-based multimedia data access tools
  - content-based manipulation and bitstream editing
  - hybrid natural and synthetic data coding
  - improved temporal random access
- compression
  - improved coding efficiency
  - coding of multiple concurrent data streams
- universal access
  - robustness in error-prone environments
  - content-based scalability

MPEG-4 achieves these goals by providing standardized ways to:

1. represent units of aural, visual or audiovisual content, called audiovisual objects (AVOs) which can be of natural or synthetic origin;
2. describe the composition of these objects to create compound AVOs that form audiovisual scenes;
3. multiplex and synchronize the data associated with AVOs, so that they can be transported over network channels providing a QoS appropriate for the specific AVOs; and
4. interact with the audiovisual scene generated at the receiver's end.

6.3.1 Coded Representation of Primitive AVOs

With reference to Fig. 6.1, the audiovisual scene is made up of AVOs which can be classified into primitive and compound AVOs. Examples the primitive AVOs are the talking person, the voice associated with the person, the background, the desk, etc. The talking person together the voice constitute a compound AVO. MPEG-4 defines the coded representation of the AVOs, be it natural or synthetic, 2- or 3-dimensional. To satisfy the above stated functionalities, the coding algorithm must be more efficient than the current standards and the coded representation allows the AVOs to be handled and accessed independently and interactively.

6.3.2 Composition of AVOs

MPEG-4 provides a standardized way to describe a scene, allowing the user to:

- place AVOs anywhere in a given coordinate system;
- apply transforms to change the geometrical or acoustical appearance of a AVO;
- group primitive AVOs in order to form compound media objects;
- apply streamed data to AVOs, in order to modify their attributes;
- change interactively the user's viewing and listening points anywhere in the scene.

Again, with reference to Fig. 6.1, for example, one can replace the person with a different person, changes her dress or hairstyle; group the desk and the globe to form a compound AVO since they are static; or change the background using a different sprite. The above features are made possible by a scene description tool which builds on several concepts from VRML in terms of both its structure and the functionality of object composition nodes.

6.3.3 Description, Synchronization and Delivery of Streaming Data for AVOs

The MPEG-4 System Layer Model is depicted in Fig. 6.2. The AVOs are delivered as streaming data in one or more *elementary streams*. Each AVO is identified by an *object descriptor* in the elementary streams, and each stream is characterised by a set of descriptors conveying the configuration and quality of service (QoS) information. These enable the decoder to determine the required resources and the precise timing for decoding, and the QoS the encoder requests for transmission.

In the *synchronization layer*, the elementary streams are packetized into access units and each access unit is timestamped for synchronization. Independent of the media type, this layer allows identification of access units in elementary streams, recovery of the AVOs or scene description's time base and enables synchronization among them.

The *delivery layer* contains a two-layer multiplexer. The first multiplexer layer known as the "FlexMux" layer, is defined according to the DMIF specification (part 6 of MPG-4 standard). This allows grouping of

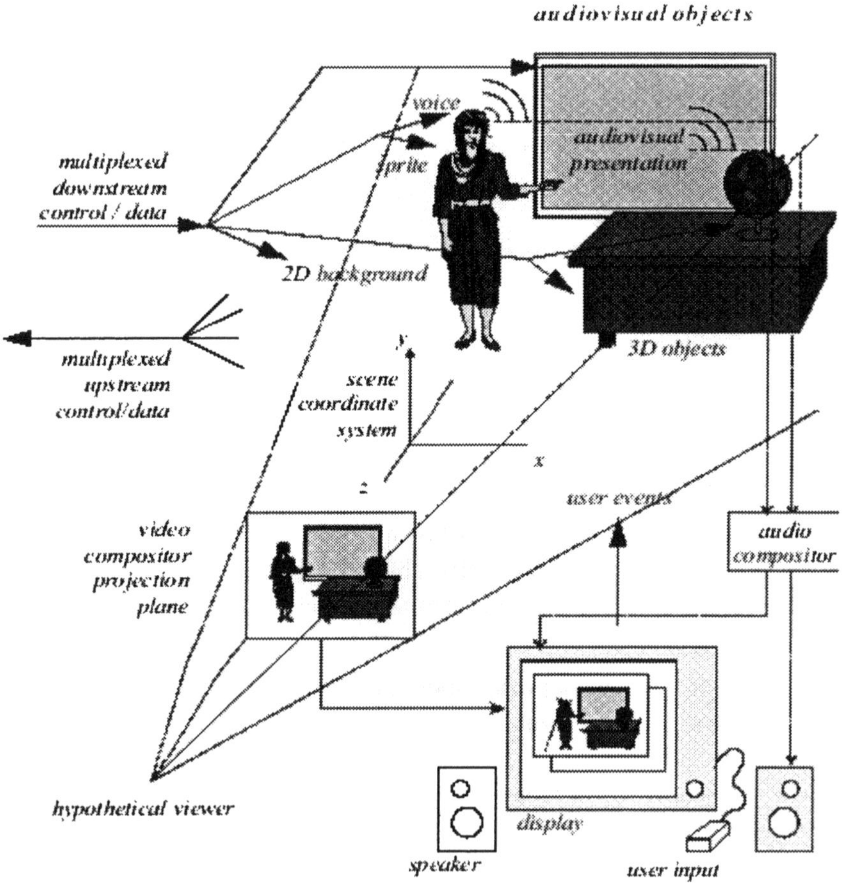

Figure 6.1: An example of an MPEG-4 audiovisual scene. ©ISO/IEC 1998

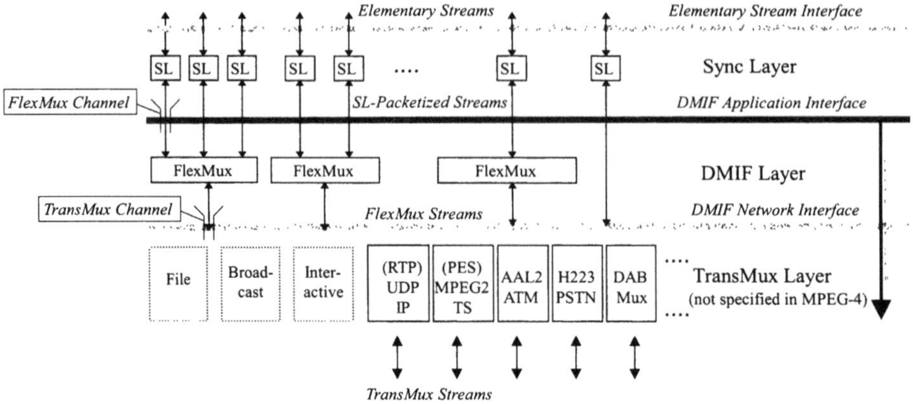

Figure 6.2: The MPEG-4 System Layer Model. ©ISO/IEC 1998

the elementary streams with a low multiplexing overhead. The "TransMux" layer offers transport services matching the requested QoS. Note that only the interface to this layer (i.e., DMIF Network Interface) is specified by MPEG-4 while the mapping of the data packets and control signalling are to be provided by the respective transport protocol offering the service.

Use of the FlexMux multiplexing tool is optional but the synchronization layer must always be present. Hence with the MPEG-4 System Layer model as shown in Fig. 6.2, the following functionalities can be provided:

1. identify access units, transport timestamps and clock reference information and identify data loss;

2. optionally interleave data from different elementary streams into FlexMux streams;

3. convey control information to

   - indicate the required QoS for each elementary stream and FlexMux stream;
   - translate such QoS requirements into actual network resources;
   - associate elementary streams to media objects;
   - convey the mapping of elementary streams to FlexMux and TransMux channels

6.3.4 Interaction with AVOs

MPEG-4 provides functionalilties for the user to interact with the audiovisual scene. Depending on the extent allowed by the author of the scene, the user has the ability to:

- change the viewing/listening point of the scene;
- drag objects in the scene to a different position;
- trigger a cascade of events by clicking on a specific object;
- select the desired language when multiple language tracks are available;
- trigger more complex kinds of behavior, e.g., answering a virtual phone.

6.3.5 Identification of Intellectual Property

It is important to have the possibility to identify intellectual property being coded into MPEG-4 media objects. Therefore, MPEG works with representatives of different creative industries in the definition of syntax and tools to support this. A full elaboration of the requirements for the identification of intellectual property can be found in MPEG97/N1918 [3], which is publicly available.

6.4 Technical Description of the MPEG-4 Standard

The MPEG-4 standard, as in the cases for MPEG-1 and MPEG-2 standards, defines an idealised decoding device together with the bitstream syntax and semantics in the form of a System Decoder Model. Fig. 6.3 shows the major components of an MPEG-4 decoder. As shown in Fig. 6.3, the MPEG-4 bitstreams are received as TransMux streams and demultiplexed into the FlexMux streams and then into the elementary streams. The elementary streams are parsed and passed to the appropriate decoders to recover the data in the AVOs and the scene description information needed for scene composition. The AVOs are composed into an audiovisual scene according to the scene description information as described by the author in the composition layer. The composed audiovisual scene is then rendered on the display or listening devices. The end user has the ability of interacting with the individual AVOs to the extend allowable by the author.

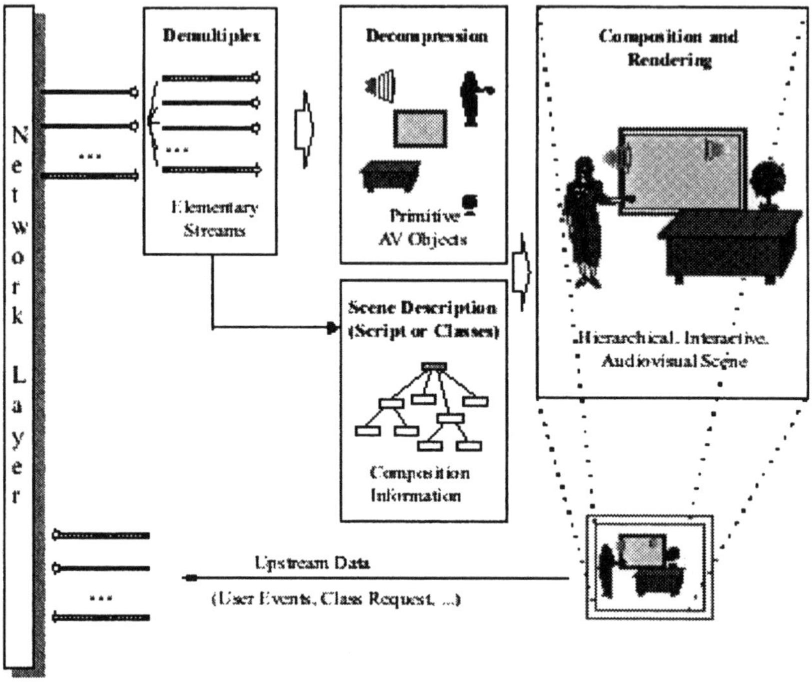

Figure 6.3: Major components of an MPEG-4 decoder. ©ISO/IEC 1998

For intellectual property rights (IPR) protection, the coded AVOs are supplemented with an optional Intellectual Property Identification (IPI) data set, carrying information about the right holders. The provision of the data sets allows the implementation of mechanisms for audit trailing, monitoring, billing and copy protection.

6.4.1 DMIF

The Delivery Multimedia Integration Framework (DMIF) is a session protocol for the management of multimedia streaming over generic networks. The DMIF architecture is such that applications which rely on DMIF for communication do not have to be concerned with the underlying communication method. The implementation of DMIF takes care of the delivery technology details presenting a simple interface to the application. Fig. 6.4 shows a conceptual model of the DMIF architecture. As shown in Fig. 6.4, the DMIF is located between the MPEG-4 application and the transport network.

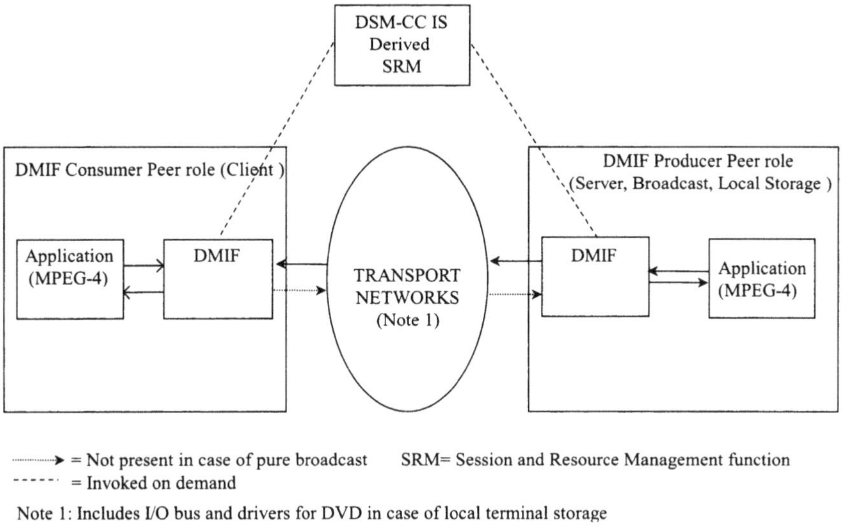

Figure 6.4: A conceptual model of the DMIF architecture. ©ISO/IEC 1998

DMIF presents a consistent interface to the application irregardless of whether MPEG-4 streams are received from a remote interactive peer over the transport networks and/or by interacting with broadcasting or storage media. An MPEG-4 application initiates a request for service through the DMIF to a another peer application or multiple peer applications. It may contain requests for certain QoS and specific channel bandwidth for that service. An interactive peer over a network may select a service, obtain a scene description and request specific streams for AVOs from the scene to be transmitted with appropriate QoS. DMIF ensures the timely establishment of channels with the specified QoSs and bandwidths over a variety of intervening networks between interactive peers. Control of DMIF spans both the FlexMux and TransMux layers as shown in Fig. 6.2 above. MPEG-4 offers a transparent interface with signalling primitive semantics at the interface to DMIF which are interpreted and translated into the appropriate protocols of each network. The exact maping for these translation are beyond the scope of MPEG-4 but left to the network provided.

The DMIF SRM functionality in Fig. 6.4 encompasses the MPEG-2 DSM-CC SRM functionality which is optional. However, DMIF provides a globally unique network session identifier which can be used to tag the resources and log their usage for subsequent billing.

6.4.2 Demultiplexing, Sychronization and Buffer Management

MPEG-4 defines a System Decoder Model along with the bitstream syntax and semantics that enables a compliant MPEG-4 decoder to decode the bitstream successfully. It provides the precise definition of the terminal's operation without making unnecessary assumptions about implementation details.

6.4.2.1 Demultiplexing

The incoming data from some network connection or a storage device have to be demultiplexed to retrieve the individual elementary streams. Demultiplexing occurs on the delivery layer that is modeled as consisting of a TransMux layer and a DMIF layer. The data retrieval consists of two tasks. Firstly, the channels must be located and opened which requires a transport control entity that manages, among others, the tables that associate transport channels to specific elementary streams. The linking of each stream to the actual channel as well as the management of the sessions and channels is handled in the DMIF layer.

Secondly, the incoming streams must be properly demultiplexed to recover SL-packetized streams from downstream channels to be passed on to the synchronization layer. In interactive applications, a corresponding multiplexing stage will multiplex upstream data in upstream channels.

The Multiplex layer is to abstract any underlying multiplex functionality that is suitable to transport MPEG-4 data streams. Note that this layer is not defined in MPEG-4. The TransMux Layer is assumed to provide protection and multiplexing functionality, indicating that this layer is responsible for offering a specific QoS.

On the other hand, the FlexMux tool is specified by MPEG-4 to optionally provide a flexible, low delay, low overhead method for interleaving data when the packet size or overhead of the underlying protocol stack is large. The FlexMux requires reliable error detection and sufficient framing of FlexMux packets (for random access and error recovery) from the underlying layer.

6.4.2.2 Sychronization

The sync layer has a minimum set of tools for consistency checking, padding, to convey time base information and to carry time stamped access units of an elementary stream. Time stamps are used to convey the nominal decoding

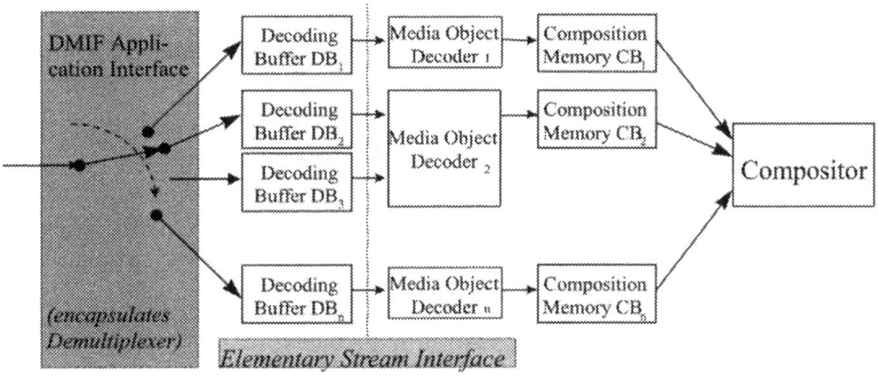

Figure 6.5: Buffer architecture of the System Decoder Model. ©ISO/IEC 1998

and composition time for an access unit. The sync layer requires reliable error detection and framing of each individual packet from the underlying FlexMux layer.

To be able to relate elementary streams to media objects within a scene, object descriptors are used which themselves are conveyed in one or more elementary streams. Object descriptors convey information about the number and properties of elementary streams that are associated to particular AVOs.

6.4.2.3 Buffer Management

To predict how the decoder will behave when it decodes the various elementary data streams that form an MPEG-4 session, the Systems Decoder Model enables the encoder to specify and monitor the minimum buffer resources that are needed to decode a session. The required buffer resources are conveyed to the decoder within object descriptors during the setup of the MPEG-4 session, so that the decoder can decide whether it is capable of handling this session. Fig. 6.5 shows the buffer architecture of the System Decoder Model.

6.4.2.4 Time Identification

For real time operation, a timing model is assumed in which the end-to-end delay from the signal output from an encoder to the signal input to a decoder is constant. Furthermore, the transmitted data streams must contain implicit or explicit timing information. There are two types of timing information. The first is used to convey the speed of the encoder clock, or time base, to the decoder. The second, consisting of time stamps attached to portions of the encoded AV data, contains the desired decoding time for access units or composition and expiration time for composition units. This information is conveyed in SL-packet headers generated in the sync layer. With this timing information, the inter-picture interval and audio sample rate can be adjusted at the decoder to match the encoder's inter-picture interval and audio sample rate for synchronized operation.

6.4.3 Syntax Description

MPEG-4 defines a syntactic description language to describe the exact binary syntax of a bit stream conveying data for a media object representation as well as that of the scene description information. This language is an extension of C++, and is used to describe the syntactic representation of objects and the overall media object class definitions and scene description information in an integrated way.

6.5 Coding of Audio Objects

MPEG-4 provides tools for coding of natural sounds (e.g. speech and music) and synthesized sounds from structured descriptions. The synthesized sounds can be derived from text data and instrument descriptions. The representations can be compressed and provide effects such as reverberation and spatialization. The audio coding tools covering bit rates from 6 kbits/s to 24 kbits/s are aimed at applications in AM digital audio broadcasting to provide improvements over the existing AM modulation services. Several codecs (e.g. CELP, Twin VQ and ACC) have been tested and compared to a reference AM system and found to give superior performance.

6.5.1 Natural Sound

For natural sound coding, MPEG-4 supports bit rates from 2 kbits/s up to 64 kbits/s. Coding at less than 2 kbits/s is also supported for variable rate coding. The inclusion of MPEG-2 AAC standard in MPEG-4 provides coding of higher quality audio at higher bit rates.

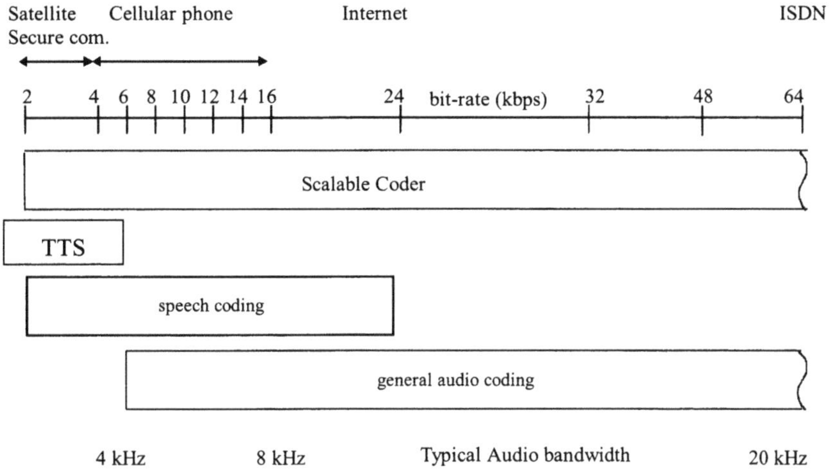

Figure 6.6: MPEG-4 audio coding framework. ©ISO/IEC 1998

6.5.1.1 Speech Coding

For bit rates between 2 to 4 kbits/s, Harmonic Vector eXcitation Coding (HVXC) is used; whilst at higher bit rates of between 4 to 24 kbits/s, Code Excited Linear Predictive (CELP) Coding is employed. HVXC can also operate at a lower bit rate of 1.2 kbits/s when coding in variable bit rate mode. In order to support narrowband and wideband speech, sampling rates of 8 and 16 kHz, respectively, are used.

6.5.1.2 Audio Coding

Transform coding techniques, such as Twin VQ and AAC, are employed to code audio at ≥ 6 kbits/s with a typical sampling rate of 8 kHz.

Fig. 6.6 shows the MPEG-4 audio coding framework.

6.5.1.3 Audio Scalability

Scalability in MPEG-4 can be in terms of:

- bit rate scalability which allows a bitstream to be parsed into a bitstream of lower bit rate to provide audio of lower quality;

- bandwidth scalability where part of the bitstream representing a part of the spectrum is discarded;

- encoder/decoder complexity scalability that allows encoder/decoder of different complexities to generate meaningful bitstreams.

Scalability tools can be applied to a combination of techniques, e.g., with CELP as a base layer and AAC as an enhancement layer.

6.5.2 Synthesized Sound

Decoders can synthesize sounds from structured inputs. Speech is converted from text input in the Text-to-Speech (TTS) coder, and more general sounds including music may be normatively synthesized with extremely low bit rate.

6.5.2.1 Text-to-Speech

MPEG-4 provides an interface for a TTS coder which allows the generation of intelligible synthetic speech from a text or a text with prosodic parameters. TTS coders operate with bit rates ranging from 200 bits/s to 1.2 kbits/s. It supports tools that allow synchronization to associated face animation, international language (for text) and international symbols (for phonemes).

6.5.2.2 Score Driven Synthesis

For audio, the decoding tools are driven by a synthesis language called Structured Audio Orchestra Language (SAOL) which defines the instruments of an orchestra downloadable in the bitstream. An instrument consists of a cluster of signal processing primitives implemented either by hardware or software that emulate some specific sounds generated by an actual acoustic instrument. Note that MPEG-4 does not standardize a method of synthesis but rather one of "describing" the synthesis.

Synthesis is controlled by the use of "scores" which are a time-sequenced set of commands that invokes various instruments at specific times to compose an overall music performance. The scores are described by a language known as Structured Audio Score Language (SASL) that can create new sounds or modify existing sounds. Careful control of the synthesis process enables the synthesis of a wide range of audio and sound effects, from footsteps, wind blowing, waterfalls, to conventional music and synthesized electronic music.

MPEG-4 also standardizes a simpler synthesis technique, the "wavetable bank format" for terminals with less functionality or applications not requiring the more sophisticated synthesis described above. This format allows the generation of sound samples from a downloadable wavetable. Simple processing such as filtering, reverberation, chorus effects are also possible.

6.6 Coding of Natural Visual Objects

Visual objects can be either of natural or of synthetic origin. First, the objects of natural origin are described.

The tools for representing natural video in the MPEG-4 visual standard aim at providing standardized core technologies allowing efficient storage, transmission and manipulation of textures, images and video data for multimedia environments. These tools will allow the decoding and representation of atomic units of image and video content, called "video objects" (VOs). An example of a VO could be a talking person (without background) which can then be composed with other AVOs (audio-visual objects) to create a scene. Conventional rectangular imagery is handled as a special case of such objects.

In order to achieve this broad goal rather than a solution for a narrow set of applications, functionalities common to several applications are clustered. Therefore, the visual part of the MPEG-4 standard [4] provides solutions in the form of tools and algorithms for:

- efficient compression of images and video
- efficient random access to all types of visual objects
- extended manipulation functionality for images and video sequences
- content-based coding of images and video
- content-based scalability of textures, images and video
- spatial, temporal and quality scalability
- error robustness and resilience in error prone environments

6.6.1 Video Object Plane (VOP)

6.6.1.1 VOP Definition

The Video Object (VO) correspond to entities in the bitstream that the user can access and manipulate (cut, paste...). Instances of Video Object in given time are called Video Object Plane (VOP). The encoder sends together with the VOP, composition information (using composition layer

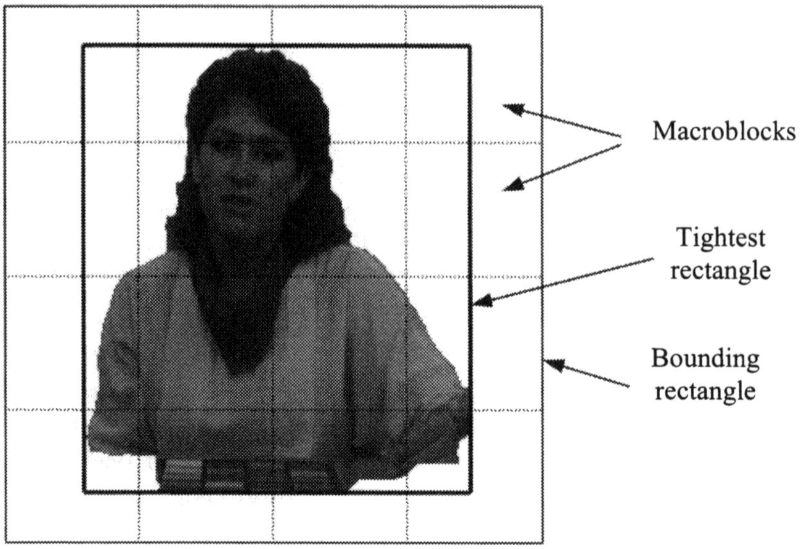

Figure 6.7: VOP Formation.

syntax) to indicate where and when each VOP is to be displayed. At the decoder side the user may be allowed to change the composition of the scene displayed by interacting on the composition information.

The VOP can be a semantic object in the scene: it is made of Y, U, V components plus shape information. When the sequence has only one rectangular VOP of fixed size displayed at fixed interval, it corresponds to the frame-based coding technique.

The exact method used to generate the VOP from the video sequences is not standardized in MPEG-4.

6.6.1.2 VOP Formation

The shape information is used to form a VOP. The VOP is formed by first drawing the tightest rectangle around the object. The rectangle is then extended to a bounding rectangle that contains a multiple of macroblocks as shown in Fig. 6.7. This ensures that the VOP contains a minimum number of macroblocks to represent the object.

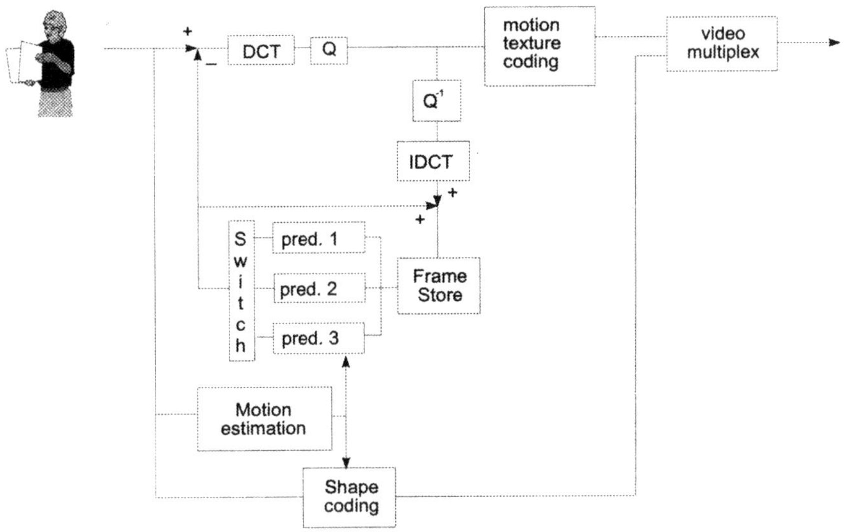

Figure 6.8: VOP encoder structure. ©ISO/IEC 1998

6.6.2 The Encoder

6.6.2.1 Overview

Fig. 6.8 presents a general overview of the VOP encoder structure. The same encoding scheme is applied when coding all the VOPs of a given session.

The encoder is mainly composed of two parts: the shape coder, and the traditional motion and texture coder applied to the same VOP. The phase between the luminance and chrominance samples of the bounding rectangle has to be correctly set according to the 4:2:0 format, as shown in Fig. 6.9. Specifically the top left coordinate of the bounding rectangle should be rounded to the nearest even number not greater than the top left coordinates of the tightest rectangle. Accordingly, the top left coordinate of the bounding rectangle in the chrominance component is that of the luminance divided by two.

The chrominance alpha plane is created from the luminance alpha plane by a conservative subsampling process. In the case of a binary alpha plane, this ensures that there is always a chroma sample where there is at least one luma sample inside the VOP.

Binary alpha plane: For each 2×2 neighborhood of luminance alpha

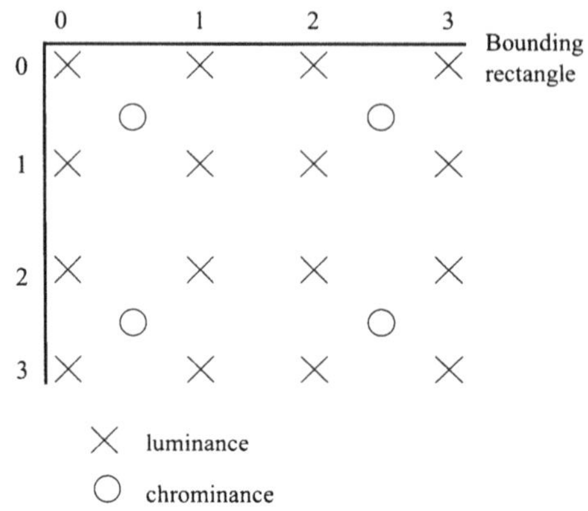

Figure 6.9: Luminance versus chrominance bounding box positioning. ©ISO/IEC 1998

pixels, the associated chroma alpha pixel is set to 255 if any of the four luminance alpha pixles are equal to 255.

Grayscale alpha plane: For each 2 × 2 neighborhood of luminance alpha pixels, the associated chroma alpha pixel is set to the rounded average of the four luminance alpha pixels.

6.6.3 Shape Coding

The shape information are in the form of binary or grey scale alpha planes as defined above. The binary alpha planes are encoded by modified context-based arithmetic encoding (CAE) whilst the grey scale alpha planes are encoded by motion compensated DCT similar to texture coding. An alpha plane is bounded by a rectangle that contains the shape of a VOP.

6.6.3.1 Binary Alpha Block Coding

The binary alpha plane contains 16 × 16 binary alpha blocks (BABs). The pixels in the BAB are raster-scanned and encoded by CAE. For I-VOPs, the BAB is encoded in INTRA mode whilst in P-VOPs, it may be encoded in INTRA or INTER mode.

The CAE Algorithm

The CAE encoding process begins by computing a context number of each pixel to be encoded. A probability table is then indexed according to the

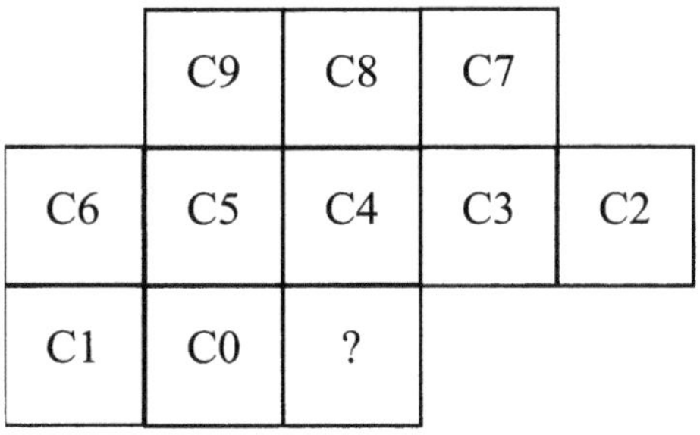

Figure 6.10: The INTRA template used in context computation. "?" denotes the pixel for which context is to be found. ©ISO/IEC 1998

computed context number. Finally the indexed probability is used to drive the arithmetic encoder. When the final pixel has been encoded, the process is terminated. The encoding process generates a single binary arithmetic codeword (BAC).

Context Computation

For INTRA coded BABs, the context for each pixel is given by

$$C = \sum_k c_k 2^k. \tag{6.1}$$

The positions of the pixels c_k are as illustrated in Fig. 6.10.

For INTER coded BABs, context computation exploits of the temporal redundancy provided by the motion-compensated BAB as depicted in Fig. 6.11.

BAB Borders

In calculating the context of each pixel of a BAB, pixels from neighboring BABs can be used. For INTRA and INTER cases, a border of 2 pixels in width is used as shown in Fig. 6.12. The top and left borders contains pixels from previously encoded and reconstructed BABs and the bottom and right borders contains zeroes which are unknown at decoding time.

BAB Encoding Decision

The BAB is encoded under all possible coding conditions both in INTRA and INTER modes. The coding condition which results in the shortest code is chosen to encode the BAB.

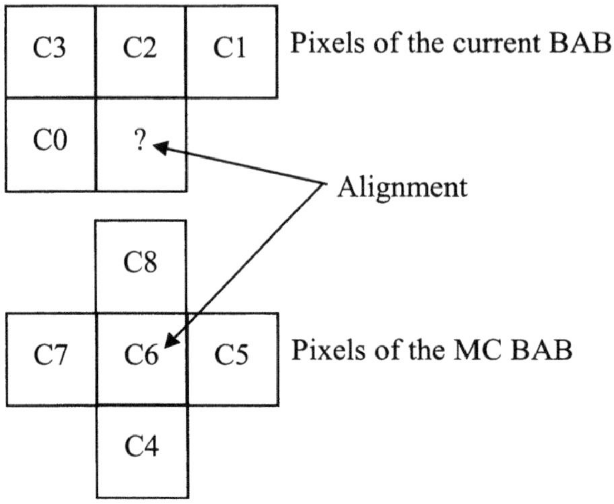

Figure 6.11: The INTER template used in context computation. "?" denotes the pixel for which context is to be found. ©ISO/IEC 1998

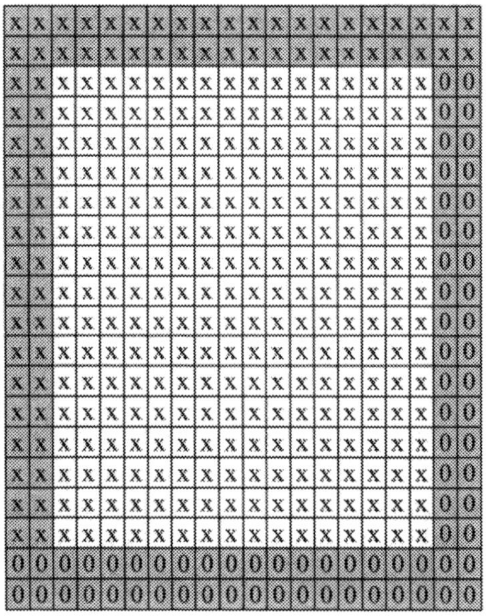

Figure 6.12: Current BAB and its borders (light grey area). ©ISO/IEC 1998

For I-VOPs, the BAB can be encoded in two ways:

- INTRA
- Transposition of the bordered BAB followed by INTRA

Similarly, for P-VOPs, the BABs can be enceode in the following ways:

- INTER
- Transposition of the bordered BAB and the MC BAB followed by INTER

The shortest codes from the INTRA and INTER modes are then compared and the coding condition which generates the shorter code is chosen.

6.6.3.2 Gray Scale Shape Coding

Gray scale alpha planes are employed for higher-quality content where each pixel belonging to a shape is assigned a value for its transparency. With this feature, transparency can differ from pixel to pixel of an object, and objects can be smoothly blended, either into a background or with other visual objects. An application of this is the superposition of a layer or layers of text or images (natural or synthetic) over a video sequence. This is particularly useful when a translucent effect is to be created, eg., subtitling, or composing synthetic objects with natural image background.

Support Function and Alpha Values Coding

Gray scale shape coding is performed by encoding the gray-level alpha plane in two separate parts: a binary alpha plane as its support function, and its alpha values as texture with arbitrary shape. The support is obtained by threshold the gray-level alpha plane by 0. The alpha values are encoded as 16×16 alpha macroblocks, the same way as luminanace in texture coding. The encoded alpha macroblock data are appended to the end of the corresponding texture macroblock for transmission.

Feathering and Translucency Coding

For most cases, when the gray-scale alpha mask is used, the texture within the alpha mask is fairly simple. It could consists of just a constant gray level or a binary alpha mask with values around the edges tapering from 255 to 0 to provide a gradual merging to the background. The process of smoothing the edge for merging to the background is known as feathering. It is done using two approaches: a feathering algorithm and a feathering filter.

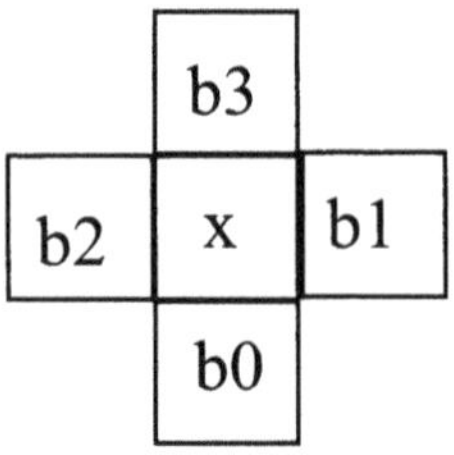

Figure 6.13: 3 × 3 filter kernel. ©ISO/IEC 1998

Feathering Algorithm

This algorithm tapers the alpha values of the pixels within a distance from the shape boundary linearly from the opaque alpha values to 0. Thus, the feathered alpha values are given by:

$$alpha = distance/(feather\_distance + 1) * opaque\_value \tag{6.2}$$

where *distance* is the distance of the pixel from the shape boundary, whereas $feature\_distance$ specifies the number of pixels to feather, and *opaque_value* is the value in the alpha mask.

6.6.3.3 Feathering Filter

A feathering filter is a 3 × 3 kernel as shown in Fig. 6.13 where x denotes the pixel to be feather and b1, b2 and b3 the surround pixels.

The filtering operation is as follows:

1. Label the alpha pixels of the input alpha mask to be 1 if it is 255, and 0 otherwise.

2. If $x = 0$, leave the pixel value intact. Otherwise, replace the pixel value with the value depending on its surrounding condition as given in Table 6.1.

3. The filtering operation is cascaded if the number of filtering iteration is more than 1. Note the the feathering filter can be different for each iteration and this is specified in the Video Object Layer.

Six modes of feathering operation can be identified in the Video Object Layer as represented by a 4-bit code as part of the VOL_descriptor:

Table 6.1: Feathering filter description table. ©ISO/IEC 1998

b3	b2	b2	b0	x'
0	0	0	0	x0
0	0	0	1	x1
0	0	1	0	x2
0	0	1	1	x3
0	1	0	0	x4
0	1	0	1	x5
0	1	1	0	x6
0	1	1	1	x7
1	0	0	0	x8
1	0	0	1	x9
1	0	1	0	x10
1	0	1	1	x11
1	1	0	0	x12
1	1	0	1	x13
1	1	1	0	x14

- no effects
- linear feathering
- constant alpha
- linear feathering and constant alpha
- feathering filter
- feathering filter and constant alpha

Linear filtering mode makes use of the feathering algorithm and feathering filter mode uses the feathering filter described above. For the linear feathering and feathering filter modes, the input can be a sequence of binary or grayscale alpha masks. For the constant alpha, linear feathering and constant alpha, and feathering filter and constant alpha modes, the input will be a sequence of grayscale alpha masks. If the mode selected is no effects, the default alpha mask compression algorithm specified will be utilized.

6.6.4 Motion Estimation and Compensation

Motion estimation is performed on a VOP basis. For the macroblocks on the VOP borders, motion estimation is modified from block matching to polygon matching. To achieve that, a special padding technique, i.e., the macroblock-based repetitive padding is used to pad the transparent region of the macroblocks in the reference VOP.

The absolute (frame) coordinate system is used for referencing all of the VOPs. At each particular time instance, a bounding rectangle is defined which includes the shape of that VOP. The left and top corner, in their absolute coordinates, of the bounding box is encoded in the VOP spatial reference. No alignment of VOP bounding boxes at different time instances is performed.

In addition to the motion estimation and compensation mode, two additional modes are supported, namely, unrestricted and advanced modes. In all three modes, the motion vector search range is $[-2^{f\_code+3}, -2^{f\_code+3} - 0.5]$ where $0 \leq f\_code \leq 7$. Unrestricted mode allows the motion vector to point outside the bounding box of the VOP. The advanced mode allows multiple motion vectors in one macroblock and overlapped motion compensation.

6.6.4.1 Padding Process

The padding process defines the values of luminance and chrominance samples outside the VOP for prediction of arbitrarily shaped objects. Fig. 6.14 shows a simplified diagram of this process.

A decoded macroblock $d[y][x]$ is padded by referring to the corresponding decoded shape block $s[y][x]$. The luminance component is padded per 16×16 samples, while the chrominance components are padded per 8×8 samples. A macroblock that lies on the VOP boundary (hereafter referred to as a *boundary macroblock*) is padded by replicating the boundary samples of the VOP towards the exterior. This process is divided into *horizontal repetitive padding* and *vertical repetitive padding.* The remaining macroblocks that are completely outside the VOP (hereafter referred to as *exterior macroblocks*) are filled by *extended padding.*

Horizontal and Vertical Padding

Horizontal repetitive padding is carried out by replicating each pixel at the boundary of a VOP horizontally to the left and/or right direction in order to fill the transparent region outside the VOP of a boundary macroblock. If there are two boundary pixels for filling a pixel position outside of a VOP,

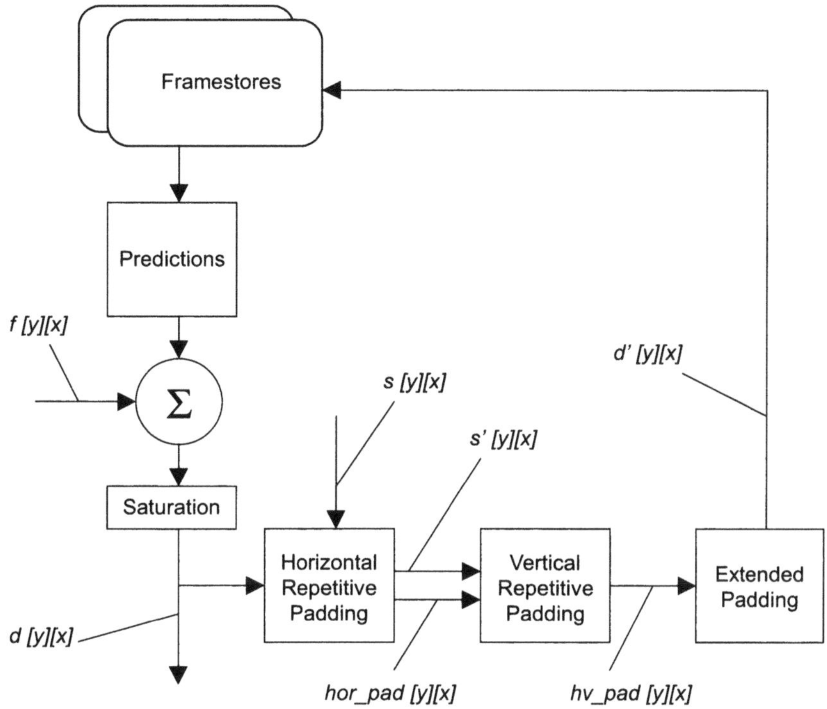

Figure 6.14: Simplified padding process. ©ISO/IEC 1998

the boundary pixel values are averaged. The remaining unfilled transparent horizontal samples are padded by a similar process as the horizontal repetitive padding but in the vertical direction. The samples already filled in are treated as if they were inside the VOP for the purpose of this vertical pass.

Extended Padding

Exterior macroblocks immediately next to boundary macroblocks are filled by replicating the samples at the border of the boundary macroblocks. If an exterior macroblock is next to more than one boundary macroblocks, If there are more one boundary macroblocks surrounding an exterior macroblock, the boundary macroblocks are numbered in priority according Fig. 6.15. The exterior macroblock is then padded by replicating upwards, downwards, leftwards, or rightwards the row of samples from the horizontal or vertical border of the boundary macroblock having the largest priority number. The remaining exterior macroblocks (not located next to any boundary macroblocks) are filled with 128.

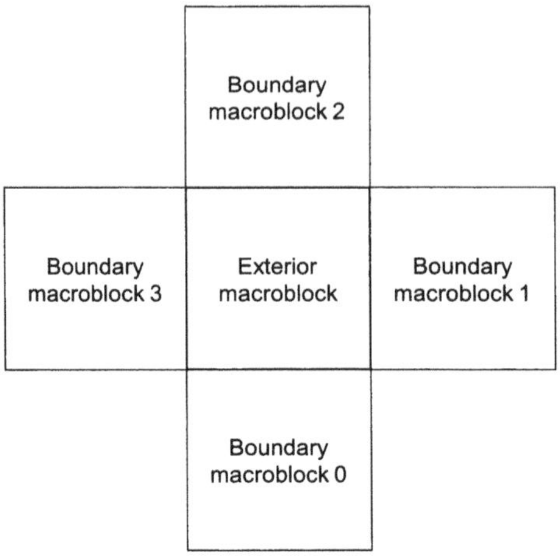

Figure 6.15: Priority of boundary macroblocks surrounding an exterior macroblock. ©ISO/IEC 1998

Padding for Chrominance Components

The above techniques for padding the luminance components are also applied to the chrominance components. The 8×8 chrominance blocks are generated by referring to a luminance shape block and decimating the shape block by a factor of 2 in the horizontal and vertical directions. The chrominance pixel is set to 1 if one of the 2×2 luminance shape samples is 1, otherwise it is set to 0.

Padding for Interlaced Macroblocks

The padding of interlaced macroblocks is the same as for the non-interlaced luminance and chrominance macroblocks except that it is performed for each field independently. A sample outside of a VOP is therefore filled with the value of the nearest boundary sample of the same field.

6.6.4.2 Basic Motion Estimation Techniques

Modified Block (Polygon) Matching

The principle of polygon matching is to use only the pixels within the VOP for motion estimation. To perform polygon matching, the bounding box of the VOP is first extended on the right-bottom corner to multiple of macroblocks, i.e., 16×16 for luminance VOPs, and 8×8 for chrominance

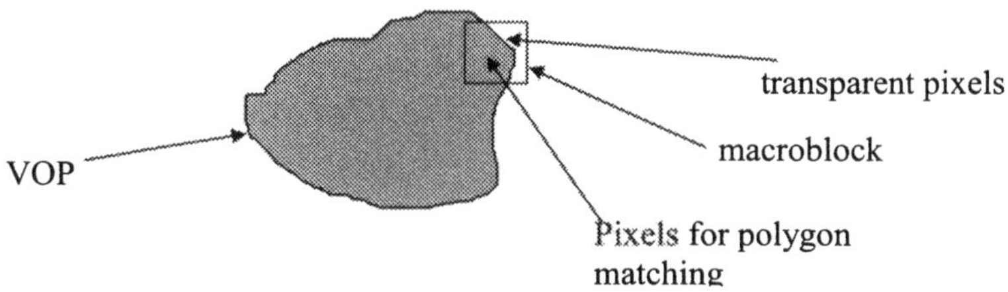

Figure 6.16: Polygon matching for an arbitrary shape VOP. ©ISO/IEC 1998

VOPs. The extended pixels and their associated alpha values are set to 0. The error criterion is the sum of absolute difference (SAD) computed using only the pixels with non-zero alpha value. The reference VOP is padded based on its own shape information. Fig. 6.16 illustrates an example.

Integer Pixel Motion Estimation

The motion estimation is performed by full search with a integer pixel accuracy within a maximum search area range specified by the *f_code*. The comparisons are made between the incoming block and the displaced block in the previous reconstructed VOP padded in a macroblock basis according to Section 6.6.4.1. The error measure is the SAD defined as:

$$SAD_N(x,y) = \sum_{i=1,j=1}^{N,N} |original - previous| * (Alpha_{original} \neq 0) \qquad (6.3)$$

where $(x, y) = [-64, 63]$, $n = 16$ or 8.

The SAD at the zero position $SAD_{16}(0,0)$ is reduced to favor the zero motion vector when there is no significant difference:

$$SAD_{16}(0,0) = SAD_{16}(0,0) - (N_B/2 + 1) \qquad (6.4)$$

N_B is the number of pixels inside the VOP multiplied by $2^{(bits\_per\_pixel-8)}$. The (x, y) pair resulting in the lowest $SAD_{16}(x, y)$ is chosen as the 16×16 integer pixel motion vector, V0.

Likewise, the (x, y) pairs resulting in the lowest $SAD_8(x, y)$ are chosen to give the four 8×8 vectors V1, V2, V3 and V4. The 8×8-based SAD for the macroblock is

$$SAD_{K\times 8} = \sum_{1}^{K} SAD_8(x,y) \tag{6.5}$$

where $0 < K \leq 4$ is the number of 8×8 blocks that do not lie outside of the VOP shape. Instead of full search, the 8×8 search is centered around 16×16 vector, with a search window of ± 2 pixels.

If interlaced video is being encoded, four field motion vectors are calculated for every macroblock (in addition to existing 16×16 and 8×8 motion vectors described above) for each reference VOP. The field motion vectors correspond to the four combinations of current field (top or bottom) and reference field (top or bottom). The top field consists of even lines (0, 2, 4, ... , $H - 2$), and the bottom field is composed of the odd lines (1, 3, 5, ldots, $H - 1$, where H is the frame height). The full-pel field motion vector $(fx_{p,q}, fy_{p,q})$ is defined by the minimum sum of absolute differences $fSAD_{p,q}$ given by

$$\begin{aligned} fSADp,q = \min_{(x,y)\in S, y \text{ even}} & \sum_{j=0}^{7}\sum_{i=0}^{15} |C[i, 2j+p] \\ & - OR[x_0 + x + i, y_o + y + 2j + q]| * (A[i, 2j+p] \neq 0) \end{aligned} \tag{6.6}$$

where

(x_0, y_0)	Upper left corner coordinates of the current macroblock
p	Field in the current frame/VOP (0 for top; 1 for bottom)
q	Field in reference frame/VOP (0 for top; 1 for bottom)
$C[x,y]$	Current macroblock luminance samples
$A[x,y]$	Current macroblock alpha values
$OR[x,y]$	Reconstructed reference VOP luminance samples
S	Search region: $\{(x,y) : -2^{f\_code+3} \leq x, y < x^{f\_code+3}\}$

The field motion vectors are unrestricted in the sense that pixels referenced above might fall outside of the VOP bounding box but within the padded extension. If any pixel is required beyond the padded reference VOPs, then the candidate motion vector is eliminated from the search region, S.

The full-pixel SAD for an interlaced P-VOP macroblock is given by:

$$\begin{aligned} SAD_{inter} = min\{ & SAD_16(x,y), SAD_{k\times 8}, min(fSAD_{0,0}, fSAD_{0,1}) \\ & + min(fSAD_{1,0}, fSAD_{1,1})\} \end{aligned} \tag{6.7}$$

INTRA/INTER Mode Decision

After the integer pixel motion estimation, the coder makes a decision on whether to use INTRA or INTER prediction in the coding. INTRA mode is chosen if:

$$A < (SAD_{inter} - 2 * N_B) \tag{6.8}$$

where

$$A = \sum_{i=1,j=1}^{16,16} |original - MB\_mean| * (!(Alpha_{original} == 0)) \tag{6.9}$$

and

$$SAD_{inter} = min(SAD_{16}(x,y), SAD_{k\times 8}) \tag{6.10}$$

$$MB_{mean} = (\sum_{i=1,j=1}^{N_c} original)/N_c \tag{6.11}$$

N_c is the number of pixels inside the VOP.

If INTRA mode is chosen, no further operations are necessary for the motion search. If INTER mode is chosen the motion search continues with half sample search around the V0 position, followed by quarter sample search if (quarter_sample $\neq$ 1).

Half Sample Search

Half sample search is performed using the previous reconstructed VOP on the luminance component of the macroblock, for 16×16 and 8×8 vectors as well as for 16×8 field motion vectors in the case of interlaced video. The search area is ± 1 half sample around the region pointed to by the motion vectors V0, V1, V2, V3, V4 or the field motion vector $(fx_{p,q}, fy_{p,q})$. The half sample values are calculated by bilinear interpolation horizontally and vertically as shown in Fig. 6.17 below:

The vector resulting in the best match during the half sample search is named MV. MV consists of horizontal and vertical components (MV_x, MV_y), both measured in half sample units.

For interlaced video, the half sample search used to refine the field motion vectors is conducted by vertically interpolating between lines of the same field. The field motion vectors are calculated in frame coordinates; that is the vertical coordinates of the integer samples differ by 2.

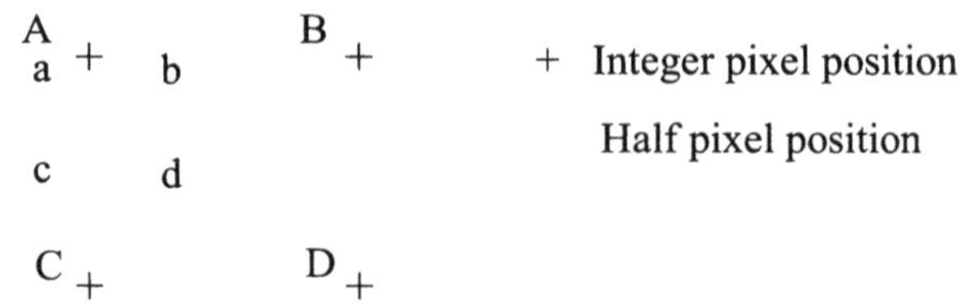

$$a = A$$
$$b = (A + B + 1 - rounding\_control)/2$$
$$c = (A + C + 1 - rounding\_control)/2$$
$$d = (A + B + C + D + 2 - rounding\_control)/4$$

Figure 6.17: Bilinear interpolation scheme. ©ISO/IEC 1998

Quarter Sample Search

For quanter sample search, an additional search step is performed for a search area of ± 1 quarter sample around the region pointed to by the best matched half sample motion vector MV. The quarter sample values are found by the same bilinear interpolation technique between surrounding half and integer samples respectively as shown in Fig. 6.18.

The vector resulting in the best match during the quarter sample search replaces the half sample vector MV. For quarter sample mode MV consists of horizontal and vertical components (MV_x, MV_y), both measured in quarter sample units.

Decision on 16×16 or 8×8 Prediction Mode

The decision to chose 16×16 or 8×8 prediction mode is based on the following criterion:

`If` $SAD_{k\times 8}(x, y) < SAD_{16}(x, y) - (N_B/2 + 1)$
  `choose` 8×8 `prediction`
`else`
  `choose` 16×16 `prediction`

Interlaced Video Prediction Mode Decision

In the case of interlaced video, the decision as to which prediction mdoe to use is by choosing the reference field that gives the smallest SAD (SAD_{top}

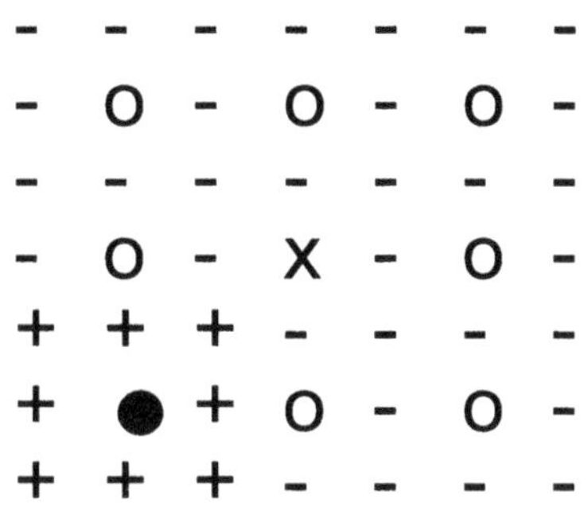

Figure 6.18: Location of best match integer (×), half (○) and quarter (−) sample; (•) marking the best match half sample, (+) the quarter sample search positions. ©ISO/IEC 1998

and SAD_{bottom}) from the field search. Therefore,

`if` $SAD_{16} \leq min(SAD_{16}, SAD_{K\times 8} + N_B/2 + 1, SAD_{top} + SAD_{bottom} + N_B/4 + 1)$
    16×16 `prediction`
`else if` $SAD_{K\times 8} + N_B/2 + 1 \leq min(SAD_{16}, SAD_{K\times 8} + N_B/2 + 1, SAD_{top} + SAD_{bottom} + N_B/4 + 1)$
    8×8 `prediction`
`else if` $SAD_{top} + SAD_{bottom} + N_B/4 + 1 \leq min(SAD_{16}, SAD_{K\times 8} + N_B/2 + 1, SAD_{top} + SAD_{bottom} + N_B/4 + 1)$
    `field based motion estimation`

Differential Coding of Motion Vectors

In inter coding, the motion vector is coded differentially by performing prediction based on the three neoghboring motion vectors as shown in Fig. 6.19. The differential coding is with reference to the reconstructed shape.

The prediction of the current motion vector is defined as:

$$\begin{aligned} P_x &= Median(MV1_x, MV2_x, MV3_x) \\ P_y &= Median(MV1_y, MV2_y, MV3_y) \end{aligned} \tag{6.12}$$

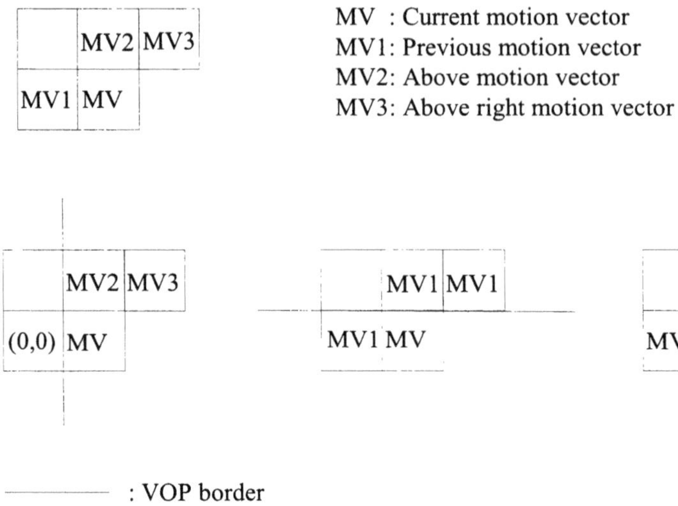

Figure 6.19: Motion vector prediction. ©ISO/IEC 1998

where $MV1$, $MV2$ and $MV3$ are the candidate predictors.

If error resilience is enabled, one dimensional prediction is used.

$$\begin{aligned} P_x &= MV1_x \\ P_y &= MV1_y \end{aligned} \tag{6.13}$$

The horizontal and vertical components of the differential motion vector are then given by:

$$\begin{aligned} MVD_x &= MV_x - P_x \\ MVD_y &= MV_y - P_y \end{aligned} \tag{6.14}$$

In the special case at the borders of the current VOP, the following decision rules are applied:

1. If only one candidate predictor is outside of the VOP, it is set to zero.

2. If only two candidate predictors are outside of the VOP, they are set to the third candidate predictor.

3. If all three candidate predictors are outside of the VOP, they are set to zero.

When interlaced video is encoded, if one or more of $MV1$, $MV2$ and $MV3$ refers to a field motion compensated macroblock, the value of MVi

is the average of the two field motion vectors. If a pixel offset finer than the motion vector resolution (half or quarter sample) is obtained by the average, it is replaced with a nearest pixel offset of the respective resolution.

If the current macroblock is a field motion compensated macroblock, then the same prediction motion vector (P_x, P_y) is used for both field motion vectors. Because the vertical component of a field motion vector has half the resolution of the horizontal component, the vertical differential motion vector encoded in the bitstream is

$$MVD_{y_{field}} = (MV_y - P_y)/2. \tag{6.15}$$

6.6.4.3 Unrestricted Motion Estimation/Compensation

Motion Estimation over VOP Boundaries

To improve the accuracy of motion estimation, the motion vector is allowed to point outside the decoded area of a reference VOP. For an arbitrarily shaped reference VOP, the decoded area refers to the area within the bounding box, padded as described in Section 6.6.4.1. The bounding box is extended multiples of 16 × 16 blocks in four directions (left, top, right and bottom) by 2^{f_code+3} pixels by repetitive padding. For the case of rectangular VOP, the rectangle is extended by the same amount in the same four directions. The target VOP remains the same except for extending to multiples of 16 × 16 blocks.

Modified block or polygon matching as described in Section 6.6.4.2 is applied to obtain the motion vectors.

Motion Compensation over VOP Boundaries

For motion compensation, when a sample referenced by a motion vector falls outside the decoded VOP area defined by the bounding box, an edge sample is used. This edge sample is the last full pixel position inside the decoded VOP area. Thus, the coordinates of the reference sample in the reference VOP are determined as follows:

$$\begin{aligned} x_{ref} &= \min[\max(x + dx, 0), x_{dim} - 1] \\ y_{ref} &= \min[\max(y + dy, 0), y_{dim} - 1] \end{aligned} \tag{6.16}$$

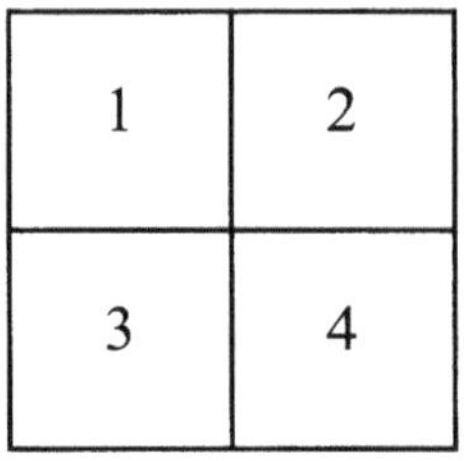

Figure 6.20: Numbering of luminance blocks in a macroblock. ©ISO/IEC 1998

where	(x, y)	are the coordinates of a sample in the current VOP,
	(x_{ref}, y_{ref})	are the coordinates of a sample in the reference VOP,
	(dx, dy)	is the motion vector, and
	(x_{dim}, y_{dim})	are the dimensions of the reference VOP.

6.6.4.4 Advanced Prediction Mode

Formation of the Motion Vectors

One, two or four motion vectors per macroblock may be used depending on the $MCBPC$ codeword and the $field\_prediction$ flag. If one motion vector MV is transmitted, all four block vectors take the same value as MV. When two field motion vectors are transmitted, each of the four block prediction motion vectors has the value equal to the average of the field motion vectors. If four motion vectors are transmitted for the current macroblock, MVD and MVD_{2-4} indicate the vector differences of luminance blocks 1 to 4 according to the block numbering in a macroblock as shown in Fig. 6.20 below. The motion vectors are obtained by adding predictors to MVD and MVD_{2-4} except that the candidate predictors $MV1$, $MV2$ and $MV3$ are redefined as illustrated in Fig. 6.21. If only one vector per macroblock is present, $MV1$, $MV2$ and $MV3$ are defined as for the 8×8 block numbered 1 in Fig. 6.20.

Motion vector MVD_{CHR} for both chrominance blocks is derived by calculating the sum of the K luminance vectors, that corresponds to K 8×8 blocks that do not lie outside the VOP shape, and dividing this sum by 2 in the case of half sample mode.

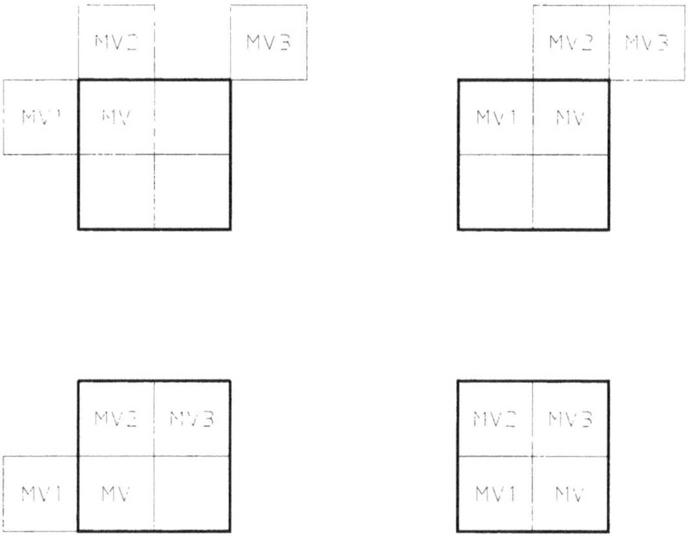

Figure 6.21: Redefinition of the candidate predictors $MV1$, $MV2$ and $MV3$ for each of the luminance blocks in a macroblock. ©ISO/IEC 1998

Overlapped Motion Compensation for Luminance Block

In overlapped motion compensation for luminance block, three motion vectors in the overlapped region (as illustrated in Fig. 6.22) are used:

- motion vector of the current block;
- motion vector of the block at the above or below the current block;
- motion vector of the block at the left or right of the current block.

For each 8×8 block in the macroblock, the motion vectors of the two nearest "remote" (left, above, right and below) blocks are used together with that of the current block. Let the motion compensated pixels of the blocks from the reference VOP be $q(i,j)$ for the current block, $r(i,j)$ for the above or bottom block, $s(i,j)$ for the left and right block. Thus

$$q(i,j) = p(i + MV_x^0, j + MV_y^0) \tag{6.17}$$

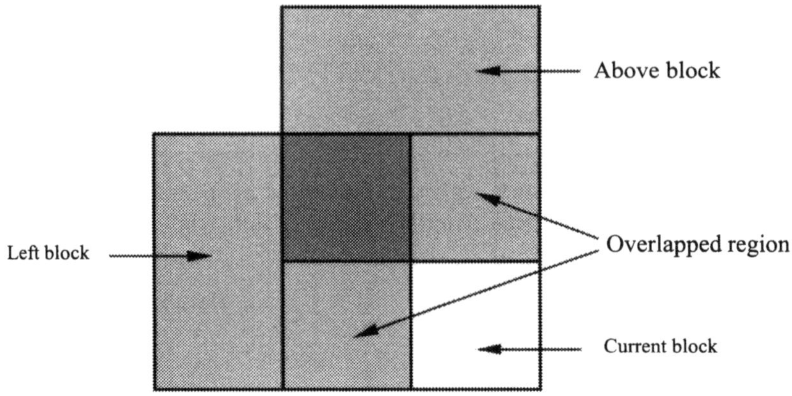

Figure 6.22: Overlapping blocks for motion compensation. ©ISO/IEC 1998

$$r(i,j) = p(i + MV_x^1, j + MV_y^1) \tag{6.18}$$

$$s(i,j) = p(i + MV_x^2, j + MV_y^2) \tag{6.19}$$

where

$p(i,j)$	is the pixel in the reference VOP;
(MV_x^0, MV_y^0)	is the motion vector of the current block;
(MV_x^1, MV_y^1)	is the motion vector of the block above or below the current block;
(MV_x^2, MV_y^2)	is the motion vector of the block at the left or right of the current block.

If MV_i^j points to a sub-sample position, the respective interpolation technique according to Section 6.6.4.2 is used.

The pixels in the prediction block $\hat{p}(i,j)$ are then given by:

$$\hat{p}(i,j) = [q(i,j) \times H_0(i,j) + r(i,j) \times H_1(i,j) + s(i,j) \times H_2(i,j) + 4]//4 \tag{6.20}$$

where $H_0(i,j)$, $H_1(i,j)$ and $H_2(i,j)$ are the weighting matrices as defined in Fig. 6.23.

Interlaced Motion Compensation

When field-based motion compensation is specified, two field motion vectors and corresponding reference fields are used to generate the prediction from

4	5	5	5	5	5	5	4
5	5	5	5	5	5	5	5
5	5	6	6	6	6	5	5
5	5	6	6	6	6	5	5
5	5	6	6	6	6	5	5
5	5	6	6	6	6	5	5
5	5	5	5	5	5	5	5
4	5	5	5	5	5	5	4

(a)

2	2	2	2	2	2	2	2
1	1	2	2	2	2	1	1
1	1	1	1	1	1	1	1
1	1	1	1	1	1	1	1
1	1	1	1	1	1	1	1
1	1	1	1	1	1	1	1
1	1	2	2	2	2	1	1
2	2	2	2	2	2	2	2

(b)

2	1	1	1	1	1	1	2
2	2	1	1	1	1	2	2
2	2	1	1	1	1	2	2
2	2	1	1	1	1	2	2
2	2	1	1	1	1	2	2
2	2	1	1	1	1	2	2
2	2	1	1	1	1	2	2
2	1	1	1	1	1	1	2

(c)

Figure 6.23: Weighting values for prediction with motion vector of current luminance block: (a)$H_0(i,j)$, (b) $H_1(i,j)$, (c) $H_2(i,j)$. ©ISO/IEC 1998 each reference VOP. For luminance motion compensation, the even lines of the macroblock are predicted from the top field motion vector using the reference field specified. The motion vector is specified in frame coordinates, i.e., full sample vertical displacements correspond to even integral values of vertical motion vector coordinate, a half sample vertical displacement is denoted by odd integral values and a quarter sample displacement by 0.5 values. The prediction of the odd luminance lines of the macroblock follows the same procedure above.

Chrominance motion compensation is perfomred field-wise. The even chrominance lines are predicted from the top field motion vector and the corresponding reference field, and the odd chrominance lines from the bottom field motion vector and the corresponding reference field. A chrominance motion vector is derived from the luminance motion vector by dividing each component by 2 followed by rounding.

6.6.5 Texture Coding

Basically the coding of the intra VOPs and the prediction error after motion compensation uses the 8×8 DCT coding scheme. In case of arbitrarily shaped VOP, the treatment of the macroblocks lying inside and on the VOP boundary is different. Those that lie completely inside the VOP boundary are coded using the H.263 coder. Whereas those intra macroblocks lying on the boundary are first padded as described in Section 6.6.5.1. For inter blocks, the region outside the VOP within the blocks are padded with zero. Padding is done for each of the luminance and chrominance blocks by using the original alpha values. The blocks are then coded same as the interior blocks. Transparent blocks are skipped and not coded. For blocks that lie outside the original shape, the intra blocks are padded with the value 128 for both the luminance and chrominance, whilst the inter blocks with the values of 0 for luminance and 128 for chrominance. Other blocks within the bounding box of a VOP are not coded at all.

6.6.5.1 Low Pass Extrapolation (LPE) Padding Technique

Before performing the DCT, for each intra block that has at least one transparent and one opaque pixel in its associated alpha information, the following block padding technique, referred to as *low-pass extrapolation* (LPE) padding, is applied to. The padding is performed in three steps.

1. Calculate the arithmetic mean value m of all block pixels $f(i, j)$ situated

within the object region R

$$m = \frac{1}{N} \sum_{(i,j)\in R} f(i,j) \tag{6.21}$$

where N is the number of pixels situated within the object region R. Division by N is done by rounding to the nearest integer.

2. Assign m to each block pixel situated outside of the object region R, i.e.

$$f(i,j) = m \qquad \text{for all}(i,j) \notin R. \tag{6.22}$$

3. Apply the following filtering operation to each block pixel $f(i,j)$ outside of the object region R, starting from the top left corner of the block and proceeding row by row to the bottom right pixel

$$f(i,j) = [f(i,j-1) + f(i-1,j) + f(i,j+1) + f(i+1,j)]/4 \tag{6.23}$$

Division is done by rounding to the nearest integer. If one or more of the four pixels used for filtering are outside the block, the corresponding pixels are not included into the filtering operation and the divisor 4 is reduced accordingly.

After this padding operation the resulting block is ready for DCT coding.

6.6.5.2 Adaptive Frame/Field DCT

MPEG-4 is capable of coding in frame or field DCT mode. The decision is based on the comparison of frame energy and field energy. That is, field DCT is used when

$$\begin{aligned} &\sum_{i=0}^{6}\sum_{i,j}^{1} 5(p_{2i,j} - p2i+1,j)^2 + (p_{2i+1,j} - p2i+2,j)^2 \\ &\quad > \sum_{i=0}^{6}\sum_{i,j}^{1} 5(p_{2i,j} - p2i+2,j)^2 + (p_{2i+1,j} - p2i+3,j)^2 \end{aligned} \tag{6.24}$$

where $p(i,j)$ is the spatial pixel value or prediction error before DCT.

When field DCT is used, the field macroblock is formed from the frame macroblock as shown in Fig. 6.24 below. The resulting macroblocks are then transformed, quantized and VLC encoded normally.

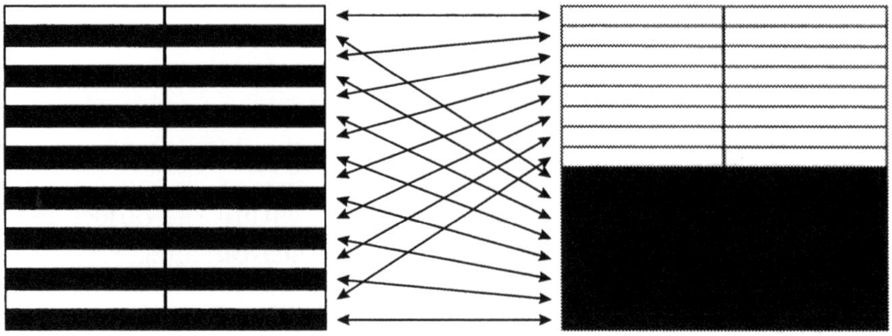

Figure 6.24: Formation of field macroblock from frame macroblock. ©ISO/IEC 1998

6.6.5.3 DCT

The $N \times N$ two dimensional (2-D) Discrete Cosine Transform (DCT) is defined as:

$$F(u,v) = \frac{2}{N}C(u)C(v)\sum_{x=0}^{N-1}\sum_{y=0}^{N-1} f(x,y)cos(\frac{(2x+1)u\pi}{2N})cos(\frac{(2y+1)v\pi}{2N}) \tag{6.25}$$

with $x, y, u, v = 0, 1, 2, \ldots, N-1$, and

$$C(u), C(v) = \begin{cases} \frac{1}{\sqrt{2}} & \text{for } u, v = 0 \\ 1 & \text{otherwise} \end{cases} \tag{6.26}$$

x, y are the spatial coordinates in the sample domain and u, v are the coordinates in the transform domain. The definition of the DCT is purely informative.

The $N \times N$ 2-D real-number inverse DCT (IDCT) is defined as:

$$F(x,y) = \frac{2}{N}\sum_{u=0}^{N-1}\sum_{v=0}^{N-1} C(u)C(v)F(u,v)cos(\frac{(2x+1)u\pi}{2N})cos(\frac{(2y+1)v\pi}{2N}) \tag{6.27}$$

The integer IDCT is defined as:

$$f'(x,y) = round[f(x,y)] \tag{6.28}$$

The saturated integer IDCT is then defined as:

$$f''(x,y) = saturate[f'(x,y)] \tag{6.29}$$

where

$$C(u), C(v) = \begin{cases} -2^n & x < -2^n \\ 2^n - 1 & x > 2^n - 1 \\ x & -2^n \leq x \leq 2^n - 1 \end{cases} \tag{6.30}$$

and n is the number of bits per pixel.

The IDCT function $f[y][x]$ used in the decoding process may be any of several approximations of the saturated integer IDCT $f''(x,y)$, provided that it meets all of the following requirements:

- The IDCT function $f[y][x]$ used in the decoding process shall have values always in the range [-256, 255].

- The IDCT function $f[y][x]$ used in the decoding process shall conform to the IEEE Standard Specification for the implementation of 8×8 Inverse Discrete Cosine Transform, Standard 1180-1990, December 6, 1990.

- This clause applies only when input blocks of DCT coefficients cause all the 64 values output of the integer IDCT $f'(x,y)$ to be in the range [-384, 383]. When $f'(x,y) > 256$, $f[y][x]$ shall be equal to 255 and when $f'(x,y) < 257$, $f[y][x]$ shall be equal to -256. For all values of $f'(x,y)$ in the range [-257, 256] the absolute difference between $f[y][x]$ and $f''(x,y)$ shall not be larger than 2.

- Let F be the set of 4096 blocks $B_i[y][x]$, $i = 0, \ldots, 4095$ defined as follows:

$$b_i[y][x] = \begin{cases} i - 2048 & y, x = 0 \\ 0 & x, y \neq 0 \end{cases} \tag{6.31}$$

For each block $B_i[y][x]$ that belongs to set F, an IDCT that conforms to this specification shall output a block $f[y][x]$ such that $f[y][x] - f''(x,y) = 0$ for all x and y.

6.6.5.4 SA-DCT & ΔDC-SA-DCT

When encoding a VOP of arbitrary shape, for the blocks which are completely within the shape, i.e. containing all opaque pixels, standard 8×8 DCT is applied. For those on the shape boundary, it is more efficient to employ DCT of arbitrary block size, known as *shape adaptive DCT* (SA-DCT) for inter-coded blocks. For intra-coded blocks, an extended version, ΔDC-SA-DCT is used.

Unlike the standard 8×8 DCT, SA-DCT and ΔDC-SA-DCT require the shape information provided by the binary alpha block. Only the opaque pixels within the shape boundary are transformed and coded thereby saving transmitted bit rate.

SA-DCT for Inter-coded Macroblocks

The SA-DCT is based on the odd or even orthonormal DCT basis functions. The procedure to calculate the SA-DCT of an arbitrary segment in a 8×8 block is illustrated in Fig. 6.25. First the segment is shifted vertically column by column to the upper edge of the block as in Fig. 6.25 (B). The length of each column N is then calculated. Depending on the length of the column, a one-dimensional N-DCT is performed on the pixels x_j of each of the columns to obtain the DCT coefficients X_j according to the following formula:

$$X_j = \sqrt{\frac{2}{N}} DCT_N x_j \tag{6.32}$$

where

$$DCT_N(p,k) = c_0 cos[p(k+0.5)\frac{\pi}{N}] \tag{6.33}$$

and

$$c_0 = \begin{cases} \sqrt{\frac{1}{2}} & p = 0; \\ 1 & \text{otherwise}; \end{cases} \quad \text{for } 0 \leq p, k \leq N-1 \tag{6.34}$$

The DCT coefficients are then shifted horizontally row by row to the left edge of the block as shown in Fig. 6.25 (E). The length of each row is found and a one-dimensional N-DCT is applied to each of the rows using the above formulae. This completes the computation of SA-DCT for an arbitrary segment in a 8×8 pixel block. Note that the number of coefficients is exactly equal to the number of pixels in the image segment. Also, the coefficients are located in a similar manner as in a standard 8×8 DCT block, i.e., the DC coefficient is located at the upper left corner with the

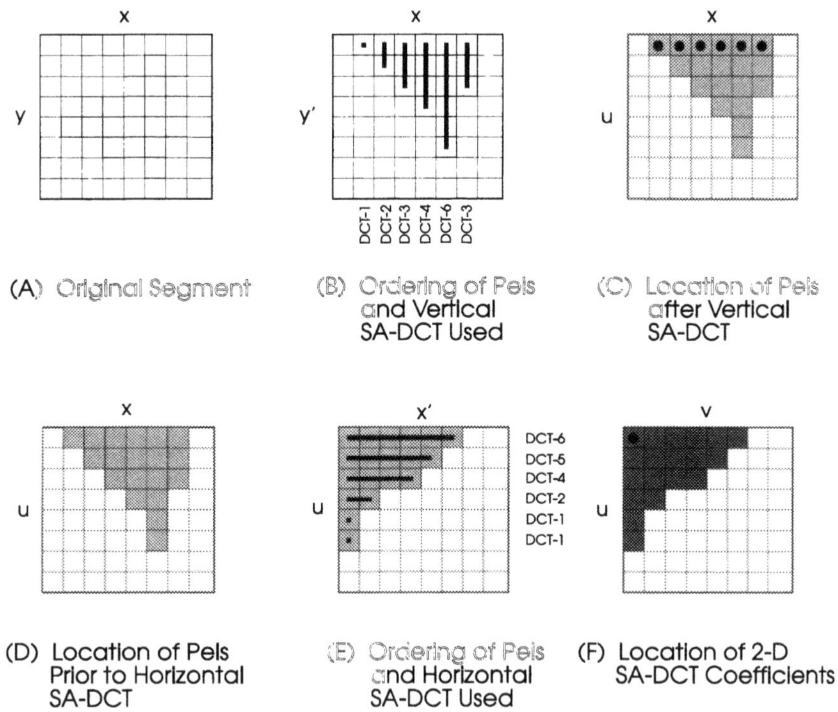

Figure 6.25: The procedure for performing a SA-DCT forward transform on a segment of arbitrary shape in a 8×8 block. ©ISO/IEC 1998

AC coefficients surrounding it, whose frequency content depends on their distance away from the DC cofficient.

Using the shape information obtained from the decoded BAB, inverse SA-DCT can be performed on the rows and columns of the transformed image segment according the following formula:

$$x_j = \sqrt{\frac{2}{N}} DCT_N^T X_j \tag{6.35}$$

where DCT_N^T is the transpose of DCT_N. The recovered pixel data together with the shape information are used to reconstruct the original image segment of arbitrary shape.

ΔDC-SA-DCT for Intra-Coded Macroblocks

For intra-coded macroblocks, ΔDC-SA-DCT, an extension of the SA-DCT is used to transform the arbitrarily shaped image segment within the blocks. Forward ΔDC-SA-DCT consists of the following steps:

1. The mean η of the pixels $x_{i,j}$ within the segment is first computed.
2. The mean is then subtracted from the pixels to generate zero-mean pixels $\hat{x}_{i,j}$ for the segment.
3. SA-DCT is applied to the zero-mean pixel data to yield the coefficients $X_{i,j}$.
4. The mean value is transmitted as the DC value of the SA-DCT coefficient matrix. For this the DC coefficient $X_{0,0}$ is redefined as

$$\hat{X}_{0,0} = 8.0\eta \tag{6.36}$$

$X_{0,0}$ which has been overwritten by $\hat{X}_{0,0}$ is called the ΔDC coefficient. Note that $X_{0,0}$ can be recomputed as a correction term during inverse ΔDC-SA-DCT in the decoder.

The inverse ΔDC-SA-DCT process follows the steps below:

1. After decoding the DC and AC coefficients, the mean value is extracted from the DC coefficient, i.e.,

$$\eta = \frac{\hat{X}_{0,0}}{8.0} \tag{6.37}$$

2. The DC coefficient is set to zero and together with the decoded AC coefficients form the coefficient matrix on which the inverse SA-DCT is performed.
3. Because of the zero-setting of the ΔDC coefficient, the sum of the inverse SA-DCT pixel values $\hat{x}_{i,j}$ is no longer zero. To correct this ΔDC error, a correction term is computed:

$$corr = \frac{sum}{\sum_{j=0}^{7} \sqrt{N_j}} \tag{6.38}$$

where

$$sum = \sum_{i=0}^{7} \sum_{j=0}^{7} a_{i,j} \hat{x}_{i,,j} \tag{6.39}$$

and

$$a_{i,j} = \begin{cases} 0 & \text{if } BAB(i,j) = 0 \\ 1 & \text{otherwise} \end{cases} \tag{6.40}$$

4. Finally, the inverse ΔDC-SA-DCT pixel values are given by:

$$x_{i,j} = a_{i,j}[\hat{x}_{i,j} + \eta - \frac{corr}{\sqrt{N_j}}] \tag{6.41}$$

1	2	6	7	15	16		
3	5	8	14	17			
4	9	13	18				
10	12						
11							
19							

Figure 6.26: Adaptive SA-DCT zig-zag scan for the active SA-DCT domain area of Fig. 6.25 (F). ©ISO/IEC 1998

Adaptive Scanning of the SA-DCT-Coefficients

Unlike the standard block-based 8×8 DCT, SA-DCT processes only a segment which forms only part of the block. Therefore, for SA-DCT we define an "active SA-DCT domain area", an example is the dark shaded area of Fig. 6.25 (F). For efficient coding of the SA-DCT coefficients using VLC, the scanning of the coefficients is modified as shown in Fig. 6.26 below for the example of Fig. 6.25 (F). Note that if the active SA-DCT domain area is rectangular, the SA-DCT degenerates into a standard 8×8 DCT and the scanning follows the standard zig-zag scanning defined in H.261, H.263, MPEG-1 and MPEG-2.

6.6.5.5 H.263 Quantization Method

The quantization parameter QP may take integer values from 1 to 31. The quantization stepsize is $2 \times QP$.

Quantization

For INTRA: $LEVEL = |COF|/(1 \times QP)$
For INTER: $LEVEL = (|COF| - QP/2)/(x \times QP)$

where COF is a transform coefficient to be quantized and $LEVEL$ is the absolute value of the quantized version of the transform coefficient. Clipping to [-127, 127] is performed for all coefficients except intra DC.

Dequantization

$$|COF'| = \begin{cases} 0 & \text{if } LEVEL = 0 \\ 2 \times QP \times LEVEL + QP & \text{if } LEVEL \neq 0, QP \text{ is odd} \\ 2 \times QP \times LEVEL + QP - 1 & \text{if } LEVEL \neq 0, QP \text{ is even} \end{cases}$$

where COF' is the reconstructed transform coefficient. The sign of COF is then added to obtain $COF' = Sign(COF) \times |COF'|$. Clipping to [-2048, 2047] is performed before IDCT.

The DC coefficient of an INTRA block is quantized as described below. 8 bits are used for the quantized DC coefficient.

Quantization: $LEVEL = COF//8$
Dequantization: $COF' = LEVEL \times 8$

6.6.5.6 MPEG Quantization Method

Quantization of Intra Macroblocks

Nonlinear Quantization of DC Coefficients

Note that his section is also valid for H.263 quantization.

The quantization strategy for intra DC coefficients is by scaling the coefficients with a non-linear function of the quantizer parameter Q_p. Separate DC Scalers for luminance (Type 1) and chrominance (Type 2) are used, with smaller value for chrominance than luminance. The characteristics of the DC Scalers as a function of Q_p is shown in Fig. 6.27.

The forward quantization is given by:

$$level = dc\_coef//dc\_scaler \tag{6.42}$$

whilst the reconstructed DC values are given by

$$dc\_rec = dc\_scaler * level \tag{6.43}$$

Quantization of AC Coefficients

The AC coefficients are first scaled by a intra quantizer matrix which by default is a flat matrix consists of 16 for all values. A user-downloaded non-flat quantizer matrix can alternatively be used.

The scaled AC coefficients are:

$$ac'[i][j] = (16 * ac[i][j])//w_1[i][j] \tag{6.44}$$

where $w_1[i][j]$ is the $[i][j]$th element of the default intra quantizer matrix. The resulting $ac'[i][j]$ is limited to the range of [-2048, 2047].

The quantized level is thus given by:

$$QAC[i][j] = \frac{ac'[i][j] + sign(ac'[i][j]) \times (3 \times Q_p//4)}{2 \times Q_p} \tag{6.45}$$

where $QAC[i][j]$ is limited to the range of [-127, 127].

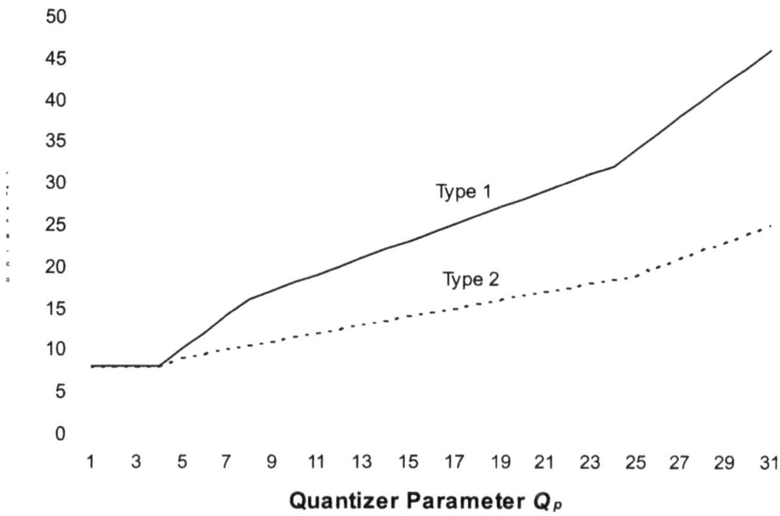

Figure 6.27: DC_Scaler characteristic as a function of quanitzer parameter Q_p. ©ISO/IEC 1998

Quantization of Non-Intra Macroblocks

Non-intra macooblocks in P- and B-VOPs are quantized with a uniform quantizer that has a dead-zone about zero. The default quantizer matrix is a flat matrix consists of 16 for all values. As in the case of quantization of intra AC coefficients, a user-downloaded non-flat quantizer matrix can alternatively be used.

The scaled AC coefficients are:

$$ac'[i][j] = (16 \times ac[i][j])//W_N[i][j] \tag{6.46}$$

where $W_N[i][j]$ is the $[i][j]$th element of the default non-intra quantizer matrix.

The quantized level is thus given by:

$$QAC[i][j] = \frac{ac'[i][j]}{2 \times Q_p} \tag{6.47}$$

where $QAC[i][j]$ is limited to the range of [-127, 127].

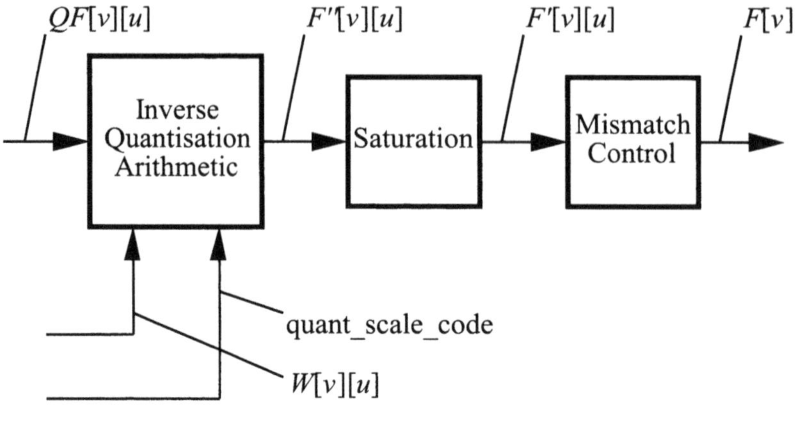

Figure 6.28: Inverse quantization process. ©ISO/IEC 1998

Inverse Quantization of Intra and Non-Intra Macroblocks

Fig. 6.28 illustrates the inverse quantization process. After the appropriate inverse quantisation arithmetic, the resulting coefficients $F''[v][u]$ are saturated to yield $F'[v][u]$ and then a mismatch control operation is performed to give the final reconstructed DCT coefficients $F[v][u]$.

Intra DC Coefficient

In intra-coded blocks, the reconstructed DC values $F''[0][0]$ are computed from the quantized values $QF[0][0]$ as follows:

$$F''[0][0] = dc\_scaler \times QF[0][0] \tag{6.48}$$

Other Coefficients

To reconstruct $F''[v][u]$ from $QF[v][u]$, the following equation is used:

$$F''[v][u] = \frac{2 \times QF[v][u] + k \times w[v][u] \times quantizer\_scale}{32} \tag{6.49}$$

In the the above expression, $k = 0, w[v][u] = w_I[v][u]$, for intra blocks; and $k = sign(QF[v][u]), w[v][u] = w_N[v][u]$, for non-intra blocks.

The coefficients resulting from the Inverse Quantisation Arithmetic are saturated to lie in the range [-2048, 2047]. Thus:

$$F''[v][u] = \begin{cases} 2047 & F''[v][u] > 2047 \\ F''[v][u] & -2048 \leq F''[v][u] \leq 2047 \\ -2048 & F''[v][u] < -2048 \end{cases} \tag{6.50}$$

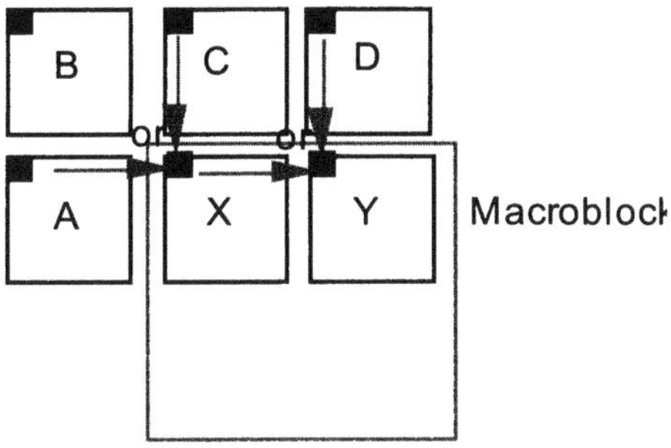

Figure 6.29: Previous neighboring blocks used in DC prediction. ©ISO/IEC 1998

Mismatch control is carried out to compensate for mismatch between DCT and IDCT. Note that only the last coefficient $F[7][7]$ is compensated. It is carried out according to the following procedure:

1. The sum of all coefficients is calculated:

$$sum = \sum_{v=0}^{v<8} \sum_{u=0}^{u<8} F'[v][u] \tag{6.51}$$

2. For all u and v except,$u = v = 7$, F[v][u] = F'[v][u].

3. If sum is even, a correction is made to $F[7][7]$, i.e.,

$$F[7][7] = \begin{cases} F'[7][7] - 1 & \text{if } F'[7][7] \text{ is odd} \\ F'[7][7] + 1 & \text{if } F'[7][7] \text{ is even} \end{cases} \tag{6.52}$$

6.6.5.7 Intra DC and AC Prediction for I-VOP and P-VOP

Adaptive DC Coefficient Prediction

The adaptive prediction of DC coefficient of the current block is based on the quantized DC (QDC) values of the immediately previous block on the same row and the block above it. Fig. 6.29 shows the neighboring blocks used in DC prediction.

DC coefficients are first divided by 8 to obtain the quantized values:

$$QDC = dc\_coef//8 \tag{6.53}$$

Let the predicted QDC value of the current block 'X' be $QPC_{X'}$, the prediction rule is as follows:

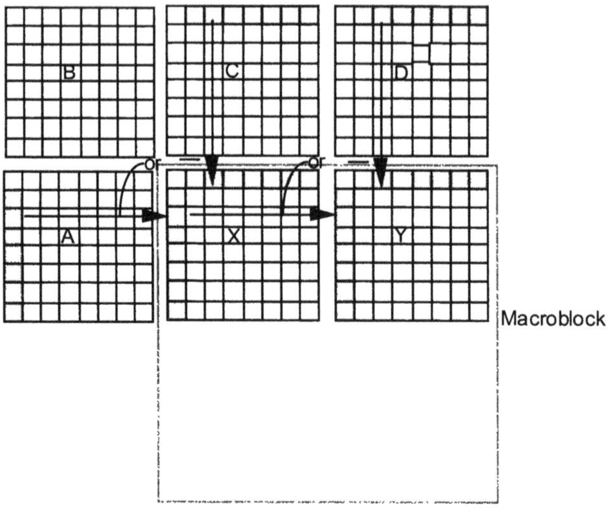

Figure 6.30: Previous neighboring blocks and coefficients used in AC prediction. ©ISO/IEC 1998

`if` $|QDC_A - QDC_B| < |QDC_B - QDC_C|$
    $QDC_{X'} = QDC_C$
`else`
    $QDC_{X'} = QDC_A$

The differential DC value is then obtained by subtracting the DC prediction, $QDC_{X'}$ from the QDC of block 'X', $QDC_{X'}$.

Adaptive AC Coefficient Prediction

Prediction of First Row or First Column of AC Coefficients

The prediction of the AC coefficients in the first row or the first column of the current block is based on the co-sited previous coded blocks. The direction of the DC coefficient prediction is used to determine the direction of the AC coefficient prediction. The AC coefficient prediction is illustrated in Fig. 6.30.

Q-step Scaling

To compensate for differences in the quantization of previous horizontally adjacent or vertically adjacent blocks used in AC prediction of the current block, the prediction is scaled is scaled by the ratio of the current quantisation stepsize and the quantisation stepsize of the predictor block. Thus, if the AC prediction is from block 'A', the horizontal AC prediction of block 'X' is given by:

$$QAC_{i0X'} = \frac{QAC_{i0A} \times QP_A}{QP_X} \tag{6.54}$$

where QAC_{i0A} is the quantized horizontal AC coefficients of block 'A', QP_A and QP_B are QP values of blocks 'A' and 'X', respectively.

If the AC prediction is from block 'C', the vertical AC prediction of block 'X' is:

$$QAC_{0jX'} = \frac{QAC_{0jC} \times QP_C}{QP_X} \tag{6.55}$$

where QAC_{0jC} is the quantized vertical AC coefficients of block 'C', and QP_C is QP value of block 'C'.

If block 'A' or block 'C' are outside of the VOP, then the corresponding QP values are assumed to be equal to QP_X.

AC Prediction Enable/Disable Mode Decision

The decision to enable or disable the AC prediction mode is based on a criterion S calculated as follows for the case if the AC prediction is from block 'A':

$$S = \sum_{i=1}^{7} |QAC_{i0X}| - \sum_{i=1}^{7} |QAC_{i0X'}| \tag{6.56}$$

If it is from block 'C', the criterion S is given as follows:

$$S = \sum_{j=1}^{7} |QAC_{0jX}| - \sum_{j=1}^{7} |QAC_{0jX'}| \tag{6.57}$$

Next for for all blocks in the macroblock for which a common decision is to be made, a single $\sum S$ is calculated and the AC prediction is enabled/disabled according to the rules below:

if $\sum S > 0$
```
   enable AC prediction
else
   disable AC prediction
```

0	1	2	3	10	11	12	13
4	5	8	9	17	16	15	14
6	7	19	18	26	27	28	29
20	21	24	25	30	31	32	33
22	23	34	35	42	43	44	45
36	37	40	41	46	47	48	49
38	39	50	51	56	57	58	59
52	53	54	55	60	61	62	63

0	4	6	20	22	36	38	52
1	5	7	21	23	37	39	53
2	8	19	24	34	40	50	54
3	9	18	25	35	41	51	55
10	17	26	30	42	46	56	60
11	16	27	31	43	47	57	61
12	15	28	32	44	48	58	62
13	14	29	33	45	49	59	63

Figure 6.31: (a) Alternate-horizontal scan (b) Alternate-vertical (MPEG-2) scan. ©ISO/IEC 1998

Other simple rules apply to AC prediction are:

- If any of the blocks A, B or C are outside of the VOP boundary or do not belong to an intra coded macroblock, their QAC values are assumed to take a value of 0 and are used to compute prediction values.
- AC predictions are performed similarly for the luminance and each of the two chrominance components using the direction identified by the corresponding direction of DC prediction.
- The process for DC/AC prediction for alpha plane is similar to that of texture as described above.

6.6.5.8 VLC Encoding of Quantized Transform Coefficients

VLC Encoding of Intra Macroblocks

In addition to the zig-zag scan, two scans are employed depending on the DC prediction direction, as shown in Fig. 6.31.

For intra blocks, if the AC prediction is disabled, the zig-zag scan is selected for all blocks in a macroblock. If the prediction is from the horizontal adjacent block, alternate vertical scan (Fig. 6.31(a)) is used for the current block, otherwise alternate horizontal scan (Fig. 6.31(b)) is used.

For non-intra blocks, the transform coefficients are scanned according to the zig-zag scan.

A three-dimensional variable length code is used to code transform coefficients. An EVENT is a combination of three parameters:

LAST 0: There are more nonzero coefficients in the block.

1	2	6	7	15	16	28	29
3	5	8	14	17	27	30	43
4	9	13	18	26	31	42	44
10	12	19	25	32	41	45	54
11	20	24	33	40	46	53	55
21	23	34	39	47	52	56	61
22	35	38	48	51	57	60	62
36	37	49	50	58	59	63	64

Figure 6.32: Zig-zag scanning pattern. ©ISO/IEC 1998

1: This is the last nonzero coefficient in the block.
RUN Number of zero coefficients preceding the current nonzero coefficient.
LEVEL Magnitude of the coefficient.

The most commonly occurring combinations of (LAST, RUN, LEVEL) are coded with variable length codes given in [4]. The remaining combinations of (LAST, RUN, LEVEL) are coded with a 22 bit word consisting of:

ESCAPE	7 bit
LAST	1 bit (0: Not last coefficient, 1: Last nonzero coefficient)
RUN	6 bit
LEVEL	8 bit

The code words for these fixed length ESCAPE codes are described in [4].

For intra macroblock chroma AC Coefficients, the VLC used is the same as that used for intra AC luminance coefficients

VLC Encoding of Inter Macroblocks

For inter blocks, the 8×8 transform coefficients are scanned with the zig-zag scanning as illustrated in Fig. 6.32. The coding of the transform coefficients is using the same VLC codetable as in the coding of intra macroblocks above.

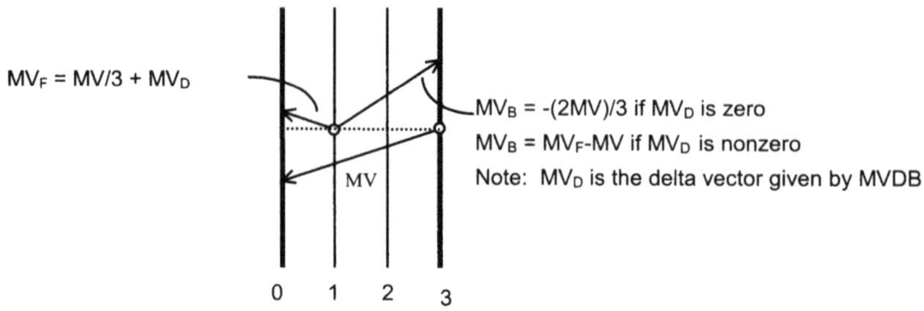

Figure 6.33: Progressive direct bidirectional prediction. ©ISO/IEC 1998

6.6.6 Prediction and Coding of B-VOPs

Prediction and coding of macroblocks in B-VOPs can use either the H.263 or the MPEG-1 approach. The main difference is in the amount of motion vector and quantization related overhead needed.

6.6.6.1 Direct Coding

Progressive Direct Coding

Progressive direct coding mode is used whenever the macroblock at the same location in the future anchor picture is coded as:

- a 16×16 (frame) macroblock,
- an intra macroblock or
- an 8×8 (advanced prediction) macroblock.

This coding mode uses direct bidirectional motion compensation derived by extending H.263 approach of employing P-picture macroblock motion vectors and scaling them to derive forward and backward motion vectors for macroblocks in B-picture. As per H.263, using B-frame syntax, only one delta motion vector is allowed per macroblock. Fig. 6.33 shows scaling of motion vectors.

The calculation of the forward and backward motion vectors follows the procedure in H.263. The only difference is that here we are dealing with VOPs instead of pictures, and instead of only a single B-picture between a pair of reference pictures, multiple B-VOPs are allowed between a pair of reference VOPs. As in H.263, the temporal referenceof the B-VOP relative to difference in the temporal reference of the pair of reference VOPs is used to determine scale factors for computing motion vectors which are corrected by the delta vector MV_δ.

The forward and the backward motion vectors, MV_F and MV_B, given in half or quarter sample units are obtained as follows:

$$MV_F = \frac{TR_B \times MV}{TR_D} + MV_\delta \tag{6.58}$$

$$MV_B = \frac{(TR_B - TR_D) \times MV}{TR_D} \quad \text{if } MV_\delta = 0 \tag{6.59}$$

$$MV_B = MV_F - MV \quad \text{if } MV_\delta \neq 0 \tag{6.60}$$

where MV is the direct motion vector of a macroblock in P-VOP with respect to a reference VOP; TR_B is the difference in temporal reference of the B-VOP and the previous reference VOP; and TR_D is the difference in temporal reference of the temporally next reference VOP with temporally previous reference VOP, assuming B-VOPs or skipped VOPs in between.

The prediction blocks are generated using computed forward and backward motion vectors to obtain appropriate blocks from reference VOPs and averaging these blocks. Motion compensation is performed individually on 8×8 blocks to generate a macroblock irregardless of whether the direct prediction motion vectors are derived by scaling a single motion vector or four 8×8 block motion vectors. It should be noted that if if the next reference VOP is an I-VOP instead of a P-VOP, then the MV vectors are by default '0'.

The direct coding mode does not allow quantizer change and thus the quantizer value for previous coded macroblock is used.

Interlaced Direct Coding

The extension to interlaced video is used whenever the corresponding macroblock of the future anchor VOP is a field motion compensated macroblock.

In interlaced direct mode, the prediction macroblock is formed separately for the top and bottom fields. The four field motion vectors of a bi-directional field motion compensated macroblock are calculated from the motion vectors of the corresponding macroblock of the future anchor picture. Only one delta motion vector (used for both fields) is used for the field predicted macroblock.

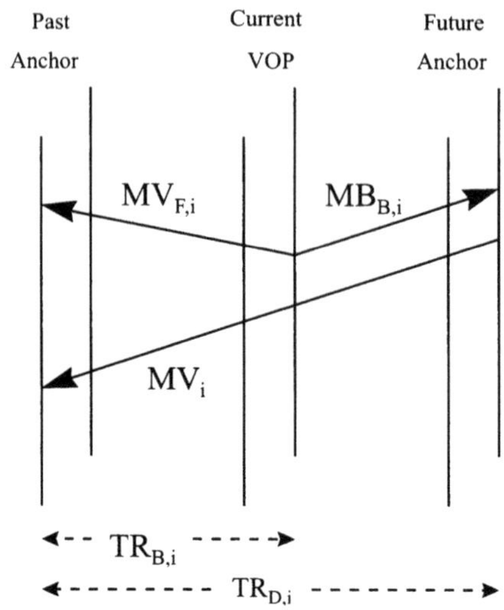

Figure 6.34: Interlaced direct bidirectional prediction. ©ISO/IEC 1998

The motion vectors for field i (top or bottom) are calculated by:

$$MV_{F,i} = \frac{TR_{B,i} \times MV_i}{TR_{D,i}} + MV_\delta \tag{6.61}$$

$$MV_{B,i} = \frac{(TR_{B,i} - TR_{D,i}) \times MV_i}{TR_{D,i}} \quad \text{if } MV_\delta = 0 \tag{6.62}$$

$$MV_{B,i} = MV_{F,i} - MV_i \quad \text{if } MV_\delta \neq 0 \tag{6.63}$$

where $MV_{F,i}$ is forward motion vector for field i whose reference field is the reference field of MV_i; $MV_{B,i}$ is backward motion vector for field i; MV_i is field motion vector for field i of the macroblock at the same location as the current macroblock in the future anchor VOP; $TR_{B,i}$ is temporal distance in fields between the past reference field for field i and field i of the current B-VOP; and $TR_{D,i}$ is temporal distance in fields between the past reference field and the future reference field for the current VOP's field i.

The calculation of $TR_{B,i}$ and TRD, i depends not only on the current field, reference field, and frame temporal references, but also on whether the current video is top field first or bottom field first, i.e.,

$$TR_{D,i} = 2(TR_{future} + TR_{past}) + \delta \tag{6.64}$$

$$TR_{B,i} = 2(TR_{current} + TR_{past}) + \delta \tag{6.65}$$

where TR_{future}, $TR_{current}$ and TR_{past} are the frame number of the future, current and past frames in display order, and δ is 0 or ± 1 depending on whether the reference is top or bottom field.

6.6.6.2 Forward, Backward and Bidirectional Coding Modes

Forward coding mode uses forward motion compensation in the same manner as in MPEG-1/2 with the difference that a VOP is used for prediction instead of a picture. Only one motion vector in half or quarter sample units is employed for a 16×16 macroblock being coded. Chrominance vectors are derived by scaling of luminance vectors as in MPEG-1/2, but are restricted to half sample accuracy.

Backward coding mode is the same as above except backward motion compensation is used. *Bidirectional coding* mode is the same as above except *bidirectional* motion compensation is used. The above coding modes allow switching of quantizer from the one previously in use.

6.6.6.3 Mode Decisions

To select the coding mode for a macroblock in B-VOP, the motion compensated prediction macroblocks difference SAD (sum of absolute differences) is calcualted for each mode. The decision is according to the following rule:

`if` $[SAD_{direct} - (0.5N_B + 1) \leq min(SAD_{forward}, SAD_{backward}, SAD_{bidirectional})]$
  `direct mode`
`else if` $[SAD_{bidirectional} \leq min(SAD_{forward}, SAD_{backward}, SAD_{bidirectional})]$
  `bidirectional mode`
`else if` $[SAD_{backward} \leq min(SAD_{forward}, SA_{backward}, SAD_{bidirectional})]$
  `backward mode`
`else`
  `forward mode`

For interlaced B-VOPs, a macroblock can be coded using:

- direct coding;
- 16×16 motion compensated (including forward, backward and bidirectional modes); or
- field motion compensation (including forward, backward and bidirectional modes).

Table 6.2: Prediction motion vectors. ©ISO/IEC 1998

Function	PMV
Top field forward	0
Bottom field forward	1
Top field backward	2
Bottom field backward	3

The decision regarding coding mode of the macroblock is based on the minimum luminance SADs with respect to the decoded anchor pictures:

$$SAD_{direct} + b_1, SAD_{forward} + b_2, SAD_{forward,field} + b_3, SAD_{backward} + b_2,$$

$$SAD_{backward,field} + b_3, SAD_{average} + b_3, SAD_{average,field} + b_4$$

where the subscripts direct, forward, backward, average and field mean forward motion prediction, backward motion prediction, average (i.e., interpolated or bidirectional) motion prediction and field mode. The field SADs above are the sums of the top and bottom field SADs each with its own reference field and motion vector. The bias terms b_i's take on the values 0, or $\pm(N_B/2+1)$ depending on the motion prediction mode.

6.6.6.4 Motion Vector Coding

Only the motion vectors used for the selected prediction mode are coded. The motion vectors are coded differentially with the left neighboring vector used as predictor in the forward, backward and bidirectional modes. In the case where the current macroblock is located on the left edge of the VOP, or no vector is present in the left neighboring macroblock, the predictor is set to zero.

For interlaced B-VOP motion vector predictors, four prediction motion vectors (PMVs) are as given in Table 6.2. These PMVs are used for the different macroblock prediction modes as indicated in Table 6.3.

The PMVs used by a macroblock are set to the value of current macroblock motion vectors after being used. The prediction motion vectors are reset to zero at the beginning of each row of macroblocks. The predictors are not zeroed by skipped macroblocks or direct mode macroblocks.

For macroblocks coded in direct bidirectional mode no vector differences are transmitted. Instead, the forward and backward motion vectors are

Table 6.3: PMVs for different macroblock prediction modes. ©ISO/IEC 1998

Macroblock mode	PMVs used
Direct	none
Frame forward	0
Frame backward	2
Frame bidirectional	0,2
Field forward	0,1
Field backward	2,3
Field bidirectional	0,1,2,3

directly computed from the temporally consecutive P-vector as described in Section 6.6.6.1.

6.6.7 Generalized Scalable Coding

Scalable coding enables MPEG-4 to encode video at different resolutions and/or quality. A scalable encoder generates a bitstream that allows decoding appropriate subset of the bitstream to generate complete pictures of resolution and/or quality commensurate with the proportion of bitstream decoded and the capability and complexity of the decoder.

Two main types of scalability are allowed in MPEG-4, namely, spatial scalability and temporal scalability. As the names imply, spatial scalability offers scalability of the spatial resolution, and temporal scalability offers that of the temporal resolution. Each type of scalability consists of two layers: the base layer and the enhancement layer, with the base layer provides the basic resolution whilst the enhancement layer enhances the resolution of the base layer. In the case of spatial scalability, the enhancement layer increases the spatial resolution of the video, while in temporal scalability, the enhancement layer provides a higher temporal resolution for the video.

Traditionally, these scalabilities are applied to rectangular frames of video, but nowadays, many applications necessitate not only traditional frame based scalabilities but also scalabilities of VOPs of arbitrary shapes. MPEG-4 supports rectangular VOPs (frames) in both the spatial and temporal scalabilities, but only temporal scalability is capable to handle arbitrarily shaped VOPs at this moment.

A generalized scalable codec structure is shown in Fig. 6.35. The input to the Scalability PreProcessor are VOPs of arbitrary shape (rectangular

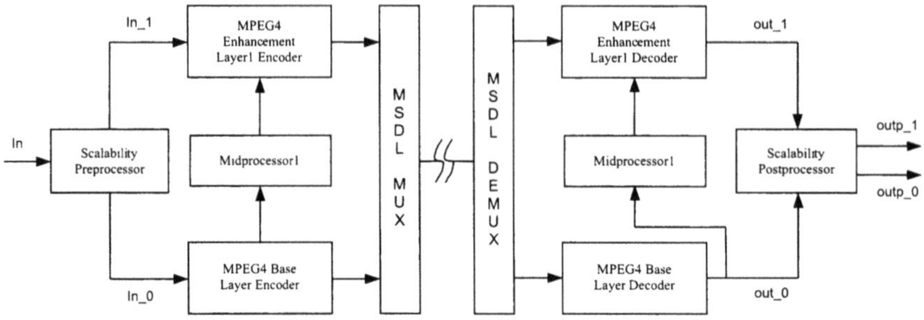

Figure 6.35: A generalized scalable codec structure. ©ISO/IEC 1998

shape is a special case). If spatial scalability is to be carried out, spatial downsampling is performed on the input VOPs to generate the base layers (in_0) as input to the Base Layer Encoder which performs a non-scalable encoding. The encoder could be one of the standard video encoders, such as H.263, H.261, MPEG-1 or MPEG-2. The reconstructed VOPs are then upsampled by the Midprocessor and used as prediction for the original VOPs (in_1) in the Enhancement Layer Encoder. The output bitstreams of the Enhancement Layer Encoder and the Base Layer Encoder are then multiplexed by the MSDL Multiplexer for transmission. The MSDL Demultiplexer, Base Layer Decoder, Enhancement Layer Decoder, Midprocessor and Scalability Postprocessor in the receiver peroforms the reverse operations to retrieve the resulting outputs for the Base Layer (outp_0) and the Enhancement Layer (out_0).

In the case of temporal scalability, the Scalability PreProcessor demultiplexes the VOP into two streams with different temporal resolutions, i.e., in_0 and in_1 as inputs to the Base Layer Encoder and the Enhancement Encoder, respectively. The MidProcessor does not perform any subsampling but merely allows the decoded base-layer VOPs to pass through and used for temporal prediction in encoding the enhancement layer. The output bitstreams of the Enhancement Layer Encoder and the Base Layer Encoder are then multiplexed for transmission. The operation of the Base Layer Decoder and Enhancement Layer Decoder are the reverse of their corresponding encoders. The PostProcessor does not carry out any conversion but temporally multiplexes the base and enhancement layers to produce higher temporal resolution enhancement layer VOPs.

6.6.7.1 Spatial Scalability Encoding

In spatial scalability, the base layer is encoded at a smaller spatial resolution than the enhancement layer. For example, if the base layer is encoded at QCIF resolution, the enhancement layer will be encoded at CIF resolution. Therefore, if QCIF resolution is required, only base layer will be decoded. However, if CIF resolution is required, both base and enhancement layers need to be decoded.

The down sampling process is performed at the Scalability PreProcessor. The down sampling process for the factor of 2 is only described here, however down sampling for any factor is allowed. The encoding process of the base layer is the same as non-scalable encoding process, such H.263, MPEG-1 or MPEG-2.

Upsampling Process

For spatial prediction of the enhancement layer, the reconstructed base-layer VOP is upsampled in the MidProcessor. The upsampling is performed in two steps, viz vertical upsampling and horizontal upsampling as described below.

Vertical Upsampling

The vertical upsampled image $vert\_pict[y][x]$ is obtained from the base-layer image $d_{lower}[y][x]$ using linear interpolation according to the following formula:

$$vert\_pic[y_h][x] = (16 - phase) \times d_{lower}[y_1][x] + phase \times d_{lower}[y_2][x] \tag{6.66}$$

where

$$\begin{aligned}
y_h &= \text{output sample coordinate in vert_pic} \\
y_1 &= y_h \times vertical_sampling_factor_m / vertical_sampling_factor_n \\
y_2 &= \begin{cases} y_1 + 1 & \text{if } y_1 < video_object_layer_height - 1 \\ y_1 & otherwise \end{cases} \\
phase &= [16 * ((y_h * vertical_sampling_factor_m)/vertical_sampling_factor_n)]
\end{aligned}$$

where $video\_layer\_height$ is the height of the reference VOL.

Samples which lie outside the lower layer reconstructed frame which are required for upsampling are obtained by border extension of the lower layer reconstructed frame.

NOTE: The calculation of phase assumes that the sample position in the enhancement layer at is spatially coincident with the first sample position of the lower layer. It is recognized that this is an approximation for the chrominance component if the chroma_format = 4:2:0.

Horizontal Upsampling

Horizontal upsampling is performed using a similar procedure as the vertical upsampling. The horizontal upsampled image $hor\_pict[y][x]$ is obtained from the vertically upsampled image $ver\_pict[y][x]$ using linear interpolation according to the following formula:

$$hor\_pic[y_h][x] = (16 - phase) \times vert\_pic[y][x_1] + phase \times vert\_pic[y][x_2]//256 \tag{6.67}$$

where

$$\begin{aligned} x_h &= \text{output sample coordinate in } hor\_pic \\ x_1 &= x_h \times horizontal\_sampling\_factor\_m / horizontal\_sampling\_factor\_n \\ x_2 &= \begin{cases} x_1 + 1 & \text{if } y_1 < video\_object\_layer\_width - 1 \\ x_1 & otherwise \end{cases} \\ phase &= [16 * ((x_h * horizontal\_sampling\_factor\_m) \\ &\qquad /horizontal\_sampling\_factor\_n)] \end{aligned}$$

where $video\_layer\_width$ is the width of the reference VOL.

Samples which lie outside the lower layer reconstructed frame which are required for upsampling are obtained by border extension of the lower layer reconstructed frame.

Encoding of Enhancement Layer

The VOP in the enhancement layer is encoded either as P-VOP or B-VOP. The relationship between VOP in the base layer and that of the enhancement layer is illustrated in Fig. 6.36. The VOP which is temporally coincident with the base-layer I-VOP is encoded as P-VOP. The VOP which is temporally coincident with the base-layer P-VOP is encoded as B-VOP. In spatial scalability, a decoded unsampled VOP in the base layer is used as a reference for prediction of the enhancement layer. The temporally coincident VOP in the reference layer (base layer) must be coded before the encoding of the VOP in the enhancement layer.

In the case of P-VOP, the INTRA/INTER mode decision is the same as that in the coding of the base layer. For Inter mode, the motion vector is always set to 0 and is not coded.

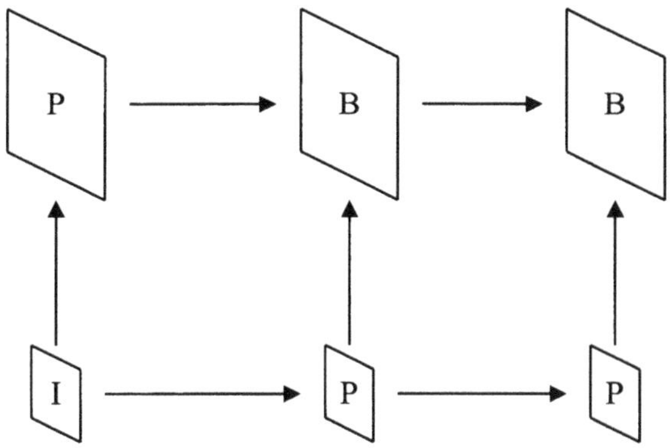

Figure 6.36: Prediction references for enhancement layer. ©ISO/IEC 1998

In the case of B-VOP, the backward prediction reference is set to P-VOP which is the temporally coincident VOP in the base layer, and the forward prediction reference is set to P-VOP or B-VOP which is the most recent decoded VOP of the enhancement layer. A macroblock in B-VOP in forward, backward or bidirectional modes and never in direct mode. The decision is which mode is best is to compare the SAD of the three modes and select the minimum as follows:

if $[SAD_{forward} \leq min(SAD_{forward}, SAD_{backward}, SAD_{bidirectional})]$
  direct mode
else if $[SAD_{bidirectional} \leq min(SAD_{forward}, SAD_{backward}, SAD_{bidirectional})]$
  bidirectional mode
else
  backward mode

6.6.7.2 Temporal Scalability Encoding

Temporal scalability can be applied to arbitrarily shaped VOPs, in addition to the traditional rectangular VOPs. In Object-based Temporal Scalability (OTS), the frame rate of the selected object in the enhancement layer is increased to provide a smoother motion than the same object in the base layer.

Type I

There are two types of temporal scalability. Fig. 6.37 shows Type 1 temporal scalability where VOL0 (Video Object Layer 0) is the entire frame containing the object and the backgroung, while VOL1 represents the object of VOL0. VOL0 is coded with a lower frame rate than VOL1. Prediction of VOL1 can be by (1) forward prediction with VOL0 as reference to form P-VOPs in the enhancement layer as shown in Fig. 6.37 (a), or (2) bidirectional prediction with VOL0 as reference to form B-VOPs in the enhancement layer as shown in Fig. 6.37 (b). In both cases, two additional shape data, a forward shape and a backward shape, are encoded to perform the background composition.

Type II

The other type of temporal scalability is Type II where the VO0 (Video Object 0) contains only the background and has no scalability layer. VO1 contains a particular object and consists of two scalability layers, VOL0 and VOL1 as shown in Fig. 6.38. VOL0 is considered the base layer and is coded at a lower frame rate than VOL1, the enhancement layer.

Enhancement Types

There are 2 types of enhancements for scalability, described by the enhancement_type flag. The base layer VOL0 for both enhancement types is coded with a lower a spatial of temporal resolution. At the enhancement layer, enhancement_type flag distinguished how it is to be enhanced. When this flag is 1, the enhancement layer increases the picture quality of a partial region of the base layer. For example, in Fig. 6.39, the temporal resolution or the spatial resolution of VOL1 (i.e., the car) is enhanced. When this flag is 0, the enhancement layer increases the picture quality of the entire region of the base layer. For example, in Fig. 6.39, the temporal or spatial resolution of VOL1 is enhanced. The region represented by VOL1 depends on the definition of VOL0.

NOTE: All scalability modes can be applied in the case of extension to 8-bit video. It is not necessary that enhancement and base layers are specified as having the same number of bits per pixel.

6.6.8 Sprite Coding

Sprite is a still image present in all scenes in a video segment. A classic example of a sprite is the background of a video sequence generated by a panning camera. The sprite generated by the panning sequence when

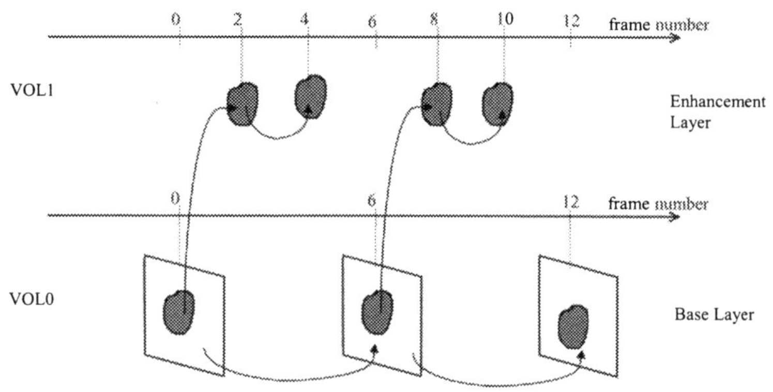

(a) Prediction of enhancement layer to form P-VOPs.

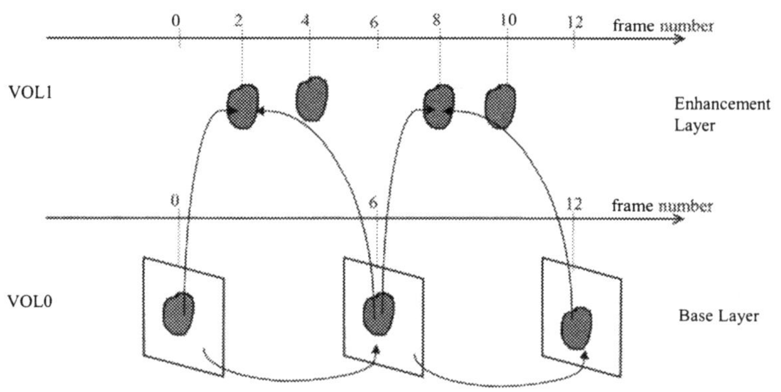

(b) Prediction of enhancement layer to form B-VOPs.

Figure 6.37: Type I temporal scalability. ©ISO/IEC 1998

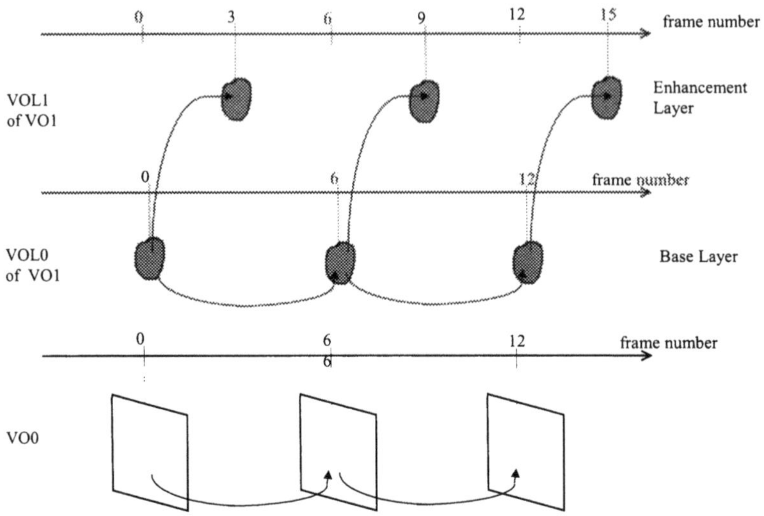

Figure 6.38: Type II temporal scalability. ©ISO/IEC 1998

composed can be transmitted as a large still image separately from the foreground object. This assumes the foreground objects can be segmented from the background and the sprite image can be extracted from the sequence prior encoding. In this way, the transmitted bit rate is reduced enormously as the sprite needs only to be transmitted once as the first frame of the sequence.

In the receiver, the background can be reconstructed based on the sprite using the global motion parameters describing the camera motion transmitted in subsequent frames. The foreground objects are transmitted separately as arbitrary-shaped video objects. Fig. 6.40 shows an exmaple of sprite coding of video sequence.

In sprite-based coding, two types of sprites are used, namely, (1) off-line static sprites, and (2) on-line dynamic sprites. The following describes them in more details.

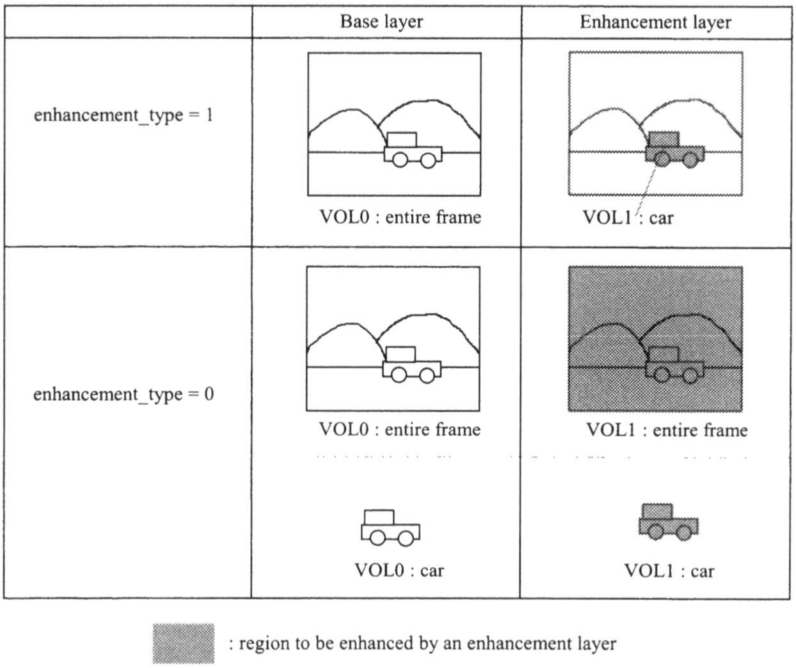

Figure 6.39: Enhancement types for scalability. ©ISO/IEC 1998

Figure 6.40: Sprite coding of video sequence. ©ISO/IEC 1998

6.6.8.1 Off-line Static Sprites

Off-line Stripe Generation

Off-line sprites, also known as static sprites are built off-line prior to encoding assuming the entire video object from which the sprite is derived is available. They can be directly copied, warped and cropped to generate a particular rendition of the sprite at a particular instant in time. For each VOP in the original video sequence, the global motion field is estimated using one of the following transform methods:

- stationary transform;
- translational transform;
- isotropic transform;
- affine transform; or
- perspective transform.

Each transformation is defined as a set of coefficients or the motion trajectories of some reference points. While the former representation is convenient for performing the transformation, the latter is required for encoding the transformations. Using the global motion parameters, the VOP is registered with the sprite by warping and blending the VOP to the sprite coordinate system. The number of reference points needed to encode the warping parameters determines the transform to be used for warping.

Off-line static sprites are particularly suitable for synthetic video objects and natural video objects undergoing rigid motion when a wallpaper-like rendering is appropriate.

Static Sprite Coding

As static sprite is a still image, the shape and texture of static sprite are treated as an I-VOP and therefore coded as such. Since sprites consists of information needed to reconstruct the background of multiple frames of a video sequence, they are typically much larger than a single frame of the video sequence. Transmitting this large amount of information as the first frame takes time and therefore a significant latency is incurred at the start of the display of a video sequence when large sprites are used.

There are two approaches one can adopt to reduce the latency incurred when large sprite are transmitted:

1. First transmit only portion of the sprite needed to reconstruct the first few frames and transmit the remaining pieces when the decoder requires them subject to the availability of bandwidth.

2. First transmit a low resolution or coarsely quantized sprite to enable the reconstruction of the first few frames and transmit the residual information to progressively build up the image quality as the bandwidth becomes available.

The above two techniques can be employed independently or in combination.

According to the sprite coding syntax, the size of sprite, the location offset of the initial piece of the sprite and the shape information for the entire sprite are transmitted at the Video Object Layer (VOL), while the transmission of the remaining portions of the sprite is done at the Video Object Plane (VOP). At the VOP, the remaining portions of the sprite are sent in small pieces along with the trajectory points. During each frame period, there may be one or more pieces of the sprite being transmitted along with size, location, and the corresponding trajectory points information where for simplicity sake, the size and location information are constrained to be of multiples of 16. The process continues until all the pieces are transmitted. Note that the encoder has the responsibility to ensure the timely delivery of pieces in a way that regions of the sprite are always present at the decoder before they are needed.

The functionality of the syntax also provides for the transmission of the sprite at a lower resolution at times of timing and bandwidth constraint and improves the quality by sending the residual information later. These residual information may be sent in place of or along with other sprite pieces at anytime subject to the bandwidth and timing constraints. The encoder can make the quality update process more efficient by determining the regions to be updated beforehand and send only the residual information when needed.

The global motion information obtained using the transformations as described above are used to represent the warping information instead of the transform coefficients. Specifically, we define a set of reference points $(x_r(n), y_r(n))$ in the current VOP to be coded. The corresponding sprite points $(x'_r(n), y'_r(n))$ in the sprite or in the reference VOP are computed using the global motion parameters estimated by global motion estimation. The sprite points $(x'_r(n), y'_r(n))$ are quantized to half-pel accuracy. The set of reference and sprite points defines the quantized transform. This process is illustrated in Fig. 6.41.

Motion vectors of the reference points which are the corner points of

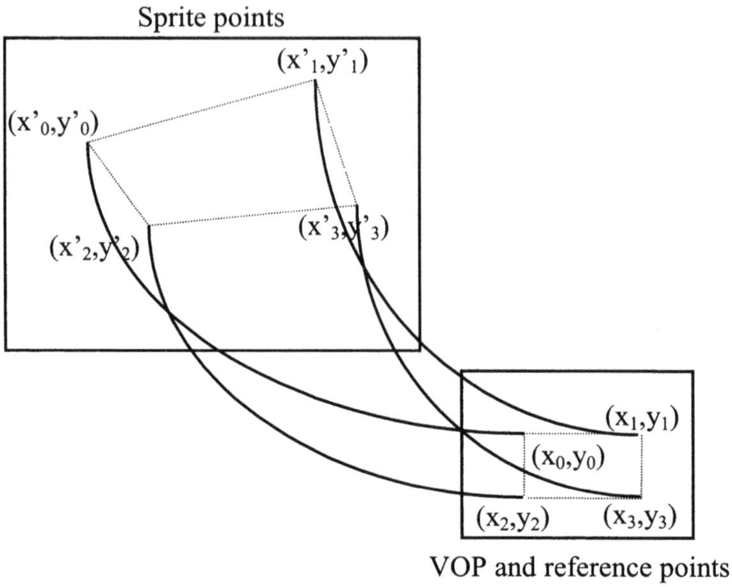

Figure 6.41: Warping of reference points to sprite points. ©ISO/IEC 1998

the bounding rectangle are coded as differential motion vectors. They are transmitted as the global motion parameters for each VOP and are at half-pixel resolution. The actual translation values are retrieved by dividing the decoded values by 2.

To reconstruct the VOP from the sprite, we scan the pixels of the current VOP and compute the corresponding location of this pixel in the sprite using the qnantized transformation described above.

6.6.8.2 On-line Dynamic Sprites

On-line Stripe Generation

On-line sprites or dynamic sprites are generated on-line during coding in both the encoder and the decoder. In on-line sprite coding, the current VOP is used as the reference, from which global motion estimation is performed between successive VOPs. The stripe is updated for each input VOP by being warped with respect to the current VOP coordinates using the estimated motion parameters between two consecutive VOPs. The current sprite is then built by blending the current VOP onto the newly aligned

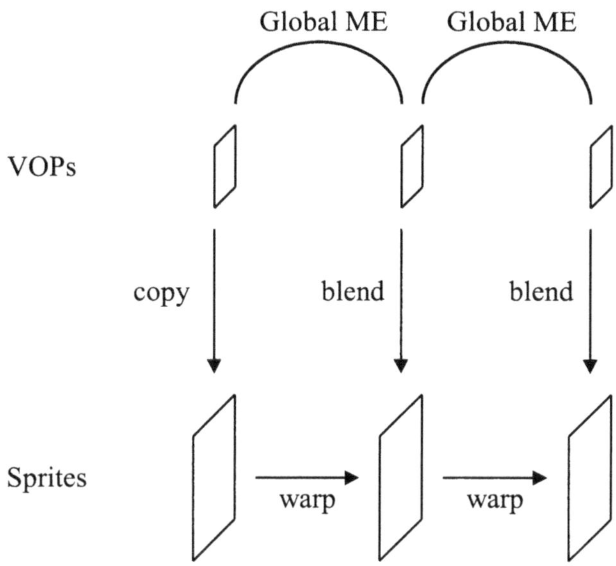

Figure 6.42: On-line dynamic sprite generation process. ©ISO/IEC 1998

sprite. Fig. 6.42 depicts the sprite generation process.

Dynamic Sprite Coding

In the case of dynamic sprites, the sprite is used for predictive coding. The prediction of a MB using the sprite is obtained using the warping parameters and a transform function. The procedure is as follows:

- the coordinates of the MB are scanned;
- using the transform function, the coordinates of the warped pixels in the sprite are found;
- the prediction of the pixel values is obtained by using bilinear transformation.

As the global motion estimation using the transformation produces pixel-wise motion vectors, the candidate motion vector predictor from the reference MB is obtained as the average value of the pixel-wise motion vectors in motion vector coding for MBs in sprite-VOPs. However, there may be regions where sprite content is undefined, therefore padding may be needed as for normal VOPs.

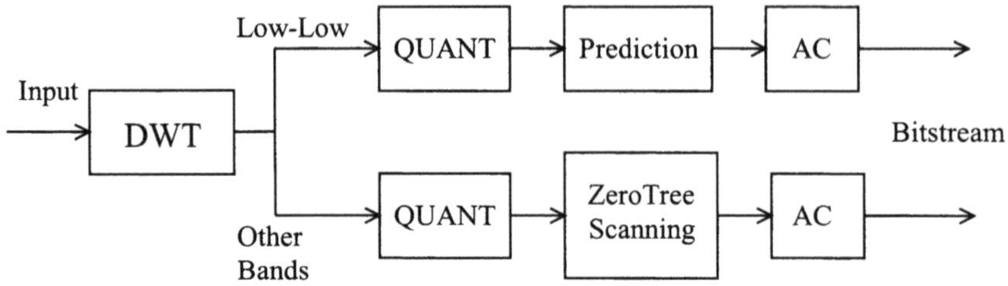

Figure 6.43: Block diagram of the wavelet encoder. ©ISO/IEC 1998

Shape coding in sprite-VOPs is the same as that in P-VOPs.

6.6.9 Still Image Texture Coding

The coding of still images employs zerotree wavelet coding technique. This technique enables the coding of still image textures with a high efficiency and spatial/SNR scalability at fine granularity which can be selected at a wide range of possible levels.

6.6.9.1 The Encoder Structure

Fig. 6.43 shows the structure of the wavelet encoder. The input is decomposed into various subbands by the discrete wavelet transform (DWT). The low-low band is quantized and coded by predictive coding scheme while the other bands are zerotree wavelet coding technique. Both the outputs of the predictive and wavelet coders are then entropy-coded by adaptive arithmetic coder (AC).

6.6.9.2 Discrete Wavelet Transform

The two-separable wavelet decomposition is performed using a Daubechies (9,3) tap biorthogonal filter with the filter coefficients given by Table 6.4. A group delay of 1 and -1 sample is applied to the highpass analysis and highpass synthesis filter, respectively.

Before applying the wavelet decomposition, symmetric extensions are performed at the leading and trailing of the texture data sequences to satisfy the perfect reconstruction criterion of wavelet filtering. Downsampling by a factor of 2 is carried out at each level of decomposition to preserve the total number of samples in the image.

Table 6.4: Coefficients of Daubechies (9,3) tap biorthogonal filter. ©ISO/IEC 1998

Lowpass filter	Highpass filter
0.03314563036812	-0.35355339059327
-0.06629126073624	0.70710678118655
-0.17677669529665	-0.35355339059327
0.41984465132952	
0.99436891104360	
0.41984465132952	
-0.17677669529665	
-0.06629126073624	
0.03314563036812	

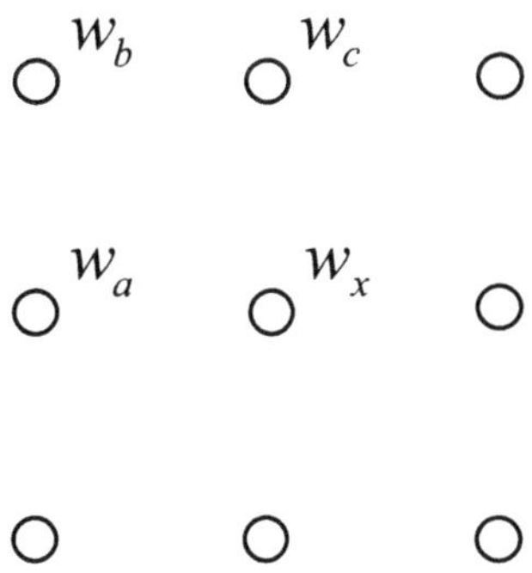

Figure 6.44: Coding of lowest subband coefficients. ©ISO/IEC 1998

6.6.9.3 Coding of the Lowest Subband

The lowest subband (i.e., low-low band) is the most important subband and is encoded independently from other subbands. The encoding technique used is a simple predictive coding scheme, the differential pulse code modulation (DPCM). Quantization of the wavelet coefficients is by an uniform midrise quantizer. The quantized coefficient w_x is predicted from its three nearest neighbors w_a, w_b, and w_c as illustrated in Fig. 6.44.

The prediction rule is as follows:

if $(|w_a - w_b| < |w_a - w_c|)$

  $\hat{w}_x = w_c$

  $w_x = w_x - \hat{w}_x$

```
else
```

$$\hat{w}_x = w_a$$
$$w_x = w_x - \hat{w}_x$$

The coefficients after DPCM are then encoded using an adaptive arithmetic coder. The minimum and maximum values of the coefficients are found. The minimum value is subtracted from all the coefficients to limit their lower bound to zero. The AC model is initiated with an uniform distribution with the maximum value as seeds. The coefficients are then scanned and encoded adaptively by the AC.

6.6.9.4 Zerotree Coding of the Higher Subbands

A multiscale zerotree coding scheme is employed to achieve a wide range of scalability levels as shown in Fig. 6.45. The wavelet coefficients of the first layer are first quantized with the quantizer Q0. The quantized coefficients are zerotree scanned and the significant maps and the coefficients are entropy coded with the AC producing output BS0. The quantized wavelet coefficients of the first layer are also reconstructed and subtracted from the original coefficients forming the coefficients of the second layer. These coefficients are quantized by the quantizer Q1, zerotree scanned and entropy coded producing the output BS1. The quantized wavelet coefficients of the second layer are also reconstructed and subtracted from the original coefficients forming the coefficients of the third layer. The process is repeated until the final Nth layer is reached where $N + 1$ defines the number scalability layers.

Zerotree Scanning

As a result of the wavelet subband decomposition, there exists a parent-child relationship, i.e., high correlation, between wavelet coefficients at the same location across different subbands. With reference to Fig. 6.46, a wavelet tree can be constructed as we scan from the parent in the lowest subband to the higher subbands as indicated by the dotted line.

Zerotree is formed at any node of the wavelet tree if the coefficient is zero and all the node's children are also zero. This is based on the principle if a wavelet coefficient in a lower subband is insignificant, because of the high correlation between parent and children, then all the coefficients in the same location in the higher will also likely to be insignificant.

The wavelet trees are coded by scanning each tree from the root in the lowest subband through the children in the higher subbands, and assigning

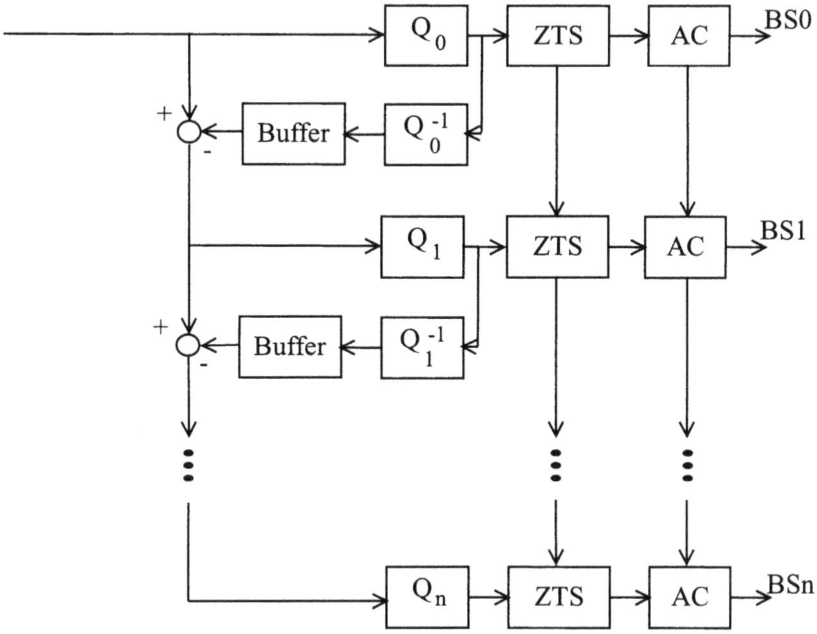

Figure 6.45: Multiscale zerotree coding scheme. ©ISO/IEC 1998

one of three symbols to each node, namely, zerotree root, valued zerotree root, or value. A zerotree root is the coefficient at the root of a zerotree. Zerotrees need not be scanned anymore since all the coefficients are zero. A valued zerotree root is a node where the coefficient has a nonzero amplitude, and all four children are zerotree roots. Scanning terminates at a valued zerotree. A value identifies a coefficient with amplitude either zero or nonzero, but also with some nonzero children. The symbols and the quantized coefficients are encoded using an adaptive arithmetic coder.

6.6.9.5 Quantization

Two quantization schemes are employed levels, i.e., multilevel quantization and bi-level quantization. To achieve a wide range of scalability levels, a multilevel quantizer is used where the quantization levels are defined by the encoder. Different quantization step sizes can be specified for each level of scalability. All higher subband quantizers are uniform midrise quantizers with a dead zone twice the quantizer step size. The multilevel quantization scheme provides a flexible tradeoff between levels and types of scalability, complexity and coding efficiency for any application.

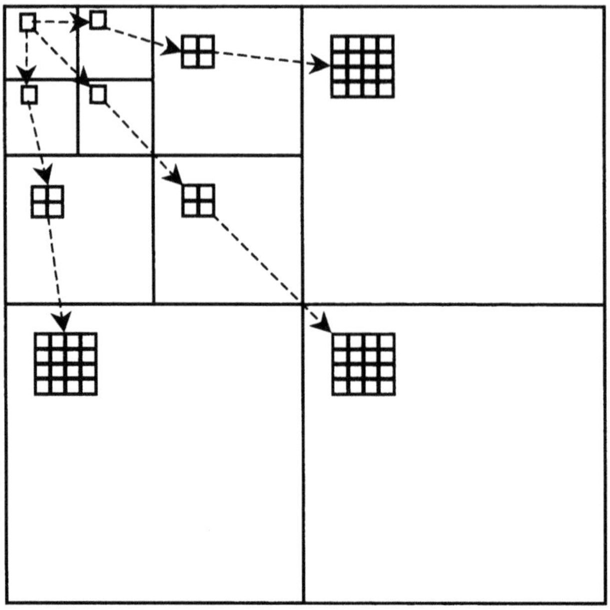

Figure 6.46: The parent-child relationship of wavelet coefficients. ©ISO/IEC 1998

In order to achieve the finest granularity of SNR scalability, a bi-level quantization scheme is used for all the quantizers. This is also a uniform midrise quantizer with a dead zone twice the quantization step size. The coefficients that are outside the dead zone are quantized with a 1-bit accuracy. The number of quantizers is equal to the maximum number of bitplanes in the wavelet coefficient representation.

6.6.9.6 Entropy Coding

The zerotree symbols and quantized wavelet coefficients are entropy-coded using an adaptive arithmetic coder with a three-symbol alphabet. Therefore, at least three different tables, namely, type, valz and valnz, must be codet at the same time. The arithmetic coder must track at least three probability models, one for each table. There may be two more models to track, one for non-zero quantized coefficients of the low-low band and one for the non-zero quantized coefficients of the other three low resolution bands. For each wavelet coefficient, first the coefficient is quantized, then its type and value are calculated, and lastly these values are arithmetic coded.

The probability model of the arithmetic coder is initialized with an uniform distribution and switched appropriately for each table.

6.7 Coding of Synthetic Objects

Synthetic objects can be generated by computer graphics, or formed from natural objects by using a parametric description of the objects. It is the latter type of synthetic objects that MPEG-4 has its focus. In its current version, MPEG-4 provides standards for:

Parametric descriptions of

- a synthetic description of human face and body
- animation streams of the face and body

Static and dynamic mesh coding with texture mapping Texture coding for view dependent applications

6.7.1 Facial Animation

Animation of the face, i.e., the shape, texture and expressions of the face, is controlled by the Facial Description Parameter (FDP) sets and/or Facial Animation Parameter (FAP) sets. The positions of the various feature points on the face as defined in MPEG-4 are shown in Fig. 6.47. Initially, the face object contains a generic face with a neutral expression. Upon receiving the animation parameters, the face can be rendered to produce animation of different facial expressions, movements and speech utterances. Together with the definition parameters, the generic face can be transformed into faces of different shapes and textures. If required, a complete face model, e,g., a wireframe model, can be downloaded via the FDP set.

Note that the face models are not normative. MPEG-4 only standardizes the coding of description and animation parameters when decoded can drive an unlimited range of models. In cases where custom models and specialized interpretation of the FAPs are needed, the Systems Binary Format for Scenes (BIFS) provides the following features to support face animation:

1. FDPs in BIFS - downloadable model data to configure a baseline face model pre-stored in the terminal into a particular face or to install a specific face model at the beginning of a session;
2. Face Animation Table (FAT) within FDPs - downloadable functional mapping from the incoming FAPs to feature control points in the face mesh to control facial movements;

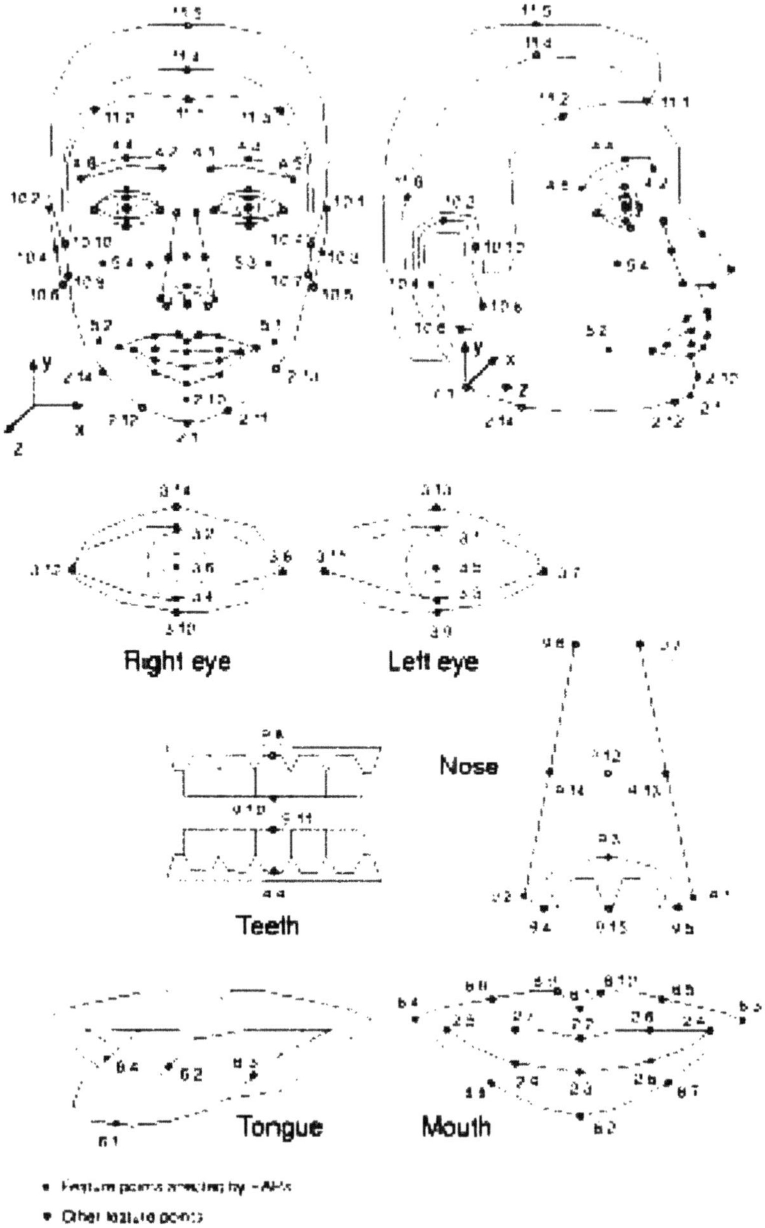

Figure 6.47: Description of the feature points in MPEG-4 [5]

Figure 6.48: An example of 2-D mesh modeling on a video object. ©ISO/IEC 1998

3. Face Interpolation Technique (FIT) in BIFS - downloadable definitions of mapping of incoming FAPs into a total set of FAPs before their applications to feature points which can be used for complex cross-coupling of FAPs to link their effects, or to interpolate FAPs using those available in the terminal.

The above BIFS features enables the customization of face models including the calibration of an established model or the downloading of a new model with its shape, texture and color.

6.7.2 Body Animation

Decoding and scene description of body animation follows the same approach in face animation described in Section 6.7.1 above and is designed to work in an integrated way with it.

6.7.3 2-D Animated Meshes

Triangular mesh has long being used for 3-D object shape modeling and rendering. When projected onto the 2-D image plane, a 2-D mesh results. A 2-D dynamic mesh refers to 2-D mesh geometry and the associate motion information of all its node points at a particular instant of time. Fig. 6.48 shows a 2-D mesh on a human face.

6.7.3.1 Mesh Tracking

Using the dynamic mesh, the image features can be tracked forward in time by non-uniform sampling of the motion field at the node points between the reference mesh and the current mesh. The 2-D mesh may be regular or adapted to the image content, which in the case is known as content-based mesh. Method of selection and tracking of the node points are not subject to standardization.

6.7.3.2 Texture Mapping

Mapping of the texture within the triangular meshes is carried out by first deforming the triangular patches in the reference frame according to the movement indicated by the motion vectors at the node points, and warping the texture within each patch in the reference frame onto the corresponding patches in the current frame. Affine mapping seems to be the popular choice for triangular meshes as it is a linear mapping which can model translation, rotation, scaling, reflection and shear, and at the same time, preserve straight lines at low computational complexity. Moreover, the degree of freedom given by the three motion vectors match well with the six parameters of the affine mapping which means that a continuous affine motion field can be reconstructed from the 2-D motion field represented by the three motion vectors at the node points.

One big advantage of the 2-D mesh modeling is it only requires a single view without range data. However, it is easily extendable to 3-D mesh if the additional information is available. In summary, 2-D mesh representation is able to model the shape and motion of a VOP in a unified framework. It also enables the following content-based functionalities:

Video Object Manipulation

- *Augmented reality*: Merging computer animated video images with natural video images to create enhanced display information. The computer-generated images must be tracked to have perfect synchronization with the moving natural images.

- *Synthetic-object-transfiguration*: Replacing a natural video object in a video clip by another video object extracted from another natural video clip or transfigured from a still image object using the motion information of the object to be replaced.

- *Spatio-temporal interpolation*: Creating intermediate frames by motion-compensated temporal interpolation e.g., in frame rate up-conversion.

Video Object Compression

- Using 2-D mesh modeling, large compression can be achieve by transmitting only texture maps of selected key frames and animate (interpolate) those in intermediate frames using the motion information of the selected frames.

Content-Based Video Indexing

Mesh representation

- enables animated key snapshots for a moving visual synopsis of objects;
- provides accurate object trajectory information that can be used to retrieve visual objects with specific motion; and
- provides vertex-based object shape representation for shape-based object retrieval.

6.8 Error Resilience

MPEG-4 provides error robustness and resilience to allow accessing image or video information over a wide range of storage and transmission media. The error resilience tools developed for MPEG-4 can be divided into three major categories: resynchronization, data recovery, and error concealment.

6.8.1 Resynchronization

The use of variable length codewords in the MPEG-4 bitstream means that there may be loss of synchronization when an error occurs in the bitstream. Resynchronization tools attempt to re-establish synchronization between the decoder and the bitstream after a residual error or errors have been detected. Generally, the data between the synchronization point prior to the error and the first point where synchronization is re-established, is discarded. If the resynchronization approach is effective at localizing the amount of data discarded by the decoder, then the ability of other types of tools which recover data and/or conceal the effects of errors is greatly enhanced.

The resynchronization technique adopted by MPEG-4 is the video packet approach based on providing periodic resynchronization markers throughout the bitstream. In other words, the length of the video packets are not based on the number of macroblocks, but instead on the number of bits contained in that packet. If the number of bits contained in the current video packet exceeds a predetermined threshold, then a new video packet is created at the start of the next macroblock and a resynchronization marker is inserted. This marker is distinguishable from all possible VLC code words as well as the VOP start code. Header information is also provided at the start of a video packet. Contained in this header is the information necessary to restart the decoding process.

It should be noted that when utilizing the error resilience tools within MPEG-4, some of the compression efficiency tools are modified. For example, all predictively encoded information must be confined within a video packet so as to prevent the propagation of errors.

In conjunction with the video packet approach to resynchronization, a second method called fixed interval synchronization has also been adopted by MPEG-4. This method requires that VOP start codes and resynchronization markers (i.e., the start of a video packet) appear only at legal fixed interval locations in the bitstream. This helps to avoid the problems associated with start codes emulations. That is, when errors are present in a bitstream it is possible for these errors to emulate a VOP start code. In this case, when fixed interval synchronization is utilized the decoder is only required to search for a VOP start code at the beginning of each fixed interval. The fixed interval synchronization method extends this approach to be any predetermined interval.

6.8.2 Data Recovery

After synchronization has been re-establish, data recovery tools attempt to recover data that in general would be lost. These tools are not simply error correcting codes, but instead techniques which encode the data in an error resilient manner. For instance, one particular tool that has been endorsed by the Video Group is Reversible Variable Length Codes (RVLC). In this approach, the variable length code words are designed such that they can be read both in the forward as well as the reverse direction, as depicted in Fig. 6.49. Obviously, this approach reduces the compression efficiency achievable by the entropy encoder. However, the improvement in error resiliency is substantial.

6.8.3 Error Concealment

Error concealment is an important component of any error robust video codec. The effectiveness of a error concealment strategy is highly dependent on the performance of the resynchronization scheme. Basically, if the

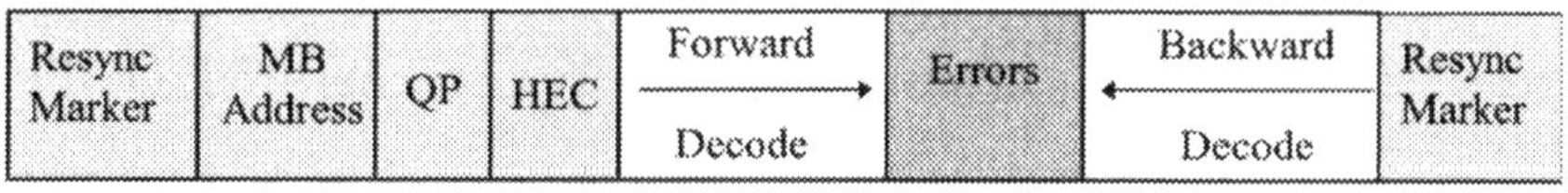

Figure 6.49: Example of Reversible Variable Length Code. ©ISO/IEC 1998

Table 6.5: Summary of error resilience modes of operation. ©ISO/IEC 1998

Mode	VOP Type	Resync Marker	RVLC	Data Partitioning (Motion Marker)
Low Complexity	I-VOP	Mandatory	Optional	Not used
	P-VOP	Mandatory	Optional	Not used
Medium Complexity	I-VOP	Mandatory	Optional	Mandatory
	P-VOP	Mandatory	Optional	Mandatory
High Complexity	I-VOP	Mandatory	Mandatory	Mandatory
	P-VOP	Mandatory	Mandatory	Mandatory

resynchronization method can effectively localize the error then the error concealment problem becomes much more tractable. For low bitrate, low delay applications the current resynchronization scheme provides very acceptable results with a simple concealment strategy, such as copying blocks from the previous frame.

An additional error resilient mode that further improves the ability of the decoder to localize an error is data partitioning. This approach requires that a second resynchronization marker be inserted between motion and texture information. If the texture information is lost, this approach utilizes the motion information to conceal these errors by motion compensating the previous decoded VOP.

6.8.4 Modes of Operation

There are several modes of opertion providing various options depending on the complexity of the scene to be encoded. Essentially there are three modes of operation, namely, low complexity mode, medium complexity mode and the high complexity mode. A summary of these various modes is given in Table 6.5.

Table 6.6: Suggested Resynchronization Marker Spacing. ©ISO/IEC 1998

Bit Rate (kbit/s)	Spacing (bits)
0-24	480
25-48	736
49-128	1500
128-512	TBD
512-1000	TBD

6.8.5 Error Resilience Encoding Tools

6.8.5.1 Resynchronization Markers

The resynchronization markers should be inserted by the encoder before the first macroblock after the number of bits output since the last resync_marker field exceeds a predetermined value. The value used for this spacing is dependent on the anticipated error conditions of the transmission channel and compressed data rate. Suggested values for this spacing are provided in Table 6.6. These values are obtained experimentally and provide good results for a variety of error conditions encountered in error-prone channels such as wireless fading channels. It is highly recommended that users of MPEG-4's error resilient tools adjust this spacing of the resynchronization markers to fit the error conditions of their particular channel.

As shown in Fig. 6.49, in addition to the *resync_marker* field, the encoder also inserts a field indicating the current macroblock address (MB address), a field indicating the current quantization parameter (QP) and a Header Extension Code (HEC). This additional information is provided to the decoder enabling it to determine which VOP a resync packet belongs to in case the VOP start_code is lost

6.8.5.2 Data Partitioning

When the data partitioning is used, then in addition to the *resync_marker* fields, MB address, QP and HEC, a *motion_marker* field is inserted after the motion data (before the beginning of the texture data). This *motion_marker* field is unique from the motion data and enables the decoder to determine when all the motion information has been received correctly.

6.8.5.3 Reversible VLCs

The use of reversible VLCs enables the decoder to recover additional texture information in the presence of errors. This is accomplished by first detecting the error and searching forward to the next *resync_marker*, once this point is determined the texture data can be read in the reverse direction until an error is detected. When errors are detected in the texture data, the decoder can use the correctly decoded motion vector information to perform motion compensation and conceal these errors.

6.8.5.4 Decoder Operation

When an error is detected in the bitstream, the decoder should resynchronize at the next suitable resynchronization point. Where a VOP header is missed or received with obvious errors, this should be the next VOP *start_code*. Otherwise, the next resynchronization point in the bitstream should be used.

Under the following error conditions, the baseline decoder should resynchronize at the next resynchronization point in the bitstream:

1. An illegal VLC is received.
2. More than 64 DCT coefficients are decoded in a single block.
3. Inconsistent resynchronization header information (i.e., QP out of range, $MBN(k) < MBN(k-1)$, etc.)
4. Resynchronization marker is corrupted.

Under the following error conditions, the decoder should resynchronize at the next VOP header:

- VOP start code corrupted.

For other resynchronization techniques, conditions for error detection and resynchronization should be as close as possible to those outlined above. Missing blocks should be replaced with the same block from the previous frame.

References

[1] ISO/IEC 14496-2, "Information technology - generic coding of audio-visual objects (final draft of international standard)," Dec. 1998.

[2] R. Koenen, "Overview of MPEG-4 standard," in *ISO/IEC JTC1/SC29/WG11 MPEG98/N2459, Atlantic City, USA*, Oct. 1998.

[3] ISO/IEC, "Managing intellectual property identification and protection within MPEG-4," in *ISO/IEC JTC1/SC29/WG11 MPEG97/N1918, Fribourg, Switzerland*, Oct. 1997.

[4] MPEG Video Group, "MPEG-4 video verification model version 11.0," in *ISO/IEC JTC1/SC29/WG11 MPEG98/N2172, Tokyo, Japan*, Mar. 1998.

[5] F. Lavagetto, R. Pockaj, and M. Costa, "MPEG-4 compliant calibration of 3d head models," in *Int. Picture Coding Symposium, PCS'99, Portland, Oregon, USA*, Apr. 1999, pp. 217–220.

Index

www.ingramcontent.com/pod-product-compliance
Ingram Content Group UK Ltd.
Pitfield, Milton Keynes, MK11 3LW, UK
UKHW050948290926
15470UKWH00005B/1714